Evolution versus Creationism:
The Public Education Controversy

Edited by J. Peter Zetterberg

ORYX PRESS
1983

The rare Arabian Oryx is believed to have inspired the myth of the unicorn. This desert antelope became virtually extinct in the early 1960s. At that time several groups of international conservationists arranged to have 9 animals sent to the Phoenix Zoo to be the nucleus of a captive breeding herd. Today the Oryx population is over 400 and herds have been returned to reserves in Israel, Jordan, and Oman.

Copyright © 1983 by The Oryx Press
2214 North Central at Encanto
Phoenix, AZ 85004

Published simultaneously in Canada

Printed and Bound in the United States of America

Library of Congress Cataloging in Publication Data

Main entry under title:

Evolution versus Creationism.

    Bibliography: p.
    Includes index.
    1. Evolution—Study and teaching—Addresses, essays, lectures. 2. Creationism—Study and teaching—Addresses, essays, lectures. I. Zetterberg, J. Peter.
QH362.E86   1983   575'.007   82–18795
ISBN 0–89774–061–0

Table of Contents

Preface

Despite the established place of evolution in modern science, the teaching of evolution remains controversial in many school districts throughout the nation. Organizations such as the Institute for Creation Research and the Creation Research Society seek to change public school curricula so that a view of creation, "scientific creationism," is taught side by side with evolution. The proponents of scientific creationism have developed a model as an alternative to the theory of evolution. They claim scientific evidence for this model, which closely parallels the Genesis account and holds that the earth is young (6,000–10,000 years), and challenge, on scientific grounds, the axioms and methods of the various sciences that support evolution.

Legislation has been introduced in many states that would mandate a two-model approach in the teaching of biology and earth science: evolution as one model, scientific creationism as the other. Such legislation was recently passed in Arkansas and Louisiana. In both states the legislation has been overturned by the courts, although the decision in the Louisiana case will probably be appealed. Whatever the final decision regarding such legislation, at the local level pressure will continue to be exerted on school boards to adopt the two-model approach.

The controversy surrounding the teaching of evolution in public schools is a confusing tangle of emotion, debate, legislation, and litigation. To help clarify the issues in this controversy and, more specifically, to examine the arguments of the proponents of scientific creationism, the University of Minnesota organized a conference entitled "Evolution and Public Education." The conference, which was held on December 5, 1981, was made possible by a grant from the Minnesota Humanities Commission, in cooperation with the National Endowment for the Humanities and the Minnesota State Legislature, and sponsored by the National Association of Biology Teachers, the Minnesota Science Teachers Association, the Minnesota Association for the Improvement of Science Education, and the University's Center for Educational Development and Department of Conferences. This book is a revised version of a resource manual compiled for the conference.

The authors of the articles in this collection, which is divided into six parts, (1) discuss the theory of evolution and its place in science education; (2) examine the creationist movement; (3) state the position of scientific creationists; (4) respond to creationists' arguments against evolution; (5) explore legal issues in the controversy; and (6) provide some perspectives on attempts to treat the Genesis creation account as science. It should be noted that Part 5, on the legal issues, includes Judge Overton's decision striking down the Arkansas Creationism Act, as well as pieces of legislation that reveal the changing tactics of creationists, who first sought to ban the teaching of evolution in the 1920s, then sought equal time for biblical creationism in the late 1960s and early 1970s, and now seek equal time for scientific creationism.

I want to thank authors and publishers for permission to reprint copyrighted material and also those authors who have developed new material for this collection. For the sake of conformity in style, I have replaced all parenthetical references occurring in original texts with end notes. The bibliography includes most, but not all, of the many works cited in the references. In compiling the bibliography, I have emphasized works that directly address the current creation/evolution controversy. As general background, I have also included some works on the philosophy of science and science and faith issues, as well as items of historical interest, including original reviews of Darwin's *On the Origin of Species* and material from the Scopes era. The scientific works on evolution in the bibliography are but a small sample of the voluminous literature on this topic.

Finally, I would like to thank the following individuals for their advice regarding the selection of material for this collection and for assistance in planning and/or participating in the conference on "Evolution and Public Education": Andrew Ahlgren, Calvin Alexander, Elving Anderson, John Bohlig, Lyle Bradley, Sr. Vera Chester, Richard Clark, Tina Egeland, Eugene Gennaro, Gina Hoffman, Samuel Kirkwood, Sr. Lucy Knoll, Malcolm Kottler, David Merrell, John A. Moore, Philip Regal, Harvey Sarles, Robert Schadewald, George Shaw, Robert Sloan, Robert Tapp, Stanley Weinberg, and Claude Welch.

J. Peter Zetterberg
Editor

Contributors

E. Calvin Alexander, Jr. is Associate Professor of Geochemistry in the Department of Geology and Geophysics at the University of Minnesota, Minneapolis, Minnesota. "Radiometric Dating and the Atmospheric Argon Correction in K/Ar Dating" is reprinted from the *Journal of the Minnesota Science Teachers Association,* vol. 1 (1979), by permission. Copyright © 1979 by the Minnesota Science Teachers Association and the Minnesota Academy of Science. All rights reserved.

Richard D. Alexander is Professor of Zoology and Curator of Insects in the Museum of Zoology at the University of Michigan, Ann Arbor, Michigan. "Evolution, Creation, and Biology Teaching" is reprinted from the *American Biology Teacher,* vol. 40 (February 1978), by permission of the National Association of Biology Teachers. Copyright © 1978 by the National Association of Biology Teachers. All rights reserved.

V. Elving Anderson is Professor of Genetics and Acting Director of the Dight Institute for Human Genetics at the University of Minnesota, Minneapolis, Minnesota. "Scientific Creationism and Its Critique of Evolution" is printed by permission of the author.

Frank T. Awbrey is Professor of Biology at San Diego State University, San Diego, California. "Defining 'Kinds'—Do Creationists Apply a Double Standard?" is reprinted from *Creation/Evolution,* Issue 5 (Summer 1981), by permission of the author. Copyright © 1981 by Frank T. Awbrey. All rights reserved.

Wendell R. Bird is an attorney for the Institute for Creation Research, El Cajon, California. "Summary of Scientific Evidence for Creation" by Duane T. Gish, Richard B. Bliss, and Wendell R. Bird is reprinted from the *ICR Impact Series,* nos. 95 and 96 (May/June 1981). Reprinted by permission of the Institute for Creation Research, 2100 Greenfield, El

Cajon, CA 92021. Copyright © 1981 by the Institute for Creation Research. All rights reserved.

Richard B. Bliss is Director of Curriculum Development for the Institute for Creation Research, El Cajon, California. "Summary of Scientific Evidence for Creation" by Duane T. Gish, Richard B. Bliss, and Wendell R. Bird is reprinted from the *ICR Impact Series,* nos. 95 and 96 (May/June 1981). Reprinted by permission of the Institute for Creation Research, 2100 Greenfield, El Cajon, CA 92021. Copyright © 1981 by the Institute for Creation Research. All rights reserved. "A Two-Model Approach to Origins: A Curriculum Imperative" by Richard B. Bliss is reprinted from the *ICR Impact Series,* no. 36 (June 1976). Reprinted by permission of the Institute for Creation Research, 2100 Greenfield, El Cajon, CA 92021. Copyright © 1976 by the Institute for Creation Research. All rights reserved.

Walter T. Brown is Director of the Institute for Creation Research— Midwest Center in Naperville, IL. "The Scientific Evidence for Creation: 108 Categories of Evidence," revised November 1982, is reprinted by permission of the author and the Institute for Creation Research—Midwest Center.

Stephen G. Brush is Professor of the History of Science in the Department of History and the Institute for Physical Science and Technology at the University of Maryland, College Park, Maryland. "Finding the Age of the Earth: By Physics or By Faith?" is reprinted from the *Journal of Geological Education,* vol. 30, no. 1 (January 1982), by permission of the National Association of Geology Teachers. Copyright © 1982 by the National Association of Geology Teachers. All rights reserved.

Neal D. Buffaloe is Professor of Biology at the University of Central Arkansas, Conway, Arkansas. "Creationism and Evolution: The Real Issues" by N. Patrick Murray and Neal D. Buffaloe is reprinted by permission of the authors. Copyright © 1981 by N. Patrick Murray and Neal D. Buffaloe. All rights reserved.

Preston Cloud is a research biologist with the U.S. Geological Survey and Professor Emeritus of Biogeology and Environmental Studies at the University of California—Santa Barbara, California. " 'Scientific Creationism'—A New Inquisition Brewing?" first appeared in the *Humanist* (January/February 1977), and is reprinted by permission. Copyright © 1977. All rights reserved.

Theodosius Dobzhansky was, before his death, Professor Emeritus at Rockefeller University, New York, New York and Adjunct Professor of Genetics at the University of California—Davis, California. "Nothing in Biology Makes Sense Except in the Light of Evolution" is reprinted from the *American Biology Teacher,* vol. 35 (March 1973), by permission of the National Association of Biology Teachers. Copyright © 1972 by the National Association of Biology Teachers. All rights reserved.

Frederick Edwords is Editor of *Creation/Evolution,* the only journal dedicated exclusively to answering the arguments of creationists. He is also Administrator of the American Humanist Association. "Decide: Evolution or Creation?," revised June 1982, is reprinted by permission of the author. Copyright © 1980 by Frederick Edwords. All rights reserved. "Why Creationism Should Not Be Taught as Science: The Legal Issues," revised, June 1982, is reprinted from *Creation/Evolution,* Issue 1 (Summer 1980), by permission of the author. Copyright © 1980 by Frederick Edwords. All rights reserved.

Stanley Freske has many years experience in industrial research and development. He currently works as a consultant in the San Diego area. "Creationist Misunderstanding, Misrepresentation, and Misuse of the Second Law of Thermodynamics" is reprinted from *Creation/Evolution,* Issue 4 (Spring 1981), by permission of the author. Copyright © 1981 by Stanley Freske. All rights reserved.

Duane T. Gish is Associate Director of the Institute for Creation Research, El Cajon, California. "Creation, Evolution, and Public Education" is printed by permission of the author. "Summary of Scientific Evidence for Creation" by Duane T. Gish, Richard B. Bliss, and Wendell R. Bird is reprinted from the *ICR Impact Series,* nos. 95 and 96 (May/June 1981). Reprinted by permission of the Institute for Creation Research, 2100 Greenfield, El Cajon, CA 92021. Copyright © 1981 by the Institute for Creation Research. All rights reserved.

Malcolm Jay Kottler is Associate Professor of the History of Science at the University of Minnesota, Minneapolis, Minnesota. "Evolution: Fact? Theory? . . . or Just a Theory?" is printed by permission of the author.

Ralph W. Lewis is Professor of Natural Science at Michigan State University, East Lansing, Michigan. "Evolution: A System of Theories" is reprinted from *Perspectives in Biology and Medicine,* vol. 23 (Summer 1980), by permission of The University of Chicago Press.

Kenneth Miller is Associate Professor of Biology at Brown University, Providence, Rhode Island. "Answers to the Standard Creationist Arguments" is reprinted from *Creation/Evolution,* Issue 7 (Winter 1982), by permission of the author. Copyright © 1982 by Kenneth Miller. All rights reserved.

John A. Moore is Professor of Biology at the University of California —Riverside, California. "Creationism" is reprinted by permission of the author from *1979 AETS Yearbook. Science Education/Society: A Guide to Interaction and Influence,* ed. by M. R. Abraham (Columbus: ERIC, 1979). "Evolution, Education, and the Nature of Science and Scientific Inquiry" is reprinted from *BioScience,* vol. 32, no. 7, copyright © 1982 by the American Institute of Biological Sciences. Reprinted with permission of the copyright holder. The article was originally prepared as a keynote address for the University of Minnesota conference on "Evolution and Public Education," December 5, 1981. "On Giving Equal Time to the Teaching of Evolution and Creation" is reprinted from *Perspectives in Biology and Medicine,* vol. 18 (March 1975), by permission of the University of Chicago Press. Copyright © 1975 by the University of Chicago Press. All rights reserved.

N. Patrick Murray is Rector of All Saints' Episcopal Church in Russellville, Arkansas. "Creationism and Evolution: The Real Issues," by N. Patrick Murray and Neal D. Buffaloe, is reprinted by permission of the authors. Copyright © 1981 by N. Patrick Murray and Neal D. Buffaloe. All rights reserved.

William R. Overton is U.S. District Court Judge, Eastern District of Arkansas, Western Division. "The Arkansas Decision: Memorandum Opinion in *Rev. Bill McLean et al.* v. *the Arkansas Board of Education, et al.* (January 5, 1982)" is in the public domain.

John W. Patterson is Professor of Material Science and Engineering at Iowa State University, Ames, Iowa. "An Engineer Looks at the Creationist Movement" is reprinted from *Proceedings of the Iowa Academy of Science* (June 1982) by permission of the Iowa Academy of Science. Copyright © 1982 by the Iowa Academy of Science. All rights reserved.

Robert J. Schadewald is a science writer who lives in Rogers, Minnesota. "Six 'Flood' Arguments Creationists Can't Answer" is reprinted

Gerald Skoog is Professor of Secondary Education in the College of Education at Texas Tech University, Lubbock, Texas. "The Topic of Evolution in Secondary School Biology Textbooks: 1900–1977" is re-printed from *Science Education,* vol. 63, no. 5 (1979), pp. 621–40, by permission. Copyright © 1979 by John Wiley & Sons, Inc. All rights reserved.

Robert E. Sloan is Professor of Paleontology in the Department of Geology and Geophysics at the University of Minnesota, Minneapolis, Minnesota. "The Association of 'Human' and Fossil Footprints" is reprinted from the *Journal of the Minnesota Science Teachers Associa-tion,* vol. 1 (1979) by permission. Copyright © 1979 by the Minnesota Science Teachers Association and the Minnesota Academy of Science. All rights reserved. "The Transition between Reptiles and Mammals" is printed by permission of the author.

Part 1
Evolution and Science Education

Evolution, Education, and the Nature of Science and Scientific Inquiry

by John A. Moore

> . . . It becomes evermore important to understand what is science and what is not. Somehow we have failed to let our students in on that secret. We find as a consequence, that we have a large and effective group of creationists who seek to scuttle the basic concept of the science of biology—the science that is essential for medicine, agriculture, and life itself; a huge majority of citizens who, in "fairness," opt for presenting as equals the "science" of creation and the science of evolutionary biology; and a president who is so poorly informed that he believes that scientists are questioning that evolution ever occurred. It is hard to think of a more terrible indictment of the way we have taught science.

Although my title is specific and formidable, my argument will develop more in relation to the general title of our conference, "Evolution and Public Education." I will attempt to present a detached view of the evolution/creationism affair; that is, to be as objective and even-handed as possible. In adopting this stance, I expect to irritate all of you—in different ways and to different degrees. I will make remarks that can be taken as insults by creationists, evolutionists, teachers, administrators, and general citizens. And since I consider myself a member of three of those five categories, I will hardly be excluded from the general and specific condemnations. I trust that each group will realize and accept that this is being done for its own good.

I am considerably more interested in the future than in the present or the past. The future is more amenable to change—we need the future to correct the present. Nevertheless, we must base the argument and the projections on the past and the present, so let's begin with some very general statements.

The argument between the evolutionists and the creationists is between those whose fundamental goal is to understand nature and those whose fundamental goal is to control thought. It is essential to remember how the argument developed.

One group, the evolutionists, has been using the available devices and procedures of science to understand the origins and changes through time of living creatures, the earth, and the cosmos. In this quest for understanding, no approaches except the supernatural are proscribed. The supernatural approach is proscribed not because it is wrong but because science has absolutely no way of dealing with it. Science is concerned with nature so, by definition, something that is above or apart from nature eludes its grasp.

The procedures of science have given us a view of origins and changes with time in the natural world that is intellectually satisfying in many respects but, in others, leaves us with only puzzles and challenges. Nevertheless, there is a steady increase in what is accepted as true beyond a reasonable doubt, a steady elimination of error and misunderstanding, and a steady increase in the sorts of problems amenable to observation and experiment. We have, therefore, a body of knowledge— the theory of evolution—that is continually growing, increasing in accuracy, and perfecting its predictive abilities.

So far as I am aware, no scientist has ever asked any sect to include the biological theory of evolution along with the sect's stories of creation. Neither am I aware of any movement by scientists to sponsor laws that mandate the teaching of evolution. It is important to remember these rather elementary points.

The creationists, however, are not concerned with understanding nature but in promoting a selection of the statements in Genesis, mainly one of the two divergent accounts of creation, and of the story of Noah and his flood. Years ago, the creationists sought to control how people think by seeking to have laws passed banning the teaching of evolution in the public schools, badgering the teachers who did, and pressuring the publishers to eliminate evolution from the textbooks. They still engage in the second and third of these activities and with considerable success. Our system of laws, however, has prevented the banning of evolution in the schools so, in the last two decades, the creationists' thrust has been to push for equal time for their sectarian views if and when the biological theory of evolution is taught. Again, realizing that they cannot be in the position of introducing religion into the classroom, they have started to call their system of beliefs creation "science," and this is done with a straight face. Yet the one thing a science cannot do is base its propositions on the supernatural—as implied by "creation." An equivalent conceptual scheme would be the Theory of Living Death.

When one inquires about the nature of creation "science," it is found to consist of two main divisions: first, notions about origins and changes that derive solely from what is said in the Book of Genesis in the Bible—and this is the lesser part—second, attempts to find flaws in the biological theory of evolution. I suppose the logic here is to pretend that there can be only two ways of looking at origins and evolution and, if one way can be obfuscated or challenged successfully, then the field is left to the other without any need for its verification.

But a field of knowledge cannot be based so heavily on this "you're wrong so I must be right" technique. Modern chemistry is far more than "what's wrong with the Theory of Alchemy" and modern astronomy is far more than "what's wrong with the Theory of Astrology."

It is exceedingly difficult to understand what it is that the creationists want us to teach as creation "science." There are no scientific data that would lead to hypotheses about the origins and diversity of living creatures that would resemble in any ways what is said in Genesis. That is, had there been no Book of Genesis, there would be no creation "scientists"—at least of the sorts associated with fundamentalist Christian religion. And, so far as I know, no other religion has sought to have its creation myths treated as science.

Thus, if due consideration is to be given to creation "science" *as science,* the science teacher and the textbooks would have to say that nearly all religions have their stories of creation, that these stories vary tremendously, and that there are no scientific data to support any one of them. The teachers and the textbooks would also have to explain the divergent procedures of science and creationism. In science the investigator seeks data in order to determine what can be believed; in creation "science," one seeks data to buttress what has already been accepted as the truth. Science seeks to understand; creationism seeks to confirm beliefs.

When the teachers and the textbooks turn to the main activities of the creationists, namely, that the theory of evolution is incomplete, all must agree. There is no disagreement, however, on the most essential fact: professional scientists who have studied the evidence accept the fact that, beyond a reasonable doubt, organic evolution has occurred. The uncertainties are in the details of evolutionary history, especially among the soft-bodied invertebrates among the animals, and the relative importance of various patterns of evolutionary change. The point to be emphasized is not our ignorance but that the procedures of science permit a steady accretion of knowledge and understanding. We know today what was a mystery yesterday and tomorrow will surely unravel some of the puzzles of our times. For example, before the 20th century, it was believed that there could never be an accurate dating of the geological past. The many varied techniques used in the 19th century for dating the geological past led to wildly different estimates and constant bickering among the proponents of one or another of the different procedures. Furthermore, it seemed improbable that there could ever be an accurate clock of the past. The discovery of radioactivity put aside that gloomy prognosis.

Today we despair of ever knowing about the evolutionary history of the soft-bodied invertebrates that left so inadequate a fossil record. Nevertheless, it is not beyond the realm of possibility that there resides in the structure of their organic molecules information that will give us a strongly probable hypothesis of their phylogenies.

Creation "science" has no procedures for adding new knowledge. This is of little moment to a creationist, however, since the final truth about the origins of the cosmos, the world, and its living creatures has been available for more than three thousand years in the writings of the Hebrew scribes of ancient Israel. The blind acceptance of that corpus is an end to inquiry.

Is this the message the creationists ask us to transmit to our students? Not likely.

Let us now turn to a second incompatibility of science and creationism: the contrasting manner in which scientists and creationists accept scientific information. In science, a hypothesis is posed and then data are sought that will support or refute deductions made from the hypothesis. The data come from observations and experiments and, to the extent that they can be verified, they allow us to accept the hypothesis as representing the state of the art. If the hypothesis deals with an important puzzle of science, there will be continuing attempts to check its accuracy as better tools and better procedures become available. Thus, our understanding of the important problems of science is characterized, not by stasis, but by evermore satisfactory explanations. Consider the history of the substance of inheritance. The dominant ideas in the late 19th Century concerned gemmules, idioplasm, and other names applied to phenomena that could not be seen or studied in a rigorous manner. Then came the beads-on-a-string genes of the early Mendelians that sufficed to provide a satisfactory explanation for the transmission of hereditary substance from one generation to another. In the early 1950's the mystery of the double helix was unwound, with the consequence that rapid progress became possible in understanding gene action and the biochemical aspects of evolution. Today the substance of inheritance is seen as evermore complex and, within that complexity, emerges an ever-increasing understanding of the basic problems of cell biology, inheritance, embryology, evolution, phylogeny, agriculture, and medicine.

In science, the new facts are used to enhance our understanding. In creationism, new facts are ignored, distorted, or ridiculed in an effort to support a never-changing, non-scientific point of view about creation that scientific data utterly refute.

This attitude of the creationists has proved to be exceedingly difficult for the scientists to deal with—especially in debates. In the first place, a public debate is an extraordinary way to decide scientific matters. Nevertheless, there are these confrontations, in pulpit and podium, between creationists and scientists. If one has a poorly informed audience, the result is usually a disaster for the scientist. How does one deal with statements that are obviously false? When a creationist claims that the Second Law of Thermodynamics proves that evolution is impossible, what is one to say? One might try to give the scientific reasons why the Second Law has nothing to do with the matter and, for this to be done adequately, a very large amount of time would be required—almost a course in science to do it right. About the only short answer that would be understood by an unsophisticated audience would be to say that those professional scientists who are familiar with the physical laws of nature and of evolution see no conflict between organic evolution and the Second Law.

I am afraid that scientists assume that facts do not disappear even if they are ignored and that constant repetition of an error does not make it a truth. In these strange times, however, the creationists are showing quite clearly that we are wrong in such assumptions. In the political arena, which is where the creationists perform, emotion can subjugate reason, and patent nonsense can be paraded as scientific fact or process. Not too surprising, I suppose, when some strange minds question that the Holocaust ever happened.

A century ago there *was* a seemingly irreconcilable conflict between the hard data of physics and the then somewhat fuzzy notions of evolutionary biology. Darwin needed eons for his process to work—about half a billion years to the beginnings of Cambrian time, for example. The British physicist, Lord Kelvin, stated with all the certainty of the hard sciences, that the entire age of the earth could be no more than 200 million years and, hence, the time to the beginning of the Cambrian very much less. He based his statement on the assumed rate of cooling of the earth and, not knowing of the heat contribution of radioactive decay, erred badly. It turned out that Mr. Darwin's guess was to be closer to the value accepted today.

But the point here is that the ever-checking, ever-repeating, and ever-testing procedures of science led eventually to agreement. Better physical methods of geological dating showed that there was plenty of time for the slow processes of evolutionary change.

The reconciling of different points of view, however, can only come to open minds. We have the ridiculous situation today where statements are made by creationists, these statements are shown to be scientifically incorrect or inadequate, but no matter, the creationists keep on repeating them. This attitude is so at variance with the accepted procedures of science that it is difficult to understand how the creationists ever confused themselves with scientists—especially when they publicly and devoutly proclaim their closed minds. Thus, the members of the Creation Research Society, the guiding force of creationism in America,

> . . . subscribe to the following statements of belief.
>
> 1. The Bible is the written Word of God, and because it is inspired throughout, all its assertions are historically and scientifically true in all the original autographs. To the student of nature this means that the account of origins in Genesis is a factual presentation of simple historical truths.
> 2. All basic types of living things, including man, were made by direct acts of God during the Creation Week described in Genesis. Whatever biological changes have occurred since Creation Week have accomplished only changes within the original created kinds.
> 3. The great Flood described in Genesis, commonly referred to as the Noachian Flood, was a historic event worldwide in its extent and effect. . . .[1]

So we have here a group of people who wish to call themselves scientists yet are committed to ignore, distort, or attempt to refute the data of science that cannot be adjusted to their supernatural point of view. Their refutations do not involve the presentation of new scientific data, which is the manner in which conflicts are resolved in science, but the device of legal or political argument, where the goal is to win the argument or win the election. No one expects a lawyer to be a seeker of truth or justice if such would be to the detriment of his client; no one expects the campaign promises of a candidate for high office to be fulfilled—much of what is promised will turn out to be impractical because of the problems "inherited from the previous administration" or "the unexpected downturn in the economy."

This behavior of the creationists, which is repugnant to professional scientists, is very effective in our society where the level of scientific literacy is low and where a scientist or a person confused with one is expected to be objective and truthful. If someone who calls himself a scientist makes a pronouncement, the public is far more likely to give credence to it than to the statements of politicians. When contrasting points of view are expressed by individuals, both claiming to be scientists, the public is thoroughly confused.

We have here a close parallel with the fluoridation controversy and a reconsideration of that sorry affair can add perspective. Social scientists have written a great deal about the extensive debates that revolved around the wisdom of adding fluorides to municipal drinking water. A nice summary is that of Harvey M. Sapolsky.[2]

The role of fluorides in preventing tooth decay was discovered quite by accident. Early in the 20th century a search was begun to determine why some individuals had discolored teeth. The condition seemed to have a geographic basis. That is, it was more frequent in the individuals living in some regions than in others. By the 1930's it was known that the discoloration was due to a higher than average amount of fluorides in drinking water. In the course of this research, a most interesting correlation was noted: dentists had a hard time making a living in areas where the discolored teeth were common—those teeth were very, very healthy. Gradually it became apparent that the fluorides were protecting the teeth from decay. Dental caries, therefore, could be considered in part as a deficiency disease to be cured by increasing the intake of fluorides. The most effective, safe, and economical way of doing this was to add minute amounts of fluorides to drinking water. The amount added was somewhat less than would cause the discoloring of the teeth but still enough to greatly reduce dental caries.

Such a program could be of enormous importance. If fluorides were added to those waters not naturally containing the necessary concentration, it was estimated that the dental bills of Americans would be halved. Since, in 1979 the national expenditure for dental care was 13.6 billion dollars, a non-trivial amount of money would be saved— to say nothing of the pain, both physical and emotional, of those with rotting teeth. Since 95 percent of Americans suffer from tooth decay, we are talking about a non-trivial number of people, too. In contrast with the huge costs of delivering some types of health care, adding fluorides to drinking water would be inexpensive—varying from a few cents to about a dollar per person per year.

Public health officials welcomed fluoridation as one of the truly great advances in protecting the public, ranking with pasteurization of milk, purification of drinking water, immunization against disease, and the enrichment of bread. Fluoridation was endorsed by the American Dental Association, the American Medical Association, the U. S. Public Health Service, the National Academy of Sciences, the American Association for the Advancement of Science, and the United Nations.

Despite the documented benefits and the endorsement by individuals and groups in a position to give authoritative opinions, there developed widespread and often virulent public opposition to the fluoridation of municipal water supplies. When voters in a community were given a choice, they rejected fluoridation in the vast majority of elections. In those communities where fluoridation was begun, the decision was usually made by mayors, city managers, or governing bodies.

The opposition to fluoridation was whipped up by a determined group of antifluoridationists who were able to engender enough doubt in the minds of the voters so that most referendums on fluoridation were rejected. There are fascinating parallels between the campaigns of the antifluoridationists and the creationists. Several will be mentioned.

1. Expert vs "expert." The creationists have been able to enlist a few scientists, far from the mainstream of research on the origins and history of life, the earth, and the cosmos, to support their cause. The list includes no scientist who has not already compromised his objectivity by accepting the notion of a supernatural creation. This is not at all surprising: one cannot be a creationist, even a creation "scientist," without believing in creation. And to a poorly informed public, the Institute for Creation Research must sound just as reliable and impressive as the National Academy of Sciences or the American Association for the Advancement of Science. In a similar vein, the antifluoridationists enlisted or began institutions such as the American Academy of Nutrition, the National Health Foundation, and the Association of American Physicians and Surgeons. These organizations said "no" to fluoridation. That being the case, should one heed a "yes" coming from the American Dental Association or the American Medical Association? The confused voter, being unable to distinguish the professional organization from the spurious, could only assume that the experts were in serious disagreement.

2. Distortion and deception. A favorite point of the antifluoridationists was that fluorides are a poison—and a rat poison at that. Clearly no American mother or father would want to give their child rat poison. There is no denying that fluorides are poisonous—even pure oxygen is for that matter—but someone figured out that a person would have to down the equivalent of 50 bathtubs of fluoridated water to obtain a lethal dose. In a similar vein, the creationists distort the situation when they claim that the Second Law of Thermodynamics makes evolution impossible or that no one has even seen one species changing into another. (I should add, parenthetically, that if a biologist ever saw one species changing into another, he should start accepting

the supernatural then and there.) Once again, the citizen observes that the authorities cannot agree among themselves.

3. The enemy. In a surprising number of instances the creationists and the antifluoridationists find that their antagonists are much the same: communists, atheists, humanists, elitists, or the powerful determined to have their way. The creationists are adding to their list of evils associated with evolution. Morris includes materialism, modernism, humanism, socialism, fascism, communism, and satanism. His associate, Gish, finds Darwinism the basis of anarchism, amoralism, racism, totalitarianism, and imperialism. Judge Dean, of Georgia, finds most of the evils of evolution associated with the 16th letter of the alphabet: pills, permissiveness, pregnancies, promiscuity, prophylactics, perversions, pornotherapy, pollution, poisoning, proliferation of crime, and abortions. If this list becomes much longer, it is obvious that the Immoral Minority will become a majority.

The purpose of the antifluoridationist campaign is to sow confusion. A confused voter is likely to opt for the status quo if a proposed course of action is questionable or possibly even dangerous—that rat poison again.

How should one deal with antifluoridationists? The remedy that Sapolsky proposes is this:

> ... for a citizen to perform adequately even in a restricted role in a scientifically advanced society, he must have at least a slight understanding of science in order to judge wisely the abilities of competing leaders to deal with complex issues. The need, then, may be for new educational programs. These programs would not train citizens to be knowledgeable in the various fields of science, but rather equip them to deal with scientific arguments and scientific experts, through understanding the limits and uses of science. This is a long-run goal.[3]

The sort of confusion that the antifluoridationists can spread is of the same sort that the creationists can spread about evolutionary biology. A standard ploy of the creationists is to select an important aspect of evolutionary biology, which requires considerable background information to understand, and then ask what appears to be a simple question. The public can be expected to understand the question but, without a sophisticated grounding in science, is hardly likely to understand the answer. An attempt by a scientist to provide an adequate answer would, more than likely, be perceived as tortuous double talk.

Let me give an example. It has been known since the early years of Drosophila genetics that essentially all new mutations produce individuals that are less fit than the wild type. Noting this fact, many evolutionists of the first part of the 20th century expressed doubt that the sorts of mutations turning up in Morgan's fly bottles had anything to do with evolutionary change. Those sorts of mutational changes could hardly be considered the advantageous variations on which natural selection could work. Genetics was seen as the study of trivial defects and aberrations, suitable to tag the movements of chromosomes throughout the life cycle, but·hardly the substance of evolutionary change.

It took a long time to work our way out of this paradox and just imagine trying to explain the following to a lay audience.

First, if we had only thought clearly about the matter we would have realized that essentially all new mutations *must* be deleterious. Had it been otherwise we would have had a problem! Consider a natural population of Drosophila. Its genes have been and are mutating constantly. Although this applies to only a few loci in each individual, in a large population every locus is at risk. Thus, the various mutated states of the loci would have occurred at some time in the past and, if any one happened to be superior, it would have been selected. The genetic state of the population at any one time, therefore, represents the best that mutation, recombination, and selection can do with the materials at hand and in the environment of the moment and the recent past.

By this stage in your argument with a creationist, the audience would surely suspect you of an elaborate coverup. First, all those brainy scientists assured us that new and beneficial mutations were required for evolution and, when the data proved them wrong, they come up with a cock-and-bull story explaining why we really knew all along that the new mutations must be bad.

Let's be honest: the explanation just given is only a hypothesis, or as the creationists would put it, "just a theory." It is up to the evolutionist to provide data to test the hypothesis. An abundance of data is now available but it is not of a sort that can be easily understood by individuals lacking training in the sciences. Nevertheless, the evolutionist has to try. Having reached the conclusion that it would be nearly impossible to observe a beneficial mutation so long as the environment remained the same, one has to look for situations where a population confronts a new environment with its genotype adapted for the old and quite different environment. If we had the lifespans of some of the patriarchs, we might transport some polar bears to the tropics and see if genes for shorter and darker hair were selected. But, being restricted to three score years and ten, it would be more practical to select organisms with ·a very short generation time. Long ago Theodosius Dobzhansky undertook such a study. He measured the fitness of a population of Drosophila under standard laboratory conditions. Then the flies were x-rayed to induce mutations—lots of them. The viability and fecundity of the flies were greatly reduced. The new mutations *were* bad—as expected. Then he allowed the flies to breed, generation after generation, and found that the various measures of fitness slowly returned to the normal level. The population became as viable and fertile as before x-raying.

We now have many other examples of genetic changes that are of adaptive significance and, to our surprise, they can occur very rapidly. I need only remind you of the many cases where insect pests have developed a genetic-based resistance to pesticides at a rate that is alarming and discouraging to us but vital for the bugs' evolutionary survival. Microorganisms are found to have the same genetic plasticity when we attempt to control the disease-causing ones with antibiotics.

Thus the data support the hypothesis that new mutations are and must be deleterious but that favorable mutations can be observed in experimental situations where we have a chance of detecting them. But let me assure you, attempts to explain this essential bit of evolutionary biology to an audience that probably isn't very sympathetic to you in the first place is unlikely to elicit many "amens."

It would be even harder to explain to a lay audience why scientists, especially biological scientists, regard evolution as the keystone of their science. The days have long passed when the prime question was whether or not biological evolution is true. We know that as well as we can ever know anything in science. The interest in the Theory of Evolution today is not that it is true but because it is so useful. Evolution provides a way of understanding a huge amount of data from all fields of biology: anatomy, embryology, systematics, physiology, genetics, distribution, behavior, and now biochemistry. It allows us to make sense out of an otherwise bewildering mass of seemingly unrelated facts.

The British biologist P. W. Medawar has expressed this notion. I quote:

> The reasons that have led professionals without exception to accept the hypothesis of evolution are in the main too subtle to be grasped by laymen. The reason is that only the evolutionary hypothesis makes sense of the natural order as it is revealed by taxonomy and the animal relationships revealed by the study of comparative anatomy, much as the notion of the roundness of the earth underlies all geodesy and navigation. In biosystematics and comparative zoology the alternative to thinking in evolutionary terms is not to think at all.[4]

Some of us older types, trained when biology was the study of organisms, will remember the grand stories of the relation of jaw bones and ear ossicles, of the derivation of the arteries of higher forms from the aortic arches of the primitive chordates, and of the recapitulation of the pronephros, mesonephros, metanephros of our own excretory system. The incredible intellectual achievements that led to these understandings are what Dobzhansky had in mind when he entitled his famous lecture "Nothing in Biology Makes Sense Except in the Light of Evolution."[5]

That title expresses why evolutionary biology is the basic concept of our field: Things make sense. Remove the concept of evolution and the conceptual framework of the field vanishes. What remains is an uncoordinated and largely meaningless mass of data—the sort of thing we often teach in beginning courses.

Having said that, one must go on and say that all too often we take evolution for granted and transmit it to our students very poorly. Some teachers feel that, with so much that is new, teaching evolutionary biology is like "Beating a Dead Horse." There is nothing dead about evolutionary biology, however, except the fossil lineages that it associates with the living.

So, we take evolutionary biology for granted and, I suppose, in parallel with taking the spherical shape of the earth and the existence of atoms for granted. I doubt that teachers of astronomy go into great detail to convince their students that our world is an oblate spheroid. Should this bit of scientific history be reviewed? I wonder, also, how many students leave first-year chemistry knowing why chemists came to accept the division of matter into atoms. Should they know? In fact, what is it that we want our students to learn about science? Enough to reach useful conclusions in the evolution/creationism debates?

I do not believe that it is possible for a person lacking a scientific background in evolutionary biology to fully understand why the creationists' arguments are usually spurious nonsense. Maybe we have to accept this situation. Of far greater concern is the fact that most American citizens, and indeed many of our leaders, are incapable of using scientific information to reach social decisions. The manner in which we deal with problems of land use, agriculture, medicine and public health, overpopulation, conservation, technology, and pollution suggest that we embrace a Cargo Cult philosophy not too different from that of those natives of the South Seas. Surely time will steer to us a bountiful tomorrow that will erase the stupidities and neglects of today.

The citizens' appalling ignorance of the nature of science—its strengths and limitations, its relation to other systems of thought, the legitimacy of its application to social problems, and the constraints it puts on our behavior—bodes ill for the future. And the more I think about this problem, the more I feel that the fault is mainly ours—we, the teachers in the schools, colleges, and universities of the nation, must accept much of the blame. And among the teachers, I believe that the main culprits are in the universities.

What is it that we are doing that is wrong? Put quite simply, I believe that we are forgetting the main goal of education: to provide information *and understanding,* not just information alone. In addition, a science course—as well as all others—should be concerned with the field as a way of knowing. The three elements—information, understanding, and ways of knowing—are closely related. The study of ways of knowing in science leads to information and, with synthesis, to conceptual schemes. It is conceptual schemes that provide us with understanding. We err seriously when we fail to reveal to students how scientific information is obtained or how this information is incorporated into conceptual schemes. It is conceptual schemes that provide us with understanding. We know enough about learning, or rather forgetting, to realize that uncoordinated facts are forgotten swiftly. The conceptual schemes last longer—especially if they continue to be used throughout life for incorporating new facts and insights. Once the scientific way of knowing is understood, a person need never confuse it with other ways of knowing.

The link between information and understanding is tenuous and the mix of the two should vary with the age of the student. For the very young, the ratio of information to understanding must be high but, as the student progresses through the grades, understanding must take on an ever-increasing importance, as does science as a way of knowing. But the facts are easy to teach and remember. Understanding is difficult to transmit and to acquire. Not surprisingly, facts become the easy option in our classrooms, especially in our age of permissiveness with its lowered standards, lower demands, and lower expectations. And fact-laden, concept-poor courses are an inevitable consequence of one of the most insidious and invidious educational innovations—the objective examination. It is easy to ask a question that requires only a simple answer that can be given by a word, a plus or minus sign, or encircling a letter from A to D. So easy for the student to answer and so easy for us to correct. Testing for understanding is, in contrast, difficult for both student and teacher.

The type of examination, whether we like it or not, sets the intellectual tone for our courses. Students learn what we want them to learn and, since this seems to be facts, that is what they learn. And there is considerable evidence that many of our students are not acquiring even enough facts and basic skills to allow them to play useful and satisfying roles in society. Seemingly, education is not regarded as important as it once was.

The fruits of our labors, the students, are turning sour in the opinion of many serious observers of American education. There is no point in elaborating the details—they are well known to all of us. The bottom line is that the educational establishment is not graduating an effective proportion of young people who can maintain our complex civilization at home nor keep America competitive in ideas and technology in the world at large. It is not unusual for high schoool graduates to be functionally illiterate—not even able to fill out a job application. And we note with concern that the genius of American discovery is more likely to enhance the GNP of Japan than our own: electronics, photographic equipment, and automobiles come to mind.

Our problems with the creationists are just one example of the inadequate job we do in educating the young. The main point I wish to make is that creationism is not the disease, merely one of its symptoms. There is no escaping the need for us to revamp the educational system in America. We lost our nerve and standards in the '60's and '70's and it becomes imperative that we regain them in the '80's. The task will be as difficult as it is important. Just think of what words come to mind when we mention schools: declining SAT scores, violence, drugs, vandalism, sex, racial problems, graduates who cannot read, write, or figure, uneducated teachers, exhausted budgets, busing, teen-age pregnancy. From time to time we do read about some splendid school where good teaching allows good learning, but these rare instances tell us only what is possible, not what is probable.

It is not too difficult to describe what effective education, especially at the university level, means. The needs of society are for a constant stream of trained minds; of individuals who are humane, versed in the skills of communication, knowledgeable of international, national, and local affairs, sensitive and appreciative of the diversity of mankind, skillful in balancing trade-offs before reaching a decision, appreciating that man is a social creature and that one's ethics must be based on the fact; who can serve as models for others; who expect and respect quality; who are disciplined and effective persons; who recognize that the preservation of society requires a constant attempt to improve what exists and to respond to what is new; that our civilization must reach a balance with the living and non-living world so that we do not destroy what is necessary and irreplaceable; to balance society's good with the individual good; and to hold a firm belief that it is important to work for a better future.

And in the last decades of the 20th century, it is more necessary than ever to come to grips with mankind's most powerful device for solving problems—science. Science controlled by the few could make slaves of the many but, harnessed as our handmaiden, it can allow us to reach the lofty humane goals of our dreams. We must never confuse humanism with science: one is all inclusive, the other derivative. Our goals should never be dictated by science. Human goals must be selected by human beings acting as and for human beings. But once the humanistic goals have been selected, scientific procedures may prove to be a powerful mechanism for reaching those goals. A desire to feed the hungry and heal the sick are humanistic goals, not dictated by any deductions from scientific principles. Nevertheless, the achievement of these goals may become possible by the application of scientific principles to agriculture and medicine.

Thus, it becomes evermore important to understand what is science and what is not. Somehow we have failed to let our students in on that secret. We find as a consequence, that we have a large and effective group of creationists who seek to scuttle the basic concept of the science of biology—the science that is essential for medicine, agriculture, and life itself; a huge majority of citizens who, in "fairness," opt for presenting as equals the "science" of creationism and the science of evolutionary biology; and a president who is so poorly informed that he believes that scientists are questioning that evolution ever occurred. It is hard to think of a more terrible indictment of the way we have taught science. But, to be fair, I should add that there is no overwhelming evidence that we have even tried. Nevertheless, we must accept the fact that, today, most Americans do not seem to know what is science and what is not.

Maybe we should try to tell them. If we succeed, creationism will become a minor problem of antiquarian interest.

REFERENCES

1. The quotation is from a leaflet, "Creation Research Society."

2. Harvey M. Sapolsky, *Science* 162 (October 25, 1968): 427–33.

3. Sapolsky, 1968, p. 433.

4. P. Medawar, *The New York Review*, February 19, 1981, p. 35.

5. Theodosius Dobzhansky, "Nothing in Biology Makes Sense Except in the Light of Evolution." Included in this volume; see following.

Nothing in Biology Makes Sense Except in the Light of Evolution

by Theodosius Dobzhansky

Seen in the light of evolution, biology is, perhaps, intellectually the most satisfying and inspiring science. Without that light it becomes a pile of sundry facts—some of them interesting or curious but making no meaningful picture as a whole.

As recently as 1966, sheik Abd el Aziz bin Baz asked the king of Saudi Arabia to suppress a heresy that was spreading in his land. Wrote the sheik:

"The Holy Koran, the Prophet's teachings, the majority of Islamic scientists, and the actual facts all prove that the sun is running in its orbit... and that the earth is fixed and stable, spread out by God for his mankind.... Anyone who professed otherwise would utter a charge of falsehood toward God, the Koran, and the Prophet."

The good sheik evidently holds the Copernican theory to be a "mere theory," not a "fact." In this he is technically correct. A theory can be verified by a mass of facts, but it becomes a proven theory, not a fact. The sheik was perhaps unaware that the Space Age had begun before he asked the king to suppress the Copernican heresy. The sphericity of the earth had been seen by astronauts, and even by many earth-bound people on their television screens. Perhaps the sheik could retort that those who venture beyond the confines of God's earth suffer hallucinations, and that the earth is really flat.

Parts of the Copernican world model, such as the contention that the earth rotates around the sun, and not vice versa, have not been verified by direct observations even to the extent the sphericity of the earth has been. Yet scientists accept the model as an accurate representation of reality. Why? Because it makes sense of a multitude of facts which are otherwise meaningless or extravagant. To nonspecialists most of these facts are unfamiliar. Why then do we accept the "mere theory" that the earth is a sphere revolving around a spherical sun? Are we simply submitting to authority? Not quite: we know that those who took time to study the evidence found it convincing.

The good sheik is probably ignorant of the evidence. Even more likely, he is so hopelessly biased that no amount of evidence would impress him. Anyway, it would be sheer waste of time to attempt to convince him. The Koran and the Bible do not contradict Copernicus, nor does Copernicus contradict them. It is ludicrous to mistake the Bible and the Koran for primers of natural science. They treat of matters even more important: the meaning of man and his relations to God. They are written in poetic symbols that were understandable to people of the age when they were written, as well as to peoples of all other ages. The king of Arabia did not comply with the sheik's demand. He knew that some people fear enlightenment, because enlightenment threatens their vested interests. Education is not to be used to promote obscurantism.

The earth is not the geometric center of the universe, although it may be its spiritual center. It is a mere speck of dust in cosmic spaces. Contrary to Bishop Ussher's calculations, the world did not appear in approximately its present state in 4004 B.C. The estimates of the age of the universe given by modern cosmologists are still only rough approximations, which are revised (usually upward) as the methods of estimation are refined. Some cosmologists take the universe to be about 10 billion years old; others suppose that it may have existed, and will continue to exist, eternally. The origin of life on earth is dated tentatively between 3 and 5 billion years ago; manlike beings appeared relatively quite recently, between 2 and 4 million years ago. The estimates of the age of the earth, of the duration of the geologic and paleontologic eras, and of the antiquity of man's ancestors are now based mainly on radiometric evidence—the proportions of isotopes of certain chemical elements in rocks suitable for such studies.

Sheik bin Baz and his like refuse to accept the radiometric evidence, because it is a "mere theory." What is the alternative? One can suppose that the Creator saw fit to play deceitful tricks on geologists and biologists. He carefully arranged to have various rocks provided with isotope ratios just right to mislead us into thinking that certain rocks are 2 billion years old, others 2 million, while in fact they are only some 6,000 years old. This kind of pseudo-explanation is not very new. One of the early antievolutionists, P.H. Gosse, published a book entitled *Omphalos* ("the Navel"). The gist of this amazing book is that Adam, though he had no mother, was created with a navel, and that fossils were placed by the Creator where we find them now—a deliberate act on His part, to give the appearance of great antiquity and geologic upheavals. It is easy to see the fatal flaw in all such notions. They are blasphemies, accusing God of absurd deceitfulness. This is as revolting as it is uncalled for.

DIVERSITY OF LIVING BEINGS

The diversity and the unity of life are equally striking and meaningful aspects of the living world. Between 1.5 and 2 million species of animals and plants have been described and studied; the number yet to be described is probably about as great. The diversity of sizes, structures, and ways of life is staggering but fascinating. Here are just a few examples.

The foot-and-mouth disease virus is a sphere 8–12 mμ in diameter. The blue whale reaches 30 m in length and 135 t in weight. The simplest viruses are parasites in cells of other organisms, reduced to barest essentials—minute amounts of DNA or RNA, which subvert the biochemical machinery of the host cells to replicate their genetic information, rather than that of the host.

It is a matter of opinion, or of definition, whether viruses are considered living organisms or peculiar chemical substances. The fact that such differences of opinion can exist is in itself highly significant. It means that the borderline between living and inanimate matter is obliterated. At the opposite end of the simplicity-complexity spectrum you have vertebrate animals, including man. The human brain has some 12 billion neurons; the synapses between the neurons are perhaps a thousand times as numerous.

Some organisms live in a great variety of environments. Man is at the top of the scale in this respect. He is not only a truly cosmopolitan species but, owing to his technologic achievements, can survive for at least a limited time on the surface of the moon and in cosmic spaces. By contrast, some organisms are amazingly specialized. Perhaps the narrowest ecologic niche of all is that of a species of the fungus family Laboulbeniaceae, which grows exclusively on the rear portion of the elytra of the beetle *Aphenops cronei,* which is found only in some limestone caves in southern France. Larvae of the fly *Psilopa petrolei* develop in seepages of crude oil in California oilfields; as far as is known they occur nowhere else. This is the only insect able to live and feed in oil, and its adult can walk on the surface of the oil only as long as no body part other than the tarsi are in contact with the oil. Larvae of the fly *Drosophila carcinophila* develop only in the nephric grooves beneath the flaps of the third maxilliped of the land crab *Geocarcinus ruricola,* which is restricted to certain islands in the Caribbean.

Is there an explanation, to make intelligible to reason this colossal diversity of living beings? Whence came these extraordinary, seemingly whimsical and superfluous creatures, like the fungus *Laboulbenia,* the beetle *Aphenops cronei,* the flies *Psilopa petrolei* and *Drosophila carcinophila,* and many, many more apparent biologic curiosities? The only explanation that makes sense is that the organic diversity has evolved in response to the diversity of environment on the planet earth. No single species, however perfect and however versatile, could exploit all the opportunities for living. Every one of the millions of species has its own way of living and of getting sustenance from the environment. There are doubtless many other possible ways of living as yet unexploited by any existing species; but one thing is clear: with less organic diversity, some opportunities for living would remain unexploited. The evolutionary process tends to fill up the available ecologic niches. It does not do so consciously or deliberately; the relations between evolution and the environment are more subtle and more interesting than that. The environment does not impose evolutionary changes on its inhabitants, as postulated by the now abandoned neo-Lamarckian theories. The best way to envisage the situation is as follows: the environment presents challenges to living species, to which the latter may respond by adaptive genetic changes.

An unoccupied ecologic niche, an unexploited opportunity for living, is a challenge. So is an environmental change, such as the Ice Age climate giving place to a warmer climate. Natural selection may cause a living species to respond to the challenge by adaptive genetic changes. These changes may enable the species to occupy the formerly empty ecologic niche as a new opportunity for living, or to resist the environmental change if it is unfavorable. But the response may or may not be successful. This depends on many factors, the chief of which is the genetic composition of the responding species at the time the response is called for. Lack of successful response may cause the species to become extinct. The evidence of fossils shows clearly that the eventual end of most evolutionary lines is extinction. Organisms now living are successful descendants of only a minority of the species that lived in the past—and of smaller and smaller minorities the farther back you look. Nevertheless, the number of living species has not dwindled; indeed, it has probably grown with time. All this is understandable in the light of evolution theory; but what a senseless operation it would have been, on God's part, to fabricate a multitude of species ex nihilo and then let most of them die out!

There is, of course, nothing conscious or intentional in the action of natural selection. A biologic species does not say to itself, "Let me try tomorrow (or a million years from now) to grow in a different soil, or use a different food, or subsist on a different body part of a different crab." Only a human being could make such conscious decisions. This is why the species *Homo sapiens* is the apex of evolution. Natural selection is at one and the same time a blind and creative process. Only a creative but blind process could produce, on the one hand, the tremendous biologic success that is the human species and, on the other, forms of adaptedness as narrow and as constraining as those of the overspecialized fungus, beetle, and flies mentioned above.

Antievolutionists fail to understand how natural selection operates. They fancy that all existing species were generated by supernatural fiat a few thousand years ago, pretty much as we find them today. But what is the sense of having as many as 2 or 3 million species living on earth? If natural selection is the main factor that brings evolution about, any number of species is understandable: natural selection does not work according to a foreordained plan, and species are produced not because they are needed for some purpose but simply because there is an environmental opportunity and genetic wherewithal to make them possible. Was the Creator in a jocular mood when he made *Psilopa petrolei* for California oil-fields and species of *Drosophila* to live exclusively on some body-parts of certain land crabs on only certain islands in the Caribbean? The organic diversity becomes, however, reasonable and understandable if the Creator has created the living world not by caprice but by evolution propelled by natural selection. It is wrong to hold creation and evolution as mutually exclusive alternatives. I am a creationist *and* an evolutionist. Evolution is God's, or Nature's, method of Creation. Creation is not an event that happened in 4004 B.C.; it is a process that began some 10 billion years ago and is still under way.

UNITY OF LIFE

The unity of life is no less remarkable than its diversity. Most forms of life are similar in many respects. The universal biologic similarities are particularly striking in the biochemical dimension. From viruses to man, heredity is coded in just two chemically related substances: DNA and RNA. The genetic code is as simple as it is universal. There are only four genetic "letters" in DNA: adenine, guanine, thymine, and cytosine. Uracil replaces thymine in RNA. The entire evolutionary development of the living world has taken place not by invention of new "letters" in the genetic "alphabet" but by elaboration of ever-new combinations of these letters.

Not only is the DNA-RNA genetic code universal, but so is the method of translation of the sequences of the "letters" in DNA-RNA into sequences of amino acids in proteins. The same 20 amino acids compose countless different proteins in all, or at least in most, organisms. Different amino acids are coded by one to six nucleotide triplets in DNA and RNA. And the biochemical universals extend beyond the genetic code and its translation into proteins: striking uniformities prevail in the cellular metabolism of the most diverse living beings. Adenosine triphosphate, biotin, riboflavin, hemes, pyridoxin, vitamins K and B_{12}, and folic acid implement metabolic processes everywhere.

What do these biochemical or biologic universals mean? They suggest that life arose from inanimate matter only once and that all organisms, no matter how diverse in other respects, conserve the basic features of the primordial life. (It is also possible that there were several, or even many, origins of life; if so, the progeny of only one of them has survived and inherited the earth.) But what if there was no evolution, and every one of the millions of species was created by separate fiat? However offensive the notion may be to religious feeling and to reason, the antievolutionists must again accuse the Creator of cheating. They must insist that He deliberately arranged things exactly as if his method of creation was evolution, intentionally to mislead sincere seekers of truth.

The remarkable advances of molecular biology in recent years have made it possible to understand how it is that diverse organisms are constructed from such monotonously similar materials: proteins composed of only 20 kinds of amino acids and coded only by DNA and RNA, each with only four kinds of nucleotides. The method is astonishingly simple. All English words, sentences, chapters, and books are made up of sequences of 26 letters of the alphabet. (They can be represented also by only three signs of the Morse code: dot, dash, and gap.) The meaning of a word or a sentence is defined not so much by what letters it contains as by the sequence of these letters. It is the same with heredity: it is coded by the sequences of the genetic "letters" — the nucleotides—in the DNA. They are translated into the sequences of amino acids in the proteins.

Molecular studies have made possible an approach to exact measurements of degrees of biochemical similarities and differences among organisms. Some kinds of enzymes and other proteins are quasiuniversal, or at any rate widespread, in the living world. They are functionally similar in different living beings, in that they catalyze similar chemical reactions. But when such proteins are isolated and their structures determined chemically, they are often found to contain more or less different sequences of amino acids in different organisms. For example, the so-called alpha chains of hemoglobin have identical sequences of amino acids in man and the chimpanzee, but they differ in a single amino acid (out of 141) in the gorilla. Alpha chains of human hemoglobin differ from cattle hemoglobin in 17 amino acid substitutions, 18 from horse, 20 from donkey, 25 from rabbit, and 71 from fish (carp).

Cytochrome C is an enzyme that plays an important role in the metabolism of aerobic cells. It is found in the most diverse organisms, from man to molds. E. Margoliash, W. M. Fitch, and others have compared the amino acid sequences in cytochrome C in different branches of the living world. Most significant similarities as well as differences have been brought to light. The cytochrome C of different orders of mammals and birds differ in 2 to 17 amino acids, classes of vertebrates in 7 to 38, and vertebrates and insects in 23 to 41; and animals differ from yeasts and molds in 56 to 72 amino acids. Fitch and Margoliash prefer to express their findings in what are called "minimal mutational distances." It has been mentioned above that different amino acids are coded by different triplets of nucleotides in DNA of the genes; this code is now known. Most mutations involve substitutions of single nucleotides somewhere in the DNA chain coding for a given protein. Therefore, one can calculate the minimum number of single mutations needed to change the cytochrome C of one organism into that of another. Minimal mutational distances between human cytochrome C and the cytochrome C of other living beings are as follows:

Monkey	1	Chicken	18
Dog	13	Penguin	18
Horse	17	Turtle	19
Donkey	16	Rattlesnake	20
Pig	13	Fish (tuna)	31
Rabbit	12	Fly	33
Kangaroo	12	Moth	36
Duck	17	Mold	63
Pigeon	16	Yeast	56

It is important to note that amino acid sequences in a given kind of protein vary within a species as well as from species to species. It is evident that the differences among proteins at the levels of species, genus, family, order, class, and phylum are compounded of elements that vary also among individuals within a species. Individual and group differences are only quantitatively, not qualitatively, different. Evidence supporting the above propositions is ample and is growing rapidly. Much work has been done in recent years on individual variations in amino acid sequences of hemoglobins of human blood. More than 100 variants have been detected. Most of them involve substitutions of single amino acids—substitutions that have arisen by genetic mutations in the persons in whom they are discovered or in their ancestors. As expected, some of these mutations are deleterious to their carriers, but others apparently are neutral or even favorable in certain environments. Some mutant hemoglobins have been found only in one person or in one family; others are discovered repeatedly among inhabitants of different parts of the world. I submit that all these remarkable findings make sense in the light of evolution; they are nonsense otherwise.

COMPARATIVE ANATOMY AND EMBRYOLOGY

The biochemical universals are the most impressive and the most recently discovered, but certainly they are not the only vestiges of creation by means of evolution. Comparative anatomy and embryology proclaim the evolutionary origins of the present inhabitants of the world. In 1555 Pierre Belon established the presence of homologous bones in the superficially very different skeletons of man and bird. Later anatomists traced the homologies in the skeletons, as well as in other organs, of all vertebrates. Homologies are also traceable in the external skeletons of arthropods as seemingly unlike as a lobster, a fly, and a butterfly. Examples of homologies can be multiplied indefinitely.

Embryos of apparently quite diverse animals often exhibit striking similarities. A century ago these similarities led some biologists (notably the German zoologist Ernst Haeckel) to be carried by their enthusiasm so far as to interpret the embryonic similarities as meaning that the embryo repeats in its development the evolutionary history of its species: it was said to pass through stages in which it resembles its remote ancestors. In other words, early-day biologists supposed that by studying embryonic development one can, as it were, read off the stages through which the evolutionary development had passed. This so-called biogenetic law is no longer credited in its original form. And yet embryonic similarities are undeniably impressive and significant.

Probably everybody knows the sedentary barnacles which seem to have no similarity to free-swimming crustaceans, such as the copepods. How remarkable that barnacles pass through a free-swimming larval stage, the nauplius! At that stage of its development a barnacle and a *Cyclops* look unmistakably similar. They are evidently relatives. The presence of gill slits in human embryos and in embryos of other terrestrial vertebrates is another famous example. Of course, at no stage of its development is a human embryo a fish, nor does it ever have functioning gills. But why should it have unmistakable gill slits unless its remote ancestors did respire with the aid of gills? Is the Creator again playing practical jokes?

ADAPTIVE RADIATION: HAWAII'S FLIES

There are about 2,000 species of drosophilid flies in the world as a whole. About a quarter of them occur in Hawaii, although the total area of the archipelago is only about that of the state of New Jersey. All but 17 of the species in Hawaii are endemic (found nowhere else). Furthermore, a great majority of the Hawaiian endemics do not occur throughout the archipelago: they are restricted to single islands or even to a part of an island. What is the explanation of this extraordinary proliferation of drosophilid species in so small a territory? Recent work of H. L. Carson, H. T. Spieth, D. E. Hardy, and others makes the situation understandable.

The Hawaiian islands are of volcanic origin; they were never parts of any continent. Their ages are between 5.6 and 0.7 million years. Before man came their inhabitants were descendants of immigrants that had been transported across the ocean by air currents and other accidental means. A single drosophilid species, which arrived in Hawaii first, before there were numerous competitors, faced the challenge of an abundance of many unoccupied ecologic niches. Its descendants responded to this challenge by evolutionary adaptive radiation, the products of which are the remarkable Hawaiian drosophilids of today. To forestall a possible misunderstanding, let it be made clear that the Hawaiian endemics are by no means so similar to each other that they could be mistaken for variants of the same species; if anything, they are more diversified than are drosophilids elsewhere. The largest and the smallest drosophilid species are both Hawaiian. They exhibit an astonishing variety of behavior patterns. Some of them have become adapted to ways of life quite extraordinary for a drosophilid fly such as being parasites in egg cocoons of spiders.

Oceanic islands other than Hawaii, scattered over the wide Pacific Ocean, are not conspicuously rich in endemic species of drosophilids. The most probable explanation of this fact is that these other islands were colonized by drosophilids after most ecologic niches had already been filled by earlier arrivals. This surely is a hypothesis, but it is a reasonable one. Antievolutionists might perhaps suggest an alternative hypothesis: in a fit of absentmindedness, the Creator went on manufacturing more and more drosophilid species for Hawaii, until there was an extravagant surfeit of them in this archipelago. I leave it to you to decide which hypothesis makes sense.

STRENGTH AND ACCEPTANCE OF THE THEORY

Seen in the light of evolution, biology is, perhaps, intellectually the most satisfying and inspiring science. Without that light it becomes a pile of sundry facts—some of them interesting or curious but making no meaningful picture as a whole.

This is not to imply that we know everything that can and should be known about biology and about evolution. Any competent biologist is aware of a multitude of problems yet unresolved and of questions yet unanswered. After all, biologic research shows no sign of approaching completion; quite the opposite is true. Disagreements and clashes of opinion are rife among biologists, as they should be in a living and growing science. Antievolutionists mistake, or pretend to mistake, these disagreements as indications of dubiousness of the entire doctrine of evolution. Their favorite sport is stringing together quotations, carefully and sometimes expertly taken out of context, to show that nothing is really established or agreed upon among evolutionists. Some of my colleagues and myself have been amused and amazed to read ourselves quoted in a way showing that we are really antievolutionists under the skin.

Let me try to make crystal clear what is established beyond reasonable doubt, and what needs further study, about evolution. Evolution as a process that has always gone on in the history of the earth can be doubted only by those who are ignorant of the evidence or are resistant to evidence, owing to emotional blocks or to plain bigotry. By contrast, the mechanisms that bring evolution about certainly need study and clarification. There are no alternatives to evolution as history that can withstand critical examination. Yet we are constantly learning new and important facts about evolutionary mechanisms.

It is remarkable that more than a century ago Darwin was able to discern so much about evolution without having available to him the key facts discovered since. The development of genetics after 1900—especially of molecular genetics, in the last two decades—has provided information essential to the understanding of evolutionary mechanisms. But much is in doubt and much remains to be learned. This is heartening and inspiring for any scientist worth his salt. Imagine that everything is completely known and that science has nothing more to discover: what a nightmare!

Does the evolutionary doctrine clash with religious faith? It does not. It is a blunder to mistake the Holy Scriptures for elementary textbooks of astronomy, geology, biology, and anthropology. Only if symbols are construed to mean what they are not intended to mean can there arise imaginary, insoluble conflicts. As pointed out above, the blunder leads to blasphemy: the Creator is accused of systematic deceitfulness.

One of the great thinkers of our age, Pierre Teilhard de Chardin, wrote the following: "Is evolution a theory, a system, or a hypothesis? It is much more—it is a general postulate to which all theories, all hypotheses, all systems must henceforward bow and which they must satisfy in order to be thinkable and true. Evolution is a light which illuminates all facts, a trajectory which all lines of thought must follow —this is what evolution is." Of course, some scientists, as well as some philosophers and theologians, disagree with some parts of Teilhard's teachings; the acceptance of his world view falls short of universal. But there is no doubt at all that Teilhard was a truly and deeply religious man and that Christianity was the cornerstone of his world view. Moreover, in his world view science and faith were not segregated in watertight compartments, as they are with so many people. They were harmoniously filling parts of his world view. Teilhard was a creationist, but one who understood that the Creation is realized in this world by means of evolution.

Evolution: Fact? Theory? ...
or Just a Theory?

prepared by Malcolm Jay Kottler

[Evolution is] theory only. In recent years it has been challenged in the world of science. If evolutionary theory is going to be taught in the schools, then I would think that also the biblical theory of creation, which is not a theory but the biblical story of creation, should also be taught.

—*Ronald Reagan,* on the campaign trail in Dallas, 1980

In the creation/evolution controversy, the terms "fact" and "theory" —as well as the phrases "just a theory" and "only a theory"—are used in a wide variety of ways. This has stimulated discussion among scientists who accept evolution, but disagree about the proper use and meaning of these words. Some of their thoughts are given below.

"Scientists are perennially aware that it is best not to trust theory until it is confirmed by evidence. It is equally true, as Eddington pointed out, that it is best not to put too much faith in facts, until they have been confirmed by theory."[1]
—*Robert MacArthur*

"... there is nothing more permanent than a theory, and there is nothing more temporary than a fact."[2]
—*Joseph S. Fruton*

"You want to know what I believe, and how I came to believe it. I probably do not believe anything as firmly as you do. For example, I am prepared to admit the possibility that I am nothing but a biologically and socially convenient fiction, that some hundreds of millions of Buddhists, in fact, are correct in referring to 'the illusion of personal identity.' In any case, our words and other symbols are so inadequate to reality that it seems likely that any statement which can be made on any subject contains at least an element of falsehood unless, perhaps, it is a purely logical statement. Certainly our ordinary ideas about space, time, matter, and so on prove meaningless when pressed too far. But, for all that, there are statements which are true enough for practical purposes. 'My wife is in the next room,' 'I own a motor-car,' and so on, the sort of statements on which an intelligent jury forms its opinion. I believe very strongly in the truth of a vast number of statements of this kind. Others, such as 'Cerdic was a leader of a Saxon invasion of southern England,' seem to me highly probable, but not certain. . . . Some propositions about the remote past—for example, the evolution of man from animal ancestors—seem to me rather more probable than the existence of Cerdic, but less so than that of Queen Anne."[3]   —*J. B. S. Haldane*

"To present evolutionary thought effectively biologists must recognize precisely the structure of scientific knowledge as it derives from scientific inquiry. We must put a stop to false claims such as 'evolution is a fact' or that science can 'prove' this or that explanation. As one who subscribes to Dobzhansky's dictum 'that nothing in biology makes sense except in the light of evolution,' and as a teacher of and researcher in evolutionary biology, I yield to no one in my respect for evolutionary thought and science in general. But that knowledge must be handled credibly and responsibly. Therefore, regarding the claim that evolution is a fact we should note the following points. In Darwinian terms, evolution refers to changes accumulated by natural selection in living things, especially regarding species formation. Furthermore, this whole process is *rarely* observed directly; for that reason alone, it is hardly factual. It is a reconstruction from facts. More significantly, evolution is often confused with natural selection. How often do we hear of 'Darwin's theory of evolution' or simply 'the theory of evolution'? Darwin's theory is that of natural selection; the operation of natural selection causes evolution. A cause and its effect can never be the same thing.

Even if we overlook the foregoing source of confusion and accept that evolution is unfortunately used in two ways—i.e., synonymous with natural selection (a causal theory) and with accumulated evolutionary change (the consequence of the theory)—we are still nowhere near justifying the statement that 'evolution is a fact.' The facts in the study of evolution come from the observed variations and differences among organisms. Those facts are to be understood as the result of changes preserved by natural selection. There has occurred here a confusion between the problem—the facts of organismic diversity—and the explanation of the problem—the theory of evolutionary change through natural selection. Facts are descriptive statements; theories are explanatory ones; they can never be equated."[4]   *—E. D. Hanson*

"Creationists have capitalized on scientific disputes among biologists on the details of the evolutionary process by pretending that serious students of the subject are themselves in doubt about evolution. Evolutionary study is a living science; as such it is rich with controversy about particular issues of detail and mechanism. Creationists have extracted published statements in these controversies and used them dishonestly to suggest that biologists are in doubt about the fact of organic evolution. Local school boards and students must clearly be impressed that scientists in nine universities seem themselves to be denying evolution.

It is time for students of the evolutionary process, especially those who have been misquoted and used by the creationists, to state clearly that evolution is *fact,* not theory, and that what is at issue within biology are questions of details of the process and the relative importance of different mechanisms of evolution. It is a *fact* that the earth, with liquid water, is more than 3.6 billion years old. It is a *fact* that cellular life has been around for at least half of that period and that organized multicellular life is at least 800 million years old. It is a *fact* that major life forms now on earth were not at all represented in the past. There were no birds or mammals 250 million years ago. It is a *fact* that major forms of the past are no longer living. There used to be dinosaurs and Pithecanthropus, and there are none now. It is a *fact* that all living forms come from previous living forms. Therefore, all present forms of life arose from ancestral forms that were different. Birds arose from nonbirds and humans from nonhumans. No person who pretends to any understanding of the natural world can deny these facts any more than she or he can deny that the earth is round, rotates on its axis, and revolves around the sun.

The controversies about evolution lie in the realm of the relative importance of various forces in molding evolution."[5]

—R. C. Lewontin

"He [Lewontin] stated that evolution is a '*fact,*' that it is a '*fact*' that the earth formed more than 3.5 billion years ago, that it is a '*fact*'. . . . In science, there must always be a margin of doubt. For the validity of evolution, that doubt margin is certainly small. I would say that the probability that evolution has and is occurring is about 99.9999 . . . 9%, but not 100%. However, if we maintain it is 'true' we are becoming dogmatic, not scientific."[6]   *—J. E. Hendrix*

"Whatever happened to the terms *probability* and *observation?* Are statements of high probability now to be deified by calling them *truths?* Does a set of consistent observations become *fact?* When I teach biology to the college student, the nature of our information mandates that the class and I preserve a healthy skepticism regarding both the broad generalizations and the specific statements of the discipline. Fact and truth are terms we almost never use. There is nothing shameful in describing what we know as having a certain probability, following from observations that have a degree of imprecision. That's the nature of science, including the science of evolution."[7]   —*R. D. Wright*

"Drs. Hendrix and Wright make, between them, three points that need to be addressed. The first concerns under what conditions something is a fact instead of simply a likely inference. I am afraid a double standard is used when evolution is considered. Dr. Hendrix thinks that "the probability that evolution has occurred is 99.999 . . . 9%," but that it is not a fact. Dr. Wright is less precise numerically, but is of approximately the same persuasion. Yet I imagine that both believe it to be a 'fact' that the earth goes around the sun, although no one, not even an astronaut, has actually seen it do so. Moreover, the 'Law of Gravitation' is only an empirical generalization. Evolution is a fact just like all other scientific facts and *intimately related to them.* To deny the fact of evolution and assert, instead, that organisms came into being fully formed in the twinkling of an eye, requires that all physics and chemistry also be thrown out. Drs. Hendrix and Wright can say that all knowledge of the physical world is in doubt if they want to, but I bet they don't think the law of combining properties should be taught as a 'reasonable inference.' "[8]   —*R. C. Lewontin*

"The fact that theories are not subject to absolute and final proof has led to a serious vulgar misapprehension. Theory is contrasted with fact as if the two had no relationship or were antitheses: 'Evolution is *only* a theory, not a fact.' Of course, theories are not facts. They are generalizations about facts and explanations of facts, based on and tested by facts. As such they may be just as certain—merit just as much confidence—as what are popularly termed 'facts.'' Belief that the sun will rise tomorrow is the confident application of a generalization. The theory that life has evolved is founded on much more evidence than supports the generalization that the sun rises every day. In the vernacular, we are justified in calling both 'facts.' ''[9]
—*G. G. Simpson and W. S. Beck*

"*Facts and theories are not separated by a magic line.* There is no magical or profound difference in what one does with these two concepts. Scientists deal in probabilities. Arbitrarily, scientists have chosen the levels of 95% probability and 99% probability as appropriate confidence levels in statistical analyses of their data. They require that the results obtained in their tests are only 5% or 1% likely to have resulted from chance alone. They call this a positive result, even though something remains unknown about the situation that somehow accounts for that last 5% or 1%. Such a result does not mean that the problem is solved. It simply means that one can proceed to the next step in the investigation with some confidence—95% or 99%, to be 'exact.'

The creationists' arguments suggest that a fact is something that, once discovered, is kept forever like a coin or a preserved butterfly. Not so. Nothing is irreversibly factual. Any fact may turn out not to be a fact at all; and in scientific investigation the only useful thing one can do with a fact is to use it to build better or more complete explanations. What researchers do with facts is establish the next line of hypotheses. And if their 'fact' proves vulnerable, they discard it and start over. It is a fact that 100% certainties are obvious only in useless tautologies such as: Hairless men have no hair. It is a fact that life insurance companies make money by operating on probabilities."[10]

—R. D. Alexander

"At this point some people will wish to object, saying that after all evolution is not a *fact*. These people are apt to be the very ones who in some other fields are the most willing to accept without question doctrines that cannot be scientifically proved. I would however ask these objectors in return: what is a fact and how is anything demonstrated to be true? The honest scientist, like the philosopher, will tell you that nothing whatever can be or has been proved with fully 100% certainty, not even that you or I exist, nor any one except himself, since he might be dreaming the whole thing. Thus there is no sharp line between speculation, hypothesis, theory, principle, and fact, but only a difference along a sliding scale, in the degree of probability of the idea. When we say a thing is a fact, then, we only mean that its probability is an extremely high one: so high that we are not bothered by doubt about it and are ready to act accordingly. Now in this use of the term fact, the only proper one, evolution is a fact. For the evidence in favor of it is as voluminous, diverse, and convincing as in the case of any other well established fact of science concerning the existence of things that cannot be directly seen, such as atoms, neutrons, or solar gravitation. . . .

So enormous, ramifying, and consistent has the evidence for evolution become that if anyone could now disprove it, I should have my conception of the orderliness of the universe so shaken as to lead me to doubt even my own existence. If you like, then, I will grant you that in an absolute sense evolution is not a fact, or rather, that it is no more a fact than that you are hearing or reading these words."[11]

—H. J. Muller

"In the American vernacular, 'theory' often means 'imperfect fact'—part of a hierarchy of confidence running downhill from fact to theory to hypothesis to guess. Thus the power of the creationist argument: evolution is 'only' a theory, and intense debate now rages about many aspects of the theory. If evolution is less than a fact, and scientists can't even make up their minds about the theory, then what confidence can we have in it? . . .

Well, evolution *is* a theory. It is also a fact. And facts and theories are different things, not rungs in a hierarchy of increasing certainty. Facts are the world's data. Theories are structures of ideas that explain and interpret facts. Facts do not go away when scientists debate rival theories to explain them. Einstein's theory of gravitation replaced Newton's, but apples did not suspend themselves in mid-air pending the outcome. And human beings evolved from apelike ancestors whether they did so by Darwin's proposed mechanism or by some other, yet to be discovered.

Moreover, 'fact' does not mean 'absolute certainty.' The final proofs of logic and mathematics flow deductively from stated premises and achieve certainty only because they are *not* about the empirical world. Evolutionists make no claim for perpetual truth, though creationists often do (and then attack us for a style of argument that they themselves favor). In science, 'fact' can only mean 'confirmed to such a degree that it would be perverse to withhold provisional assent.' "[12]

—S. J. Gould

"The evolutionists' cause is not going to be helped in the long run by self-contradiction. We cannot go on telling the public, as Harris and several other evolutionists have been doing of late, that evolution is a fact rather than a mere theory—and then in the next breath admonish our colleagues and students that science can never hope to attain certainty about anything and that facts don't exist apart from theories anyway. The two claims are not only in conflict; they are probably both wrong.

To say that evolution is a fact may be politic at present, but it seems wrong in several ways, and its wrongness points up some other peculiarities of Darwinism as a scientific theory. Well-constructed theories, basically those in the physical sciences, consist of a mathematical model—a system of equations, for example—and a set of rules for

translating the math into real-life terms that can be checked out. If the model yields equations that don't check out when translated, then it's wrong. There are parts of evolutionary biology that come close to this ideal, especially the mathematical theories of population biology and genetics. Some of the predictions of these theories can be checked out, and they tell us a good deal about how evolution can occur and under what circumstances. But they don't say anything about whether all organisms have descended from one or a few ancestral species, which is the supposed fact of evolution. That fact is really a very lofty hypothesis. There are compelling reasons for believing it, but it takes a rather thick book to hammer them home. For example, the facts of the fossil record are much better explained by evolution than by any of the alternative theories about the history of living things. Reptiles come before mammals and birds in the sequence of sedimentary rocks, and the oldest mammals and birds are more like reptiles than later mammals and birds are. These facts support the hypothesis of evolution, but only if we make certain assumptions about geology. By the time we have finished marshaling all the evidence that is relevant to the hypothesis of evolution, we have wound up dragging most of modern science into the argument, from quantum mechanics to molecular biology. Calling the resultant edifice a fact is like calling Bach's Mass in B Minor a tune."[13]
—M. Cartmill

"Then he saw it. He brought out the knife and excised the one word that created the entire angering effect of that sentence. The word was 'just.' Why should Quality be *just* what you like? Why should 'what you like' be 'just'? What did 'just' *mean* in this case? When separated out like this for independent examination it became apparent that 'just' in this case really didn't mean a damn thing. It was a purely pejorative term, whose logical contribution to the sentence was nil. Now, with that word removed, the sentence became 'Quality is what you like,' and its meaning was entirely changed. It had become an innocuous truism."[14]
—Robert Pirsig

REFERENCES

1. R. MacArthur, "Coexistence of Species" in *Challenging Biological Problems,* ed. by John A. Behnke (New York: Oxford University Press, 1972), p. 253.

2. J. S. Fruton. Source unknown.

3. J. B. S. Haldane, *Science and the Supernatural* (New York: Sheed and Ward, Inc., 1935), pp. 16–17.

4. E. D. Hanson, "Evolution/Creation Debate," *Bioscience* 30 (1980): 5.

5. R. C. Lewontin, "Evolution/Creation Debate: A Time for Truth," *Bioscience* 31 (1981): 559.

6. J. E. Hendrix, "Response to R. C. Lewontin," *Bioscience* 31 (1981): 788.

7. R. D. Wright, "Response to R. C. Lewontin," *Bioscience* 31 (1981): 788.

8. R. C. Lewontin, "Reply to Hendrix and Wright," *Bioscience* 31 (1981): 789.

9. G. G. Simpson and W. S. Beck, *Life: An Introduction to Biology,* 2d ed. (New York: Harcourt, Brace, and World, 1965), p. 16.

10. R. D. Alexander, "Evolution, Creation, and Biology Teaching." Included in this volume.

11. H. J. Muller, "One Hundred Years without Darwin Are Enough," *School Science and Mathematics* 59 (1959): 304–05.

12. S. J. Gould, "Evolution as Fact and Theory," *Discover* (May 1981): 34–35.

13. M. Cartmill, "An Ill-Timed Modesty," *Natural History* (June 1982): 61–62.

14. R. M. Pirsig, *Zen and the Art of Motorcycle Maintenance* (New York: Bantam Books, 1974), pp. 226–27.

Evolution: A System of Theories *

by Ralph W. Lewis

Many of the arguments over falsification of theories, prediction in evolution theories, circular reasoning, and even other things outside of biology like creationism can be greatly clarified if they are based on an understanding of evolution in terms of its structured theories.

The many theories in biological evolution are usually presented in a narrative form that fails to give adequate attention to the structure of the theories discussed and to the relations between them. This paper is a preliminary attempt to describe some aspects of theory structure and some kinds of relations between theories found in the study of evolution. The paper does not attempt to give a complete or well-rounded view of evolution, but I do assume that the patterns discussed and the postulate lists given in the Appendix are an initial step in the presentation of the whole of evolution in a more concise, complete, and logical order.

Knowledge in most disciplines is grouped into areas of thought called theories that are built on the pattern of Euclidean geometry. When theories are partially formalized to show some of this pattern explicitly, the intra- and interworkings of theories become more clearly visible, and the total structure of the discipline becomes more evident.

When this view is applied to the study of evolution, we find hundreds of theories which have the typical geometric form, albeit the form is often obscured. And we find relations between theories that permit them to be characterized by one or more of these terms: major theory,

*The author thanks Rollin H. Baker, Paul H. Barrett, Alain F. Corcos, Anton E. Lawson, William S. Moore, and a reviewer for their many suggestions and corrections. Reprinted from *Perspectives in Biology and Medicine,* vol. 23 (Summer 1980), by permission of The University of Chicago Press. Copyright © 1980 by The University of Chicago Press. All rights reserved.

subtheory, accessory theory, parallel theory, competing theory, and subsumed theory. In this paper only the first three will be discussed.

The two major theories in evolution as developed by Darwin are the theory of descent with modification and the theory of natural selection. By the use of the gene theory as an accessory theory, modern biologists have refined and modified natural selection theory to produce what is called the synthetic theory.

The postulates of the major theories and several subtheories are given to support the views of the author and to permit the reader to judge them. The views of both author and reader are bound to remain tentative until the postulate lists are corroborated or corrected by the best scholars and until the lines of reasoning used by biologists to construct and test the theories have been studied in greater detail. But even in the present tentative state some readers may agree with me that partial formalization of theories can simplify and clarify our knowledge of evolution.

The theses in this essay were initiated or supported by the views of many scholars. The words of three of them are a good introduction for some of the arguments that follow.

"The progress made in recent decades in the development of unifying concepts has been so great, however, that the presentation of chemistry to students of the present generation can be made in a more simple, straightforward, and logical way than formerly" (Linus Pauling).[1]

"[Ancient Greece] for the first time created the intellectual miracle of a logical system. . . . ,—Euclid's geometry. This marvelous accomplishment of reason gave the human spirit the confidence it needed for its future achievements" (Albert Einstein).[2]

"Our understanding of evolution depends on a combination of clearly formulated theories and wide comparative knowledge" (John Maynard Smith).[3]

The pattern of the Euclidean logical system pervades every theory in every discipline, but only in mathematics and physics is the pattern often made explicit. With some exceptions, the other disciplines expand their theories discursively to such a degree that it is often difficult to identify the basic premises, the postulates, of each theory. And when these are not clearly stated, the reasoning that flows to and from the postulates is cloudy and the subject is needlessly obscure. Only when each discipline makes explicit its Euclidean logical systems, as far as this is practicable, will man be able to teach, learn, use, and enlarge his knowledge with maximum efficiency. Suppes has argued convincingly for the formalization of theories as the means of achieving this end by, among other things, clarifying the concepts, the relations of parts, and the total structure of a discipline; by forcing completeness of thought;

by admitting easier critical examination; and by displaying the common aspects of the intellectual enterprise.[4]

Stimulated by Pauling's view given above as applied to biology and led by a search for the meaning of "biological principles" and by a search for what is general to "general education," I have been outlining undergraduate biological knowledge by starting to partially formalize each theory found therein. Since the study of evolution subsumes most of our knowledge in biology, evolution is the major topic in this work. And since evolution is a combination of theories, a degree of formalization of the theories and a beginning knowledge of the relations between them are essential for a clearer view of the subject. This is especially true for beginners and for nonspecialists in evolution.

MEANING OF THEORY

To view evolution as a system of theories, some agreement is necessary on the meaning of "theory," a term that has different meanings even in the natural sciences. If one goes to the philosophers for help with this term one finds it, but one also finds complications that are unnecessary for present purposes.[5] To shape my understanding of theory, I have gone directly to the works of men who create and use theories. From these works I have collected postulate lists of more than 300 biological theories. While compiling these lists I have made preliminary notes on the facts included in each theory and on the reasoning statements made by the authors. Postulate lists compiled from recent theoretical research and review papers have been sent to the authors of the papers for corrections and comments. The responses to date from 27 authors lead me to think that my view of theory is essentially the view of active scholars in biology.

A theory consists of a set of ideas, a collection of facts, many lines of reasoning, and often some definitions. A single idea is not a theory. Nor is a set of ideas standing alone, although it is sometimes convenient to speak as though it is. A theory is a Euclidean logical system, a hypothetico-deductive system, that includes ideas, facts, and lines of reasoning. Often "model" is used synonymously with "theory" or "tentative theory," but at present it is best to follow those who use "model" to mean a rather tightly structured subtheory.[6] The term "hypothesis" has often been used to mean a tentative theory, but since hypothesis is also used to mean postulate or deduction or supposition, it is best to say "tentative theory" when that is the meaning desired.

Facts play four roles in theories—they may support or fail to support a postulate, they may be explained, they may be predicted, and they may enter into lines of reasoning used to support, to explain, or to predict. The great bulk of a theory, viewed in toto, consists of facts and reasoning, but theory, by its nature and role, unburdens us of this bulk. That theory does this is attested to by the words of Pauling in the quotation above and by these words from Medawar: "As a science advances, particular facts are comprehended within, and therefore in a sense annihilated by, general statements of steadily increasing explanatory power and compass—. . . ."[7] Thus theories are in one sense implosions of knowledge into manageable systems of thought and as such are the most important units for thinking, teaching, and learning.

Postulates (basic premises, fundamental assumptions, hypotheses, axioms) are the statements of the central ideas of a theory. Blanché discusses the shifts in the meaning of "axiom" from the Greeks to the present so one can see why the different names for postulates might be used.[8] Ideally the postulates of a theory are few in number, as simple as possible, and not deducible from parts of the theory. In some theories one or more of these ideals may not hold. The theory of natural selection is a notable example because three of its postulates, according to Wallace's "demonstration,"[9] are logically derived from the others (Appendix, C). Despite this, biologists always include these among the fundamental propositions of the theory.

Postulate lists for theories are often incomplete or imperfect. This is understandable in a newly developing theory where the central ideas are being tested and modified. In established theories, difficulties are also present and will always remain because biological theories are not tight logical systems, and even if they were, mathematicians tell us that the best axiomatic systems have their limitations.[10] The imperfections that may appear in a concisely stated set of postulates are small when compared with the large advantages of clarity and completeness that accrue to both the author and the reader when postulates are explicitly stated and identified.[11]

Many modern authors in biology are explicit in their reasoning steps where these might be novel to the reader, whereas others, especially earlier biologists, are "context-dependent" reasoners,[12] that is, some premises necessary to a line of reasoning are left implicit.[13] In this kind of reasoning, which is common in biology and in the discursive disciplines, it is presumed that both the author and the reader share these premises, so it is unnecessary to include them. To attempt to do so "would be cumbersome and not worth the effort,"[14] and in some cases would be well-nigh impossible because the "reasoning" is wholly or partly intuitive.[15] This shorthand method of context-dependent reasoning works well for those well versed in the subject but creates many difficulties for the novice. When, however, new ideas, or a new grouping of ideas, are presented in a tentative theory or when such a theory is being tested, an explicit statement of the ideas is essential because every reader is a novice. When concisely stated, the new ideas in the postulates can more quickly become a part of the reasoning context. In pedagogy these considerations are especially important because we expect students to do context-dependent thinking, and often we have not made sure that the implicit premises are part of the context in the student's mind.

DARWIN'S TWO MAJOR THEORIES

In the *Origin of Species,* Darwin gives us two major theories of evolution: the kinematic theory of descent with modification and the dynamic theory of natural selection.[16] He names these theories over and over again in the *Origin,* yet for 120 years most biologists have failed to recognize the descent theory explicitly. Rather, they have spoken of it in such terms as "the *story* of evolution"[17] or the "fact of evolution."[18] Now that biology has become overtly a hypothetico-deductive science,[19] we can no longer neglect the realities of our discipline. It is time to stop repeating "evolution is a fact" when in reality it is an unshaken theory of descent with modification.

Some scholars have recognized Darwin's two major theories, but usually they did not point out that the two theories often function as separate theories in guiding the thinking of biologists. Lovejoy spoke of ". . . the theory of organic evolution—as distinct from the hypothesis of natural selection—. . . ."[20] Fisher said: "Natural Selection is not Evolution. Yet, ever since the two words have been in common use, the theory of Natural Selection has been employed as a convenient abbreviation for the Theory of Evolution by means of Natural Selection, ... This has had the unfortunate consequences that the theory of Natural Selection itself has scarcely ever, if ever, received separate consideration."[21] (Today, of course, it is not true that natural selection is neglected, rather the emphasis is reversed.) Haldane's view is similar: "We must therefore carefully distinguish between two quite different doctrines which Darwin popularized, the doctrine of evolution and that of natural selection."[22] Textbooks usually separate the subject matters of Darwin's two theories, but almost never do they state that they are discussing two theories and point out the postulates of each. The textbook by Simpson, Pittendrigh, and Tiffany is an exception. They say: "First, there is a the *theory of evolution* in the strict sense, ... Second, there is the *theory of natural selection.* . . ."[23]

Dobzhansky was fully aware of the two areas in evolution. In the preface to the first edition of *Genetics and the Origin of Species* he said: "The problem of evolution may be approached in two different ways. First, the sequence of the evolutionary events as they have actually taken place. . . . Second, the mechanisms that bring about evolutionary changes. . . ." In a more recent book he called the theory of descent with modification "the classical theory of evolution" and gives four postulates.[24] But since most of his work was on the mechanisms of evolution, it aided the development of the view that the study of evolution is largely limited to the mechanism of evolution.

Thoday's view is close to Darwin's and to the reality of the subject:

> The theory of evolution has two major components, the concept of evolution itself and mechanism of change. . . . The concept of evolution explains the classifiability of organisms, the facts of plant and animal geography, the common behavioral, morphological, embryological, anatomical, physiological, biochemical, cytological and genetical properties of diverse organisms, the facts of microevolution in "nature" and in the laboratory, and the results of plant and animal breeders. It does so by postulating that diversity is the consequence of modification over the generations of differing lines descended from common ancestors. It does not explain adaptations as such but explains the diversity of adaptations.[25]

Thoday goes on to say that the concepts of the mechanisms of evolution seek to explain this diversity. Since the "concept of evolution itself" really consists of a set of ideas, and since these ideas are used in various ways to encompass the different classes of facts listed by Thoday, it is more reasonable to stay with Darwin's work and call this part of evolution the "theory of descent with modification." Although the theory of descent and the theory of natural selection are interlocked in some explanations, most of the time they function as separate theories. And when they do interlock, usually one is the major theory and the other serves as an accessory theory.

Since there are two distinct major theories in evolution, we should be able to list their postulates and to show the range of applicability of each theory. I have compiled the postulates of the two theories from the *Origin.* They are listed in Appendix, A and B with the page numbers of the first edition where they can be found. The table of contents in the *Origin* gives an outline of the classes of facts included in each theory. Of course, today the number of classes of facts has increased, especially in cellular and molecular biology, and the theory of natural selection has been modified to become the synthetic theory, discussed in a later section.

THEORY OF DESCENT WITH MODIFICATION

The postulates of the descent theory (Appendix, A), as with most theories, tell the range of applicability of the theory and its major limitations. This theory applies to living and fossil organisms. It does not include cosmic, inorganic, or cultural evolution. I stress this because in some popular and semipopular literature biological evolution is discussed as though it is part of a general theory of evolution rather than as a separate field of learning. The fields of cosmic, inorganic, and cultural evolution possess their own theories of evolution each with its own postulates and ranges of applicability, and none of these leads to explanations or predictions about the evolution of living or fossil organisms. If authors who discuss evolution as though it were a single cosmic-to-culture theory were better schooled in the structure of theories they would see how the theory of biological evolution is separate from the evolution in other areas of knowledge, and they would differentiate between the scientific ideas of evolution and the metaphysical idea of evolution.

By presenting discussions of the origin of life adjacent to discussions of biological evolution, biologists often convey the notion that the origin of life is part of the theory of evolution. The first postulate of the descent theory says clearly that this is not so. The theory starts with a simple form of life and deals with its descendants. In his *Pencil Outline* written about 1838, Darwin noted: "Extent of my theory— having nothing to do with first origin of life...."[26] Dobzhansky thought it desirable to stress this limitation of biological evolution theory: "... the problem of the origin of life is quite distinct from that of subsequent evolution."[27] The study of the origin of life has produced its own theories that are not part of the theory of biological evolution.

The theory of descent with modifications is a kinematic theory in the way that the plate tectonic theory is kinematic.[28] These kinds of theories deal with the relations between things without considering causes. Darwin was not content with his kinematic theory, which was well developed early in his work,[29] until he had formulated the dynamic theory of natural selection to accompany the descent theory.

Despite the major attention given to the theory of natural selection today, the theory of descent with modification is still very active as a separate theory in guiding the thinking and research of many biologists. For those who work directly on natural selection theory, the synthetic theory, or their subtheories, the descent theory acts as an accessory theory in their thinking. These persons often draw upon one or more of the descent theory postulates directly in their reasoning or indirectly as a part of the context of their reasoning. The descent theory functions independently most of the time in studies of those topics listed in Thoday's quotation above. The authors of modern papers in these areas are usually applying only the theory of descent, but in some papers natural selection theory also comes into play as the authors consider interactions with the environment. The Floridean theory of the origin of the true fungi as discussed by Dennison and Carroll (Appendix, E) is a theory that combines the two.[30]

Ball's paper on the geographic distribution of the freshwater planarians is an example of a paper in which the descent theory functions independently of natural selection theory.[31] The postulates of Ball's theory as I extracted them are in Appendix, F. If you compare these with the postulates of the descent theory, you will see that postulates 2, 3, 5, and 6 from Ball are directly related to one or more of the descent theory postulates and that there is no relation to any of the natural selection postulates. Because of this Ball's theory is a subtheory of descent theory. As a subtheory, it develops a specific, limited part of the general theme of the descent theory. Hundreds of theories found in the divisions of biology noted in the quotation from Thoday above are also subtheories to the descent theory. They accept as fully established the postulates of the descent theory and use them or imply them in their postulates.

Plate tectonic theory is necessary to Ball's theory and to other recent theories of geographic distribution, but this was not always so. Many earlier evolutionary theories of geographic distribution were set forth in the absence of plate tectonics, and the more limited geographic theories do not need plate tectonics. Both the recent and earlier theories are all directly dependent upon the theory of descent; in fact, they are direct outgrowths of that theory, hence I class them as subtheories. Whereas the descent theory is the direct superior of Ball's theory, the plate tectonic theory contributes a different set of ideas. Since this kind of a relationship between theories is not uncommon, it is convenient to speak of the plate tectonic theory, and others that serve in this way, as accessory theories.

THEORY OF NATURAL SELECTION

The postulates of the theory of natural selection as I extracted them from the *Origin,* along with Wallace's[32] treatment of them and with the "informal axioms" from Williams's[33] formalized treatment of the theory, are in Appendix B, C, and D. The first three postulates of Williams's list illustrate, in one way, what is meant by context-dependent reasoning. Biologists never give these three as postulates of this theory despite the fact that they permeate their thinking in the theory. They are part of the context brought to the theory by both the biologist author and his reader so they need not be explicitly stated. But when one tightens the reasoning so that it can be put in symbolic form, as Williams does, this part of the biology context must be made explicit. Whether the tight logic of Williams and of Woodger[34] will play an important role in the growth of biology will remain to be seen. Only one biology research paper to my knowledge has made extensive use of symbolic logic.[35]

Wallace's "demonstration" (Appendix, C) was popularized by Huxley[36] and has appeared in some textbooks. If we accept Wallace's demonstration, and it is hard not to, then the postulates of this theory as given in Appendix, B violate the rule that postulates are independent and not deducible from parts of the theory. Since natural selection theory has withstood many attacks and has been very fruitful, and since biologists generally accepted the postulates in essentially the form given, we must conclude that postulates may deviate greatly from the ideal and still function successfully.

Darwin often used the phrases "the laws of inheritance" and "the principle of inheritance" as a part of his reasoning in the theory of natural selection.[37] Should the postulates of this theory include a statement about inheritance or are the "laws of inheritance" part of the context of the reasoning? I did not include a postulate about inheritance because I was inclined to think of the ideas of inheritance as a part of the context of the theory. Darwin scholars may prove me wrong in this view, as they may object to my omissions of Darwin's ideas on migration and isolation. According to Vorzimmer these last two are properly omitted, but further study of Darwin's works may require their inclusion.[38]

SYNTHETIC THEORY

The synthetic theory, according to Hull,[39] is a synthesis of the classical natural selection theory and the genetical theory of evolution produced by the works of Fisher, Haldane, and Wright. As can be seen by examining the postulate lists in Appendix, B and G, the synthetic theory is a refinement and slight modification of natural selection theory made possible mainly by an application of the gene theory acting as an accessory theory. (Because of this, would it not be more informative to call this theory the genetic theory of evolution?)

The postulates of the synthetic theory as I extracted them are largely from Stebbins[40] and Hamilton.[41] I doubt if experts in the field will agree fully with my list, but in this form it can be readily examined and corrected. By using the gene theory as an accessory theory, the theory of natural selection has been greatly refined. The problems with "variation" that many biologists found in natural selection theory have been much clarified.[42] Darwin was aware that "mere chance . . . might cause one variety to differ in some character from its parents . . ." but he did not emphasize the role of chance.[43] The synthetic theory, on the other hand, gives chance a definite position in the theory, albeit the extent of its effects is much debated. And the synthetic theory includes isolation and migration as important factors in the mechanism of evolution.

SUBTHEORIES OF THE THEORY OF DESCENT WITH MODIFICATION

Modern general theories of classification are subtheories of the descent theory because they are structured with the aid of ideas present in its postulates. The earliest and most concise statement of the assumptions of a general classification theory that I have found is by Bessey.[44] His list of postulates, which he called "dicta," are in Appendix, H along with a set of postulates (Appendix, I) which I extracted from three more recent papers.[45] The authors of these papers may disagree with parts of my list, but they, along with Bessey, would probably agree that their theories are based on the theory of descent, and, because of this, classification theories are subtheories of descent theory. It is obvious that classification theories do not belong with the theory of natural selection because they do not embody a single idea from it.

Most of the hundreds of modern classifications of different taxa are also subtheories of the descent theory. By making a choice of characteristics and by making assumptions about the relative importance of characteristics which are assumed to be derived from a common ancestor, the authors assemble the subtaxa into a classification scheme. Unfortunately, authors do not often list their assumptions, but usually these can be extracted, especially if the classifier is also building a phylogeny. Since phylogeny and classification are often intimately intertwined,[46] a theory of the phylogeny of the Lopezieae,[47] a tribe of the evening primrose family, is a good example (Appendix, J). The postulates for this phylogeny theory illustrate again the characteristics of a subtheory. They include ideas from the descent theory and include specific assumptions applicable only to the range of facts considered.

In addition to the rather tightly knit phylogenies developed by taxonomists, there are many theories about the evolution of different groups of organisms. The postulates of such a theory on the evolution of mammals taken from Dawson[48] are in Appendix, K. As is easily recognized by reading these postulates, this is a subtheory of the descent theory.

The studies of geographic distribution have produced many subtheories of the theory of descent. One of these was discussed in an earlier section. Other subdivisions of biology, those discussed by Darwin in chapters 9 through 13 in the *Origin* and a few additional ones noted in the above quotation from Thoday, contain possibly hundreds of subtheories of the descent theory. How many there are, just in the various kinds of undergraduate biology courses, I have not yet determined. Rapid surveys of textbooks do not help much because authors usually do not identify all the theories they discuss, and often they seem to be unaware that they are discussing theory. Since biology has moved into the explicit hypothetico-deductive era of its development, a new generation of textbook writers can clearly structure the theories in their books and can identify the assumptions that are today implicit in many explanations and interpretations. The explicit structuring of theories in biology will do much to eliminate the "authoritative facts and dogma" decried by Schwab[49] and the "just-so" explanations abhorred by Gould.[50]

SUBTHEORIES OF NATURAL SELECTION THEORY

The theory of sexual selection is the best known subtheory of natural selection because Darwin discussed it in the *Origin* and in the *Descent of Man and Selection in Relation to Sex*. I found difficulties in extracting Darwin's postulates of this theory so I have listed those from Wilson, (Appendix, L).[51] A comparison of these postulates with those of natural selection shows clearly why sexual selection is a subtheory of natural selection. While reading the postulates of sexual selection, we carry into our reading the thoughts about favorable and unfavorable traits drawn from natural selection theory. Only with these thoughts in mind does sexual selection make sense, thus it is a subtheory of natural selection.

There are very many subtheories to the theory of natural selection, such as the theories of convergent, divergent, and parallel evolution; mimicry theories; theories of the origin of certain groups of organisms; and theories of the origin of certain traits. This last group is illustrated by Packard's theory of the origin of air breathing in jawed fish (Appendix, M).[52] Wilson's three theories of the evolution of man clearly belong with this group.[53] In all these theories the authors make assumptions

about characteristics and their interactions with factors in the environment, and these interactions lead to natural selection. These theories,
and others like them, are subtheories to the theory of natural selection
because they use ideas from natural selection theory and apply them to
limited parts of biology.

Ecological theories that reach beyond ecological time into evolutionary time are also subtheories of natural selection. Three of these are
given in the Appendix. Seven of the eight niche theory postulates,
which Levins (Appendix, N) lists as "conclusions . . . common to"
"several different models," embody or imply ideas from natural selection theory.[54] All of the statements from May, which I have called
postulates of the r- and K-selection theory (Appendix, O), embody
natural selection ideas. May speaks of these statements as "the deliberately oversimplified concept of r selection and K selection."[55] Oversim-
plification is a characteristic of any set of postulates, but a characteristic
that is rapidly overcome as the ideas in the postulates are used to
develop the lines of reasoning found in explanations and predictions. A
concise listing of the postulates of a theory does offer the possibility of
misinterpretation, but on the other hand it furnishes an efficient starting
position for those who wish to learn and it provides a useful summary
for those who wish to compare competing theories. An example of the
latter is found in a paper by Moore on narrow hybrid zones.[56] Moore
weighs three natural selection subtheories: the ephemeral-zone theory,
the dynamic-equilibrium theory, and the hybrid-superiority theory. He
decides in favor of the last one whose postulates are in Appendix, P.

SUBTHEORIES OF THE SYNTHETIC THEORY

When subtheories of natural selection theory are developed so that
the traits considered can be assigned a genetic pattern, they become
members of this subgroup. Theories on the evolution of mimicry when
first described were simply subtheories of natural selection, but recent
identification of the genes involved has shifted some of them into the
range of the synthetic theory. The postulates of Fisher's theory of the
evolution of mimicry as taken from Ford[57] are in Appendix, Q. Also,
the postulates of the theory of balanced polymorphism are in Appendix,
R. Since these theories are concerned with genes and natural selection,
they are subtheories of the synthetic theory.

Speciation theories as a group are a mixed bag, as are molecular
evolution theories. Two speciation theories whose postulates are copied
from Eldridge and Gould (Appendix, S and T)[58] are clearly subtheories
of the descent theory, but that from Carson (Appendix, U)[59] clearly

belongs with the synthetic theory. Molecular evolution phylogeny theories can be considered as descent subtheories, but because they deal directly with gene products they are closely related to the synthetic theory. And when the gene products can be shown to have selective significance, as with the globins,[60] they are clearly subtheories of the synthetic theory.

ROLE OF PARTIAL FORMALIZATION AND CLASSIFICATION OF THEORIES

Many of the arguments over falsification of theories,[61] prediction in evolution theories,[62] circular reasoning,[63] and even other things outside of biology like creationism[64] can be greatly clarified if they are based on an understanding of evolution in terms of its structured theories. Most of the testing of evolution consists not of testing its major theories directly but of testing its subtheories, so any evaluation of falsification and prediction must first be done in the context of a particular subtheory. An evaluation of any one of the major theories will depend upon a study of its many subtheories, their fruitfulness in explaining and predicting facts, in enlarging knowledge, and in spawning new subtheories. Anyone who examines evolution in this manner will look upon the arguments over the supposed circularity existing in the postulates of natural selection theory as poorly founded. And the "evidence" used by creationists to attempt to discredit evolution is so small compared with hundreds of successful subtheories and the vast array of evidence supporting them that only ignorance and blocking metaphysical assumptions permit creationists to cling to their view of creation as a biological theory.

Partial formalization of theories in pedagogy is poorly developed in biology and the other discursive disciplines, as judged by perusing undergraduate textbooks. Fortunately, formalization is present in some classrooms where teachers, because of orderly minds and thorough training in their discipline, give explicit intellectual order to their subjects. One can start to assess the degree of formalization present in a textbook or a course by asking a two-part question of the book or the teacher: What are the names of the theories being taught and what are the postulates of each theory? With answers to these questions at hand, one can then begin to see the structure of each theory, the relations between theories, and the structure of the knowledge. From years of teaching in the tradition of textbooks followed by years of teaching partially formalized theories, I can attest to the great advantages of formalization. A set of ideas correctly presented in the appropriate context almost forces students into a pattern of intellectual activity that greatly enhances learning. In part it does this by leading the teacher to organize his materials effectively and to ask significant questions.

The role of formalization in the thinking of creative scholars is unknown to me, but if Suppes's analysis is correct,[65] I do not see how an awareness of formalization could fail to aid them. As I read a wide range of research and review papers in biology, I find that the papers from persons who have earned an outstanding reputation almost always have their material organized in a way that makes it easy to extract their fundamental assumptions and to follow their lines of reasoning. From this I conclude that an awareness of the geometric structure of knowledge exists in work of creative scholars, but the degree to which the scholars give it conscious attention I have not yet learned.

SUMMARY

This paper argues first, for the partial formalization of theories and, second, for the overt recognition of relations between theories. The first permits a more concise and more logical management of theories in pedagogy and in practice and the second permits a more orderly and logical grasp of a body of knowledge that contains a system of theories. To aid the reader in evaluating the author's arguments and conclusions, postulate lists from 21 evolution theories are included in the Appendix.

APPENDIX

The numbers following the postulates of theories (A) and (B) are the pages in the first edition of *On the Origin of Species* on which the essence of these postulates are found. The origin of the other postulate lists are as follows.

Copied Verbatim

(C) A. R. Wallace, *Natural Selection and Tropical Nature* (London: Macmillan, 1981).

(D) M. B. Williams, "Falsifiable Predictions of Evolutionary Theory," *Philosophy of Science* 40 (1973): 518–37.

(E) W. C. Denison and G. C. Carroll, "The Primitive Ascomycete: A New Look at an Old Problem," *Mycologia* 58 (1966): 249–69.

(H) C. E. Bessey, "Phylogenetic Taxonomy of Flowering Plants," *Annals, Missouri Botanical Garden* 2 (1915): 109–64.

(U) H. L. Carlson, "The Genetics of Speciation at the Diploid Level," *American Naturalist* 109 (1975): 83–92.

(S) N. Eldredge and S. J. Gould, "Punctuated Equilibria: An Alterna-
(&) tive to Phyletic Gradualism," in *Models in Paleobiology,* ed. by T. J. M.

(T) Schopf (San Francisco: Freeman, Cooper, 1972), pp. 82–115.

Copied Nearly Verbatim

(F) I. R. Ball, "Nature and Formulation of Biogeographical Hypotheses," *Systematic Zoology* 24 (1975): 407–30.

(N) R. Levins, "Toward an Evolutionary Theory of the Niche," in *Evolution and Environment,* ed. by E. T. Drake (New Haven: Yale University Press, 1968), p. 337.

Listed by Me and Corrected by the Authors of the References

(L) E. O. Wilson, *Sociobiology* (Cambridge, MA: Harvard University Press, 1975).

(O) R. M. May, "The Tropical Rainforest," *Nature* 257 (1975): 737–38.

(P) W. S. Moore, "An Evaluation of Narrow Hybrid Zones in Vertebrates," *Quarterly Review of Biology* 52 (1977): 263–77.

Extracted by Me

(G) G. L. Stebbins, *Process of Organic Evolution,* 2d ed. (Englewood Cliffs, NJ: Prentice-Hall, 1971).

T. H. Hamilton, *Process and Pattern in Evolution* (New York: Macmillan, 1967).

(I) M. Beckner, *The Biological Way of Thought* (New York: Columbia University Press, 1959).

E. Mayr, "Theory of Biological Classification," *Nature* 220 (1968): 545.

S. Lovtrup, *Epigenetics: A Treatise on Theoretical Biology* (London: Wiley, 1974).

(J) R. H. Eyde and J. T. Morgan, "Floral Structure and Evolution in Lopezieae (Onagraceae)," *American Journal of Botany* 60 (1973): 771–87.

(K) T. J. Dawson, "Kangaroos," *Scientific American* 237 (1977): 78.

(M) G. C. Packard, "The Evolution of Air-Breathing in Palezoie Gnathostome Fish," *Evolution* 28 (1974): 320–25.

(Q) E. B. Ford, *Ecological Genetics,* 4th ed. (London: Chapman & Hall, 1975).

(R) is extracted by me from various sources.

Postulate Lists

A. Theory of descent with modification

1. All life evolved from one simple kind of organism or from a few simple kinds. 484.
2. Each species, fossil or living, arose from another species that preceded it in time. 6, 306, 316, 321, 341, 351, 356, 385, 389, 405, 461, 481, 486.
3. Evolutionary changes were gradual and of long duration. 84, 102, 287, 302, 314, 317, 343, 429, 459, 462, 463, 471, 475, 479.
4. Over long periods of time new genera, new families, new orders, new classes, and new phyla arose by a continuation of the kind of evolution that produced new species. 125, 126, 128, 316, 351, 427, 462, 471, 474, 483.
5. Each species originated in a single geographic location. 352, 356, 407, 427, 461, 487.
6. The greater the similarity between two groups of organisms the closer is their relationship and the closer in geologic time is their common ancestral group. 321, 412, 413, 420, 425, 426, 476, 477, 479, 485.
7. Extinction of old forms (species, etc.) is a consequence of the production of new forms or of environmental change. 126, 344, 463, 471, 475.
8. Once a species or other group has become extinct it never reappears. 127, 313, 316, 343, 344, 475.
9. Evolution continues today in generally the same manner as during preceding geologic eras. 409, 480.
10. The geologic record is very incomplete. 342, 345, 464, 475, 487.

B. The theory of natural selection

1. A population of organisms has the tendency and the potential to increase at a geometric rate. 63, 64, 78, 109, 186, 322, 467, 470.
2. In the short run the number of individuals in a population remains fairly constant. 65, 67, 69.
3. The conditions of life are limited. 63, 64, 67, 68, 140, 319, 322.
4. The environments of most organisms have been in constant change throughout geologic time. 81, 107, 108, 126, 201, 314, 356, 382, 462, 468, 476.
5. Only a fraction of the offspring in a population will live to produce offspring. 61, 63, 65, 66.

6. Individuals in a population are not all the same: some have heritable variations (variable traits). 60, 61, 102, 108, 127, 130–170, 459, 466, 474, 479, 481.
7. Life activities ("struggle for existence") determine which traits are favorable or unfavorable by determining the success of the individuals who possess the traits. 53, 61, 62, 63, 79, 102, 109, 127, 459, 467.
8. Individuals having favorable traits (favorable variations) will, on the average, produce more offspring and those with unfavorable traits will produce fewer offspring. 61, 81, 82, 83, 84, 320, 344, 459, 476. ("Natural selection" is the term used to encompass statements 7 and 8.)
9. Natural selection causes the accumulation of new variations and the loss of unfavorable variations to the extent that a new species may arise. 53, 470, 490.

C. Wallace's presentation of natural selection

A Demonstration of the Origin of Species by Natural Selection

PROVED FACTS	NECESSARY CONSEQUENCES (afterwards taken as Proved Facts)
RAPID INCREASE OF ORGANISMS, pp. 23, 142 (*Origin of Species,* p. 75, 5th ed.) TOTAL NUMBER OF INDIVIDUALS STATIONARY, p. 23.	STRUGGLE FOR EXISTENCE, the deaths equalling the births on the average, p. 24 (*Origin of Species,* chap. iii).
STRUGGLE FOR EXISTENCE. HEREDITY WITH VARIATION, or general likeness with individual differences of parents and offsprings, pp. 142, 156, 179 (*Origin of Species,* chaps. i, ii, v).	SURVIVAL OF THE FITTEST, or Natural Selection; meaning, simply, that on the whole those die who are least fitted to maintain their existence (*Origin of Species,* chap. iv).
SURVIVAL OF THE FITTEST. CHANGE OF EXTERNAL CONDITIONS, universal and unceasing.—See Lyell's *Principles of Geology.*	CHANGES OF ORGANIC FORMS, to keep them in harmony with the Changed Conditions; and as the changes of conditions are permanent changes, in the sense of not reverting back to identical previous conditions, the changes of organic forms must be in the same sense permanent, and thus originate SPECIES.

D. Williams's informal axioms of Darwin's theory of evolution

1. No biological entity is a parent of itself.
2. If B_1 is an ancestor of B_2, then B_2 is not an ancestor of B_1.
3. Every Darwinian subclan is a subclan of a clan of some biocosm.
4. There is an upper limit to the number of organisms in any generation of a Darwinian subclan.
5. For each organism there is a positive real number which describes its fitness in its particular environment.
6. Consider a subclan D_1 of D. If D_1 is superior in fitness to the rest of D for sufficiently many generations (. . .) then the proportion of D_1 in D will increase.
7. In every generation m of a Darwinian subclan D which is not on the verge of extinction, there is a subclan D_1 such that D_1 is superior to the rest of D for long enough to insure that D_1 will increase relative to D: and as long as D_1 is not fixed in D it remains sufficiently superior to insure further increases relative to D.

E. Floridean theory of the origin of the fungi

1. The Ascomycetes evolved from autotrophic marine ancestors with many of the morphological and cytological features of modern Rhodophyta.
2. The primitive Ascomycetes evolved as saprophytes on driftwood in oceans and estuaries, possibly as early as the Devonian era, and spread, after the development of airborne ascospores, to dead and dying wood ashore and thence to other saprophytic and parasitic niches.
3. The primitive Ascomycetes were pyrenomycetes with membranaceous to carbonaceous, unilocular perithecia which were not embedded in stromata. They were monecious and heterothallic and had functional spermatia and trichogynes, dicaryotic ascogenous hyphae, deliquescent asci, and two-celled ascospores. Asexual spores were either poorly developed or lacking.
4. The major evolutionary trends with the Ascomycetes incorporate adaptive changes associated with emergence from the sea, together with subsequent specialization to exploit diverse and discontinuous habitats on the land. At least two groups, the ascolocular bitunicates (Loculascomycetes) and the ascohymenial unitunicates, developed mechanisms for discharge of ascospores into air and, in some instances, multilocular stromata. On land, the passive, vulnerable spermatia formed a

bottleneck in the sexual cycle which prompted widespread evolutionary experimentation with the mechanics of plasmogamy, stimulated the development of asexual spores, and led to the development of parasexuality.

5. From the ancestral Rhodophyta several heterotrophic lines emerged to give rise to at least four modern classes of fungi. These are, in addition to the Ascomycetes, the Laboulbeniomycetes, the Basidiomycetes, and the Zygomycetes. Together these groups constitute a division, the Eumycophyta.

F. **Theory of the geographic distribution of the flatworm genera in the family Dugesiidae**

1. The center of origin and dispersal of the Dugesiidae was south of the present-day equator.
2. This group arose in Gondwanaland in what is now Antarctica.
3. By the beginning of the Mesozoic (220 million years B.P.) the early diversification of the Dugesiidae was complete, with a main massing of *Girardia* in the west and *Neppia* and *Sptahula* in the east.
4. The northward dispersal of these elements coincided with the early stages of Gondwanaland breakup, leading to a concentration of *Girardia* in the Americas with outliers across the southern hemisphere to Australasia, and of *Neppia* in Africa with a few in Australasia with connections to South America.
5. After separation was well under way, the *Dugesia gonocephala* group arose in Africa, and, after closure of the Tethys Sea, dispersed northward into Palaearctic Africa and eastward to India, populated entirely from the north, and southeastern Asia.
6. *Schmidtea* arose later in Europe.

G. **Synthetic theory of evolution**

1. Evolution is the change of gene frequencies in the gene pool of a species or a subspecies population.
2. Each species is an isolated pool of genes possessing regional (racial, populational) gene complexes which are connected by gene flow.
3. An individual contains only a portion of the genes in the gene pool of the species to which it belongs, and the portions are different for each individual.
4. The kinds of genes and gene combinations in an individual of a species that reproduces sexually are due to the transmissible

halves of the genomes of the parents, to recombination, and to mutation.

5. An individual with a phenotype that favors the production of more offspring will contribute a larger proportion of genes and gene combinations to its gene pool.
6. Isolation that restricts gene flow between a subpopulation and its parent population is essential if the subpopulation is to evolve into a new species.
7. Changes of gene frequencies come about by natural selection, migration, gene flow, and mutation and other random genetic changes. Natural selection is the most important cause of changes in gene frequency.
8. Evolution of a species may result in a temporal sequence of species without an increase in the number of species (phyletic evolution), in a group of new species (adaptive radiation), or in variation on these two possibilities.
9. Speciation is completed when variations have accumulated in a species subpopulation such that genetic exchanges with the parent population, or with "sister" populations, cannot occur even though the two populations meet.
10. Mutation is the ultimate source of new genes in a gene pool.

H. General theory of plant classification

1. Evolution is not always upward, but often it involves degradation and degeneration.
2. In general, homogeneous structures (with many and similar parts) are lower and heterogeneous structures (with fewer and dissimilar parts) are higher.
3. Evolution does not necessarily involve all organs of the plant equally in any particular period, and one organ may be advancing while another is retrograding.
4. Upward development is sometimes through an increase in complexity and sometimes by a simplification of any organ or a set of organs.
5. Evolution has generally been consistent, and when a particular progression or retrogression has set in it is persisted in to the end of the phylum.
6. In any phylum the holophytic (chlorophyll-green) plants precede the colorless (hysterophytic) plants, and the latter are derived from the former.
7. Plant relationship are up and down the genetic lines, and these must constitute the frame work of phylogenetic taxonomy.

I. **General theory of biological classification**
1. Ideally a biological classification represents the evolutionary development of the taxa considered (i.e., classifications are phylogenetic).
2. Species populations are the basic taxonomic units, the basic taxa.
3. Taxa can be arranged in a phylogenetic hierarchy with species (sometimes subspecies) populations at the base of the hierarchy.
4. Each taxon is assigned to the lowest status to which it can reasonably be assigned.
5. Each taxon is polytypic with respect to a set of characters.
6. Each taxon is monophyletic.
7. The value of a character is determined primarily by the size of the group which exhibits it.
8. The more characteristics shared by two taxa, the more closely are they related and the closer they are to their common ancestor.

J. **Theory of evolution of the Lopezieae**
1. The ancestral Lopezieae were bird-pollinated, woody perennials with regular flowers, two fertile stamens, and no floral tube distal to the ovary.
2. During the evolution of the modern taxa the following changes occurred: abortion of the abaxial stamen, development of an epignyous floral tube, decrease in floral symmetry without conversion to insect pollination, and decrease in floral symmetry with conversion to insect pollination.
3. Tubercles on upper petals and an associated snapping mechanism of the stamens and pistil evolved as an adaptation to fly pollination.

K. **Theory of the evolution of mammals**
1. All mammals descended from a mammal-like reptile that lived late in the Triassic period, more than 200 million years ago.
2. About 180 million years ago, the prototherians (egg-laying mammals) and the therians (marsupials and placental mammals) evolved from the mammal-like reptiles.
3. The initial radiation of the therians early in the Cretaceous period stemmed from tiny insect-eating animals and was based on adaptations to the newly developing flowering plants and their pollinating insects.

4. The evolution that gave rise to the marsupials and placentals took place at this time 130 million years ago.
5. The marsupials developed in North America and were dominant therians in that region for most of the Cretaceous, or until about 70 million years ago.
6. The placentals developed initially in Asia and reached North America late in the Cretaceous.
7. The invasion of North America by the placentals was followed by a major period of adaptive radiation by the placentals.
8. This radiation coincided with the extermination of all of but one of the numerous marsupial species.
9. By the end of the Cretaceous the marsupials had spread widely through South America.
10. The marsupials spread still further into Australia via Antarctica before these three continents were separated by continental drift.
11. North and South America were separated for much of the Cenozoic. When they were rejoined, placentals migrated into South America and replaced most of the marsupials.

L. **Theory of sexual selection**
  1. Epigamic selection is determined by choices among courting partners or differences in breeding time.
  2. Choice among different types of suitors is dependent upon their relative frequencies, but choice itself is not necessarily frequency-dependent.
  3. Differences in breeding time offer superior suitors a greater chance of breeding and leaving offspring.
  4. Competition for mating partners between members of the same sex permits some individuals to leave more offspring (intrasexual selection). This is especially true among males.
  5. Intrasexual selection may find expression in precopulatory and/or postcopulatory competition.

M. **Theory of origin of air breathing in jawed fish**
  1. Gnathostome fishes originated in tropical marine environments during the Silurian period.
  2. In the late Silurian-early Devonian these fishes occupied shallow waters at the continental margins.
  3. During periodic droughts the reduced surface flow of rivers and streams contributed to the creation of hypersaline conditions.
  4. The hypersalinity led to further reduction in the solubility of oxygen in these waters.

5. Low oxygen placed severe constraints on metabolic scope and therefore on activity.
6. An aberrant behavior pattern involving gulping air at the water surface was indirectly subjected to positive selection.
7. Air gulping permitted oxygen to be absorbed by bucco-pharyngeal surfaces.
8. Individuals securing more oxygen by gulping could be more active in gathering food, escaping enemies, and other pursuits including the production of offspring.

N. Evolutionary niche theory
1. Insofar as the same phenotype is not optimal in all environments, niche spread involves some fitness loss in each habitat.
2. In a completely certain stable environment narrow specialization would evolve.
3. It is the uncertainty of the environment that creates the selective pressure toward a broad niche.
4. This uncertainty may arise from (1) temporal availability of the environment, (2) coarse grained habitat, and (3) low productivity.
5. The final niche breadth that evolves will be an increasing function of uncertainty.
6. In an uncertain environment there is a loss of fitness at the optimum niche structure. This loss of fitness is roughly proportional to the uncertainty of the environment.
7. When the uncertainty of the environment exceeds the upper limit of the niche breadth, habitat selection can reduce the uncertainty.
8. Both the lower and upper limits to the niche breadth depends upon the uncertainty of the environment compared to the tolerance of the individual.

O. r- and K-selection theory
1. At low population densities, there is essentially pure exponential growth, at the rate r.
2. At high densities the population stabilizes at a value of K which is set by the environmental carrying capacity.
3. An r-selected organism sees its environment as unstable and unpredictable, and this produces episodes of boom and bust in population growth.
4. The evolutionary pressures on the r-selected organism are for opportunism—to produce many offspring rapidly in good times.

5. A K-selected organism sees its environment as stable and pre-dictable, and its population usually remains at an equilibrium level.
6. The evolutionary pressures on the K-selected organism are to produce fewer offspring but with more time and energy spent in raising them.
7. All organisms participate to some degree in both r- and K-selection, but they vary greatly as to which kind of selection has the major effect on their life and evolution.

P. Hybrid-superiority theory of narrow hybrid zones
1. The hybrids that form at the zone of contact between two species or subspecies are more fit than the parental phenotypes in the hybrid zone and possibly in other environments.
2. The range of the hybrid population is determined by the range of environmental conditions within which the hybrids are superior.
3. Hybrids, in some cases, can succeed in environments such as ecotones where competition from parental phenotypes is weak.
4. The breadth of a hybrid zone is determined by the geographic range of ecological conditions to which the hybrid is adapted, that is, to which the parental phenotypes are less adapted.
5. Narrow hybrid zones are associated with ecotones.

Q. Fisher's theory of the evolution of mimicry
1. A mutant gene appeared in the to-be mimic species that gave a slight resemblance to a protected species.
2. The effect of the mutant gene was modified by selection operating upon segregation taking place within the gene complex, leaving the mutant gene unchanged.
3. Owing to the advantage conferred by the mutant gene it spread, and its effects were further modified by other genes to improve the mimicry.
4. If polymorphism evolved, the original mutant would remain as the switch-gene controlling alternative forms.

R. Theory of balanced polymorphism
1. A mutant gene may be detrimental when homozygous and beneficial when heterozygous.
2. The mutant and normal alleles determine three phenotypes in appreciable numbers.
3. The mutant and normal alleles will reach an equilibrium in the gene pool.

4. At least two opposing selective forces act upon the phenotypes to determine the equilibrium.

S. Theory of allopatric speciation
1. New species arise by the splitting of lineages.
2. New species develop rapidly.
3. A small subpopulation of the ancestral form gives rise to the new species.
4. The new species originates in a very small part of the ancestral species' geographic extent—in an isolated area at the periphery of the range.

T. Theory of phyletic gradualism
1. New species arise by the transformation of an ancestral population into modified descendants.
2. The transformation is even and slow.
3. The transformation involves large numbers, usually the entire ancestral population.
4. The transformation occurs over all or a large part of the ancestral species' geographic range.

U. Theory of speciation in diploid species
1. A diploid species has two differing systems of variability.
2. The open system consists of polymorphic gene loci which recombine freely without drastic effects on viability.
3. The closed system consists of blocks of genes forming coadapted, internally balanced gene complexes with or without the presence of inversions as a stabilizing mechanism.
4. Perturbation of these blocks (supergenes) by crossing-over results in greatly reduced viability under normal natural selection.
5. These blocks vary between but not within species.
6. When natural selection is relaxed, as during a population flush-crash-founder cycle, the coadaptive balances of the closed system may become disorganized, and one or more discordant individuals may survive.
7. Speciation may occur as selection operates on the perturbed genetic system to organize new coadapted closed systems which come to characterize the new species.

ADDENDUM FOR "EVOLUTION: A SYSTEM OF THEORIES"

My views on structure have changed a little since the above paper was written. Today they are as follows: The structure of knowledge in biology, and in all the intellectual disciplines, consists of the structure of its theories plus the relations between theories. The relations are such that each theory can be categorized as major theory, over-theory, sub-theory, accessory theory, parallel theory, competing theory, subsumed theory, or sequential theory. A theory may belong to more than one category depending on the context in which it is considered. The relations between theories are determined by the content of postulates and by the components found in the lines of reasoning within theories.

Ralph W. Lewis, May 1982

REFERENCES

1. Linus Pauling, *College Chemistry* (San Francisco: W. H. Freeman, 1950), p. viii.

2. Albert Einstein, "The Method of Science," in *The Structure of Scientific Thought,* ed. by E. H. Madden (Boston: Houghton Mifflin, 1960), p. 81.

3. J. Maynard Smith, "The Limits of Evolutionary Theory," in *The Encyclopedia of Ignorance,* ed. by R. Duncan and M. Weston-Smith (Oxford: Pergamon, 1977), p. 241.

4. Patrick Suppes, "The Desirability of Formalization in Science," *Journal of Philosophy* 65 (20) (1968): 651–64.

5. See, for example, F. Suppe, *The Structure of Scientific Theories* (Urbana, IL: University of Illinois Press, 1974).

6. For a discussion of the many meanings of "model," see R. C. Lewontin, "Biological Models," in *Dictionary of the History of Ideas,* ed. by P. R. Wiener (New York: Scribners, 1973), vol. 1, pp. 242–46.

7. P. W. Medawar, *The Art of the Soluble* (London: Methuen, 1976), p. 114.

8. R. Blanché, "Axiomatization," in *Dictionary of the History of Ideas,* vol. 1, pp. 162–72.

9. A. R. Wallace, *Natural Selection and Tropical Nature* (London: Macmillan, 1891), p. 166.

10. See, for example, L. A. Steen, "Foundations of Mathematics: Unsolvable Problems," *Science* 189 (1975): 209–10.

11. See, for example, P. Suppes, 1974.

12. Morton Beckner, "Aspects of Explanation in Biological Theory," in *Philosophy of Science Today,* ed. by S. Morgenbesser (New York: Basic Books, 1967), p. 150.

13. See E. Nagel, *The Structure of Science* (New York: Harcourt, Brace and World, 1961), p. 30.

14. W. J. Van der Steen and W. Boontje, "Phylogenetic versus Phenetic Taxonomy: A Reappraisal," *Systematic Zoology* 22 (1973): 55–63.

15. Mary B. Williams, "Falsifiable Predictions of Evolutionary Theory," *Philosophy of Science* 40 (1973): 518–37.

16. C. Darwin, *On the Origin of Species,* a facsimile of the 1st edition with an introduction by E. Mayr and added index (Cambridge, MA: Harvard University Press, 1964), p. 111.

17. D. Hull, *Philosophy of Biological Science* (Englewood Cliffs, NJ: Prentice-Hall, 1974), pp. 45, 46, 50.

18. J. Huxley, *Evolution, the Modern Synthesis* (New York: Harper, 1943), pp. 13–14.

19. See Stephen D. Fretwell, "The Impact of Robert MacArthur on Ecology," *Annual Review of Ecology and Systematics* 6 (1975): 1–13.

20. A. O. Lovejoy, "The Argument for Organic Evolution before the 'Origin of Species'," in *Forerunners of Darwin, 1745–1859,* ed. by B. Glass, O. Temkin, and W. L. Straus, Jr. (Baltimore, MD: The Johns Hopkins Press, 1968), p. 356.

21. R. A. Fisher, *The Genetical Theory of Natural Selection,* 2d ed. (New York: Dover, 1958), p. vii.

22. J. B. S. Haldane, *The Causes of Evolution* (Ithaca, NY: Cornell University Press, 1966), p. 2.

23. G. G. Simpson, C. S. Pittendrigh, and L. H. Tiffany, *Life: An Introduction to Biology* (New York: Harcourt Brace, 1957), p. 25.

24. T. Dobzhansky, *The Genetics of the Evolutionary Process* (New York: Columbia University Press, 1970), p. 28.

25. J. M. Thoday, "Non-Darwinian 'Evolution' and Biological Progress," *Nature* 225 (1975): 675.

26. See S. S. Schweber, "Genesis of Natural Selection—1838: Some Further Insights," *Bioscience* 28 (1978): 325.

27. T. Dobzhansky, "Darwinian Evolution and the Problem of Extraterrestrial Life," *Perspectives in Biology and Medicine* 15 (1972): 157–75.

28. See D. P. McKenzie, "Plate Tectonics and Its Relationship to the Evolution of Ideas in the Geological Sciences," *Daedalus* 106 (1977): 97–124.

29. See M. T. Ghiselin, *The Triumph of the Darwinian Method* (Berkeley, CA: University of California Press, 1969), p. 38.

30. W. C. Denison and G. C. Carroll, "The Primitive Ascomycete: A New Look at an Old Problem," *Mycologia* 58 (1966): 249–69.

31. Ian R. Ball, "Nature and Formulation of Biogeographical Hypotheses," *Systematic Zoology* 24 (1975): 407–30.

32. A. R. Wallace, 1891.

33. Mary B. Williams, "Deducing the Consequences of Evolution: A Mathematical Model," *Journal of Theoretical Biology* 29 (1970): 343–85.

34. J. H. Woodger, *Biology and Language* (Cambridge: Cambridge University Press, 1952).

35. J. P. Changeux, P. Courrege, and A. Danchin, "The Theory of the Epigenesis of Neuronal Networks by Selective Stabilization of Synapses," *Proceedings, National Academy of Sciences USA* 70 (1973): 2974–78.

36. J. Huxley, 1943.

37. C. Darwin [*On the Origin of Species*], 1964, index.

38. P. J. Vorzimmer, *Charles Darwin: The Years of Controversy* (Philadelphia: Temple University Press, 1970), chs. 3, 4, 7.

39. D. Hull, 1974.

40. G. L. Stebbins, *Process of Organic Evolution,* 2d ed. (Englewood Cliffs, NJ: Prentice-Hall, 1971).

41. T. H. Hamilton, *Process and Pattern in Evolution* (New York: Macmillan, 1967).

42. See Vorzimmer, 1970.

43. C. Darwin [*On the Origin of Species*], 1964.

44. Charles E. Bessey, "Phylogenetic Taxonomy of Flowering Plants," *Annals, Missouri Botanical Garden* 2 (1915): 109–64.

45. M. Beckner, *The Biological Way of Thought* (New York: Columbia University Press, 1959); E. Mayr, "Theory of Biological Classification," *Nature* 220 (1968): 545–48; S. Lovtrup, *Epigenetics: A Treatise on Theoretical Biology* (London: Wiley, 1974).

46. See Walter J. Bock, "Philosophical Foundations of Classical Evolutionary Classification," *Systematic Zoology* 22 (1973): 375–92.

47. R. H. Eyde and J. T. Morgan, "Floral Structure and Evolution in Lopezieae (Onagraceae)," *American Journal of Botany* 60 (1973): 771–87.

48. T. J. Dawson, "Kangaroos," *Scientific American* 237 (1977): 78.

49. J. J. Schwab, *Biology Teachers' Handbook* (New York: Wiley, 1967), p. 45.

50. S. J. Gould, "Sociobiology: The Art of Storytelling," *New Scientist* 80 (1978): 530–33.

51. E. O. Wilson, *Sociobiology* (Cambridge, MA: Harvard University Press, 1975), ch. 27.

52. Gary C. Packard, "The Evolution of Air-Breathing in Palezoie Gnathostome Fish," *Evolution* 28 (1974): 320–25.

53. E. O. Wilson, 1975, ch. 27.

54. Richard Levins, "Toward an Evolutionary Theory of the Niche," in *Evolution and Environment,* ed. by E. T. Drake (New Haven: Yale University Press, 1968), p. 337.

55. Robert M. May, "The Tropical Rainforest," *Nature* 257 (1975): 737–38.

56. William S. Moore, "An Evaluation of Narrow Hybrid Zones in Vertebrates," *Quarterly Review of Biology* 52 (1977): 263–77.

57. E. B. Ford, *Ecological Genetics,* 4th ed. (London: Chapman and Hall, 1975), pp. 111, 295, 314.

58. N. Eldridge and S. J. Gould, "Punctuated Equilibria: An Alternative to Phyletic Gradualism," in *Models in Paleobiology,* ed. by T. J. M. Schopf (San Francisco: Freeman, Cooper, 1972), pp. 82–115.

59. Hampton L. Carson, "The Genetics of Speciation at the Diploid Level," *American Naturalist* 109 (1975): 83–92.

60. See M. Goodman, G. W. Moore, and G. Matsuda, "Darwinian Evolution in the Genealogy of Haemoglobin," *Nature* 253 (1975): 603–08.

61. See E. O. Wiley, "Karl R. Popper, Systematics, and Classification: A Reply to Walter Bock and Other Evolutionary Taxonomists," *Systematic Zoology* 24 (1975): 233–43.

62. See H. J. Ferguson, *American Naturalist,* 110 (1976): 110.

63. See R. H. Peters, "Predictable Problems with Tautology in Evolution and Biology," *American Naturalist* 112 (1978): 759–62.

64. R. D. Alexander, "Evolution, Creation, and Biology Teaching." Included in this volume.

65. P. Suppes, 1968.

The Topic of Evolution in Secondary School Biology Textbooks: 1900–1977

by Gerald Skoog

Other textbooks, in a retreat from the 1960s, now have deemphasized evolution. The deemphasis has not been a result of a diminishing in the power of evolution to explain and make sense out of the natural world. Instead, the deemphasis has been the result of publishers, authors, educators, and politicians responding to the strenuous efforts of antievolutionists and creationists to suppress and diminish the study of evolution.

INTRODUCTION

For most biologists, evolution is a central and unifying concept of their discipline. Their position is summarized well by Mayfield's statement that "it has become trite but not less true to say that evolution is the greatest unifying concept in biology."[1] The public's acceptance of evolution as a natural process and as a valid area of study in schools has not paralleled the views of biologists and other scientists. Considerable hostility and public disagreement has surrounded evolution ever since the publication of *The Origin of Species* by Charles Darwin in 1859. Public schools have not been able to escape the conflict and, as a result, many critics have argued that evolution has not been taught in a free and unfettered manner in the past or present despite its legitimacy as an important biological concept.

PROBLEM AND PROCEDURE

In recognition of the importance of the textbook in determining what is studied, one approach to determining whether evolution has been neglected or given minor treatment in the past is to analyze the content of secondary school biology textbooks. In this study, 93 secondary school biology textbooks, which are identified in the Appendix bibliography, were analyzed to determine the extent 44 topics concerned with the study of evolution were emphasized. Word counts were used as relative indicators of emphasis and trends. In the first phase of the study, 83 textbooks published between 1900–1968 were analyzed.[2] To update the earlier study, 10 additional secondary school biology textbooks published between 1970 and 1977 were analyzed, and data from the two phases of the study were combined.

The textbooks were selected on the basis of their availability and authorship. The textbooks published between 1900 and 1919 were selected solely on the basis of availability. An attempt was made to obtain textbooks that had been revised over an extended period of time so changes from edition to edition could be noted. Also, the continued publication of a textbook was interpreted as an indication of sustained popularity. No claim has been made that the sample of textbooks analyzed in this study included all of the best sellers of this century. However, because of the large sample of textbooks analyzed for each time period, popular textbooks as well as others that were representative of the various choices available to the schools undoubtedly were included in the study.

Individual words were not counted. The average number of words in a complete line of a textbook was determined by counting the number of words in the complete lines of 10 full or nearly full pages from various parts of the textbook and dividing the sum by the total number of complete lines. The number of words for a specific topic was calculated by multiplying the average number of words per line with the number of lines devoted to that topic. The number of words in the incomplete lines were added to the product to determine the final total. When a textbook had less than 10 pages devoted to the topics related to evolution, the words on the specific topics were counted separately.

The 44 topics selected for analysis were centered around the evidences of evolution, the mechanisms of evolution, the process of evolution, the evolution of various organisms, theories pertaining to evolution, and other topics that would be expected to provide additional understanding of the process of evolution. Topics such as "defense for the study of evolution" and "science and religion" were included as they indicated procedures and practices of various authors. Minor segments of subject matter not related to any of the 43 specific topics were classified as miscellaneous. Among topics classified as miscellaneous were evolution of behavior, miscellaneous evidences of evolution, discontinuous distribution, the stability factor, adaptations, evolution of insect societies, evolution of bacteria, evolution of multicellularity and evolution of invertebrates.

SUMMARY OF DATA AND TRENDS, 1900–1977

Analysis of the 93 biology textbooks revealed that prior to 1960, evolution was treated in a cursory and generally noncontroversial manner. However, there was a continued increase in the emphasis on evolution in the textbooks from 1900 to 1950. This trend was reversed in the 1950s when the concept was deemphasized slightly. In the 1960s the activities and influence of the Biological Sciences Curriculum Study (BSCS) resulted in several textbooks that gave unprecedented emphasis to evolution. Accordingly, 51% of the total words written on the topics concerned with the study of evolution in the 83 textbooks published between 1900–1968 appeared in 17 textbooks published in the 1960s.

The analysis of the textbooks published in the 1970s revealed a reversal of the trends of the 1960s. There was a decrease in the overall coverage of the 44 topics. In certain textbooks the emphasis on selected topics concerned with evolution was drastically reduced or eliminated. Word changes in the textbooks of the 1970s resulted in many statements becoming less definite, more cautious, and thus less controversial than those appearing in earlier editions.

Data in Table I are evidence of the aforementioned trends. Thirty-seven of the 44 topics concerned with the study of evolution received more emphasis in the textbooks of the 1960s than in any other prior decade. Fifty or more percent of what was written on 20 of the 44 topics in the 83 textbooks published before 1970 was in the 17 textbooks published in the 1960s. Ninety-two percent of the 33,498 total words written concerning the hypotheses on the origin of life prior to 1970 was in the textbooks of the 1960s. Sixty to 70% of the material written in the textbooks published prior to 1970 on geological eras, cultural evolution, the fossil record of man, and the life of Darwin was in the 17 textbooks published in the 1960s. Other topics such as reproductive isolation, genetic drift, convergent and adaptive evolution, and polyploidy did not receive much emphasis until the 1960s.

Evolution was defined or explained in 87 of the 93 textbooks. No other topic had a higher frequency of appearance. Nineteen topics were covered in some manner in 50% or more of the 93 textbooks. Among the 24 topics found in fewer than half the textbooks were hypotheses concerned with the evolution of life, theses concerned with inorganic evolution, the age of earth, and the evolution of plants, birds, mammals,

TABLE I

Comparison of the Number of Words Written on the Topics Related to the Study of Evolution in 93 Secondary School Biology Textbooks Published during Different Periods between 1900 and 1977.

Topic	1900–1919	1920–1929	1930–1939	1940–1949	1950–1959	1960–1969	1970–1977	Total
Evolution Defined	534	3,233	4,421	6,670	5,796	14,671	5.994	41,319
Evolution of								
Fish	0	0	143	407	240	1,187	483	2,460
Amphibians	195	217	105	339	334	1,079	1,708	3,977
Birds	258	170	772	830	1,203	1,298	3,984	8,515
Reptiles	81	301	704	1,474	1,621	3,528	2,377	10,086
Mammals	0	210	243	372	736	2,288	3,310	7,159
Plants	730	458	874	1,436	2,036	19,985	5,497	31,016
Age of Earth	0	96	576	1,796	2,322	4,475	1,597	10,862
Geologic Eras	0	1,749	4,959	7,976	10,243	38,976	10,846	74,749
Geologic Change	0	221	1,884	6,286	5,923	4,045	1,520	19,879
Fossil Formation	0	1,141	3,319	3,045	3,891	9,799	3,121	24,316
Theses of Evolution								
Lamarckian	647	1,296	2,939	3,690	1,806	7,581	5,346	23,305
Darwinian	3,949	4,278	7,051	7,884	5,747	9,425	7,343	45,677
Life of Darwin	97	2,862	2,352	1,480	2,024	15,337	6,660	30,812
Inorganic Evolution	0	0	312	269	529	948	505	2,563
Origin of Life	0	144	108	1,984	354	30,868	13,698	47,156
Evidences of Evolution								
Fossils, etc.	474	1,869	6,402	7,572	5,465	6,194	5,360	33,336
Homologous Organs	578	1,731	1,757	2,298	1,811	6,205	2,915	17,295
Physiological	0	171	674	677	427	1,319	811	4,079
Biochemical	0	0	0	0	0	630	1,484	2,114
Vestigial Organs	260	696	2,075	2,381	878	2,378	973	9,641
Embryological	816	1,846	2,576	3,121	2,306	2,714	1,390	14,769
Domestic Breeding	595	1,523	1,626	1,288	1,122	1,827	1,848	9,829
Transformation	1,130	1,800	2,793	2,533	3,259	6,703	3,174	21,392
Convergent Evolution	702	0	0	0	0	1,064	1,464	3,230
Adaptive Radiation	0	0	0	0	0	2,973	3,282	6,255
Relationships	541	705	1,609	1,670	1,190	7,300	5,516	18,531
Religion & Science	0	498	404	323	382	184	81	1,872
Defense for the Study of Evolution	207	0	0	155	63	0	0	425
Special Creation	98	0	0	911	0	245	1,502	2,756
Mechanisms of Evolution								
Mutations	611	1,392	5,156	6,898	4,823	14,680	4,028	37,588
Recombination	0	597	1,016	227	305	11,614	1,246	15,005
Geographic Isolation	309	1,080	5,923	2,763	2,530	18,142	6,419	37,166
Reprod. Isolation	0	0	242	455	369	4,809	1,689	7,564
Genetic Drift	0	0	0	0	0	4,363	1,375	5,738
Polyploidy	0	0	87	314	204	3,760	912	5,277
Natural Selection	2,092	3,657	3,092	4,956	5,395	20,038	10,617	49,847
Evolution of Man	0	1,339	439	1,004	614	8,977	7,382	19,755

TABLE I (Continued)

Topic	1900– 1919	1920– 1929	1930– 1939	1940– 1949	1950– 1959	1960– 1969	1970– 1977	Total
				Number of Words				
Fossil Record of Man	0	798	2,355	4,764	6,010	21,049	17,189	52,165
Uniqueness of Man	69	3,475	3,329	6,696	2,317	8,131	5,032	29,049
Cultural Evolution	130	5,240	4,903	2,356	2,802	27,680	8,495	51,606
Race Differentiation	88	551	809	9,786	6,340	10,187	4,341	32,102
Future of Man	0	0	160	201	397	1,883	1,252	3,893
Miscellaneous	3,307	3,981	3,841	2,085	3,008	7,554	13,140	36,916
Total	18,498	49,325	82,030	111,372	96,822	368,093	186,906	913,046
Number of Textbooks	8	14	15	15	14	17	10	93
Average Words	2,312	3,523	5,469	7,425	6,916	21,653	18,690	9,818

amphibians, and fish. Special creation was discussed in six textbooks during the entire period.

During the period 1930–1939, an estimated 3% of the total words in the texts of the 15 textbooks was devoted to evolution. This percentage was 3.3, 3, 8.1, and 7.4% for the periods 1940–1949, 1950–1959, 1960–1969, and 1970–1977, respectively. Variation in the use of illustrations, photographs, and charts plus the presence of laboratory activities in some textbooks made it difficult to determine the average number of words per page and the total number of words in the textbooks. Because of this problem, these percentages were not calculated in the first phase of this study and the data needed to determine the percentage for the periods 1900–1920 and 1920–1929 were not available for the present study. However, an earlier study by Cubie (1958) indicated that 2% of the pages in six popular biology, botany, and zoology textbooks used during the period 1915–1925 were devoted to evidences of changes in living things and theories of evolution.[3]

REPRESENTATIVE DATA FROM SEVERAL EDITIONS OF ONE TEXTBOOK

Data from biology textbooks published and revised by a single publisher several times over a long period of years also revealed trends. *Modern Biology* (Otto & Towle, 1977), a widely used textbook today and during the past 50 years, traces its origin back to Moon's (1921) *Biology for Beginners.* An analysis of Moon's original textbook and 13 subsequent editions or revisions produced the data in Table II. These data paralleled the trends established by the data in Table I. A gradual increase in emphasis on evolution occurred from 1921–1947. The coverage of evolution was reduced in the 1950s, expanded in each of the next five editions, and reduced again in 1977.

Data also indicate a reluctance to use the word evolution in many of the textbooks of this popular series. Evolution was not used in the text, glossary, or index in the 1947, 1951, 1956, and 1960 editions. In 1933, 1938, and 1941 the word evolution appeared once in the text but not in the glossary or index. In 1921 and 1926, the word evolution was used in the text and index but not in the glossary.

DATA ON THE USE OF THE WORD EVOLUTION IN TEXTBOOKS

Data also revealed trends in the use of the term evolution in the texts, glossaries, and indexes of biology textbooks. As noted on Table III, the term evolution was seldom used in textbooks during the period 1900–1959. Prior to 1960, evolution was listed in the glossaries of 15 of the 53 textbooks having glossaries. During this same period, evolution was listed in the indexes of 24 of the 64 textbooks with indexes. Overall, only 15 of the 53 textbooks published between 1920 and 1959 included the word evolution in the text, glossary, and index.

Textbooks were grouped by publication date into seven time periods. Following are summaries of certain of the data and conclusions concerning the coverage of evolution in the textbooks of each time period.

Summary: 1900–1919

The eight biology textbooks analyzed for this period stressed both practical biology and the study of human physiology, botany, and zoology. However, regardless of the orientation, evolution was not stressed in the textbooks of this period. Only three (Bigelow & Bigelow, 1911; Abbott, 1914; Hodge & Dawson, 1918) of the eight textbooks reviewed had specific chapters on evolution. In three textbooks (Hunter, 1907; Peabody & Hunt, 1912; 1913) the term evolution was not used whereas another textbook (Hodge & Dawson, 1918) used the term twice. Three of the seven textbooks with indexes listed evolution. The data in Table I indicated that few topics were discussed and that the depth of the presentations was shallow.

Hunter's (1914) textbook, *Civic Biology,* was identified in a 1921 study by Richards as the most widely used biology textbook in 59 cities having a population of 100,000 or more.[4] This textbook was not analyzed in this study but Grabiner and Miller (1974) reported that the book had a three-page section on evolution and a brief discussion of natural selection.[5] Hunter's (1911) *Essentials of Biology* and Peabody and Hunt's (1913) *Elementary Biology* were reported as the only two biology textbooks used in 25 midwestern high schools.[6] These two textbooks also were used widely in 1921.[7] They had 1,908 and 1,035 words on evolution, respectively. In comparison, *The Elementary Principles of General Biology* by Abbott (1914) contained 8,291 words on topics related to the study of evolution. There was no evidence it was a popular textbook.

Authors of this period may not have emphasized evolution because they considered it too difficult for high school students. Peabody and Hunt (1924), authors of biology textbooks in 1912, 1913, 1924, and 1933, stated that theories of evolution were among the topics that were more suitable for discussion in advanced courses. Hunter (1911), in defining the content of his textbook, indicated that "abstractions are not a part of the thought of a first-year pupil."[8] In his coverage of heredity and variation, Hunter (1911), stated "The part played by Mendel's law is too difficult to explain to high school pupils."[9] Hunter and other authors may have considered evolution as abstract and difficult as Mendel's law and thus inappropriate for study in high school biology textbooks. Yet, evolution would not seem to be any more abstract than much of the content studied in Greek, Latin, geometry, ancient history, and other courses included in the classical curriculum that was still popular during this time period. Evolution may have been excluded for reasons other than its abstract nature.

Summary: 1920–1929

Despite an increase in emphasis in the textbooks of the 1920s, evolution still was not treated as a major concept in biology. All 14 textbooks analyzed had some material on evolution. The difference in coverage was extreme however. A 1924 textbook (Peabody & Hunt, 1924), contained 94 words concerned with evolution whereas a 1926 textbook (Kinsey, 1926) had 8,757 words on the various topics. As noted in Table III, authors and publishers were reluctant to use the word evolution in the text, glossary, and index of the textbooks during this time. Smallwood et al. (1920) evidently attempted to bootleg in a discussion of evolution by including it within the section dealing with the metamorphosis of a frog. Other authors discussed evolution openly and attested to its validity. Atwood (1922) wrote:

> ... the mass of evidence which we have outlined constitutes so complete a case that all well-read biologists accept the doctrine of evolution as the explanation of how the life on the earth came to exist in its present form.[10]

TABLE II
Comparison of the Number of Words Written on the Topics Related to the Study of Evolution in 14 Secondary School Biology Textbooks Published by Henry Holt and Holt, Rinehart & Winston from 1921–1977.

Topic	1921	1926	1933	1938	1941	1947	1951	1956	1960	1963	1965	1969	1973	1977
Evolution Defined:	343	343	329	329	329	520	140	103	135	135	420	512	512	77
Evolution of:														
Fish	0	0	0	0	0	0	0	0	0	0	45	188	188	163
Amphibians	0	0	0	0	0	0	0	0	77	72	524	795	795	728
Reptiles	0	0	0	0	0	403	291	291	709	709	1,443	1,712	1,712	1,601
Birds	0	0	0	0	0	0	0	94	72	72	539	834	834	279
Mammals	0	0	0	0	0	0	0	58	389	389	751	1,621	1,600	874
Plant Life	0	0	0	0	0	430	394	493	373	373	445	1,232	1,232	890
Age of Earth	0	0	0	0	0	182	140	140	329	329	0	50	50	46
Geologic Eras	0	0	0	0	0	0	0	0	0	0	0	376	376	225
Geologic Change	0	0	0	0	0	61	62	62	34	34	0	121	121	108
Fossil Formation	0	0	0	0	0	0	0	0	744	744	0	394	0	123
Lamarckian Theses of Evolution	119	119	487	487	487	484	181	181	243	243	519	519	519	412
Darwinian Theses of Evolution	694	694	764	764	764	738	574	593	853	853	594	594	594	535
Life of Darwin	0	0	76	76	76	0	0	0	0	0	0	0	0	0
Inorganic Evolution	0	0	0	0	0	0	0	0	0	0	0	0	0	0
Origin of Life	0	0	0	0	0	0	0	0	0	0	0	0	0	366
Evidences of Evolution:														
Fossils, etc.	73	73	314	314	314	613	382	382	0	0	0	0	0	72
Homologous Organs	288	288	147	147	147	156	156	156	180	180	96	129	129	95
Physiological	0	0	190	190	190	185	108	108	114	114	0	0	0	0

TABLE II (Continued)

Topic	1921	1926	1933	1938	1941	1947	1951	1956	1960	1963	1965	1969	1973	1977
Biochemical	0	0	0	0	0	0	0	0	0	0	40	72	72	60
Vestigial Organs	218	218	494	567	567	420	134	118	246	246	127	172	172	103
Embryological	180	180	264	404	404	256	140	140	178	178	43	87	87	97
Domestic Breeding	79	79	76	76	76	151	117	117	129	129	67	67	67	0
Transformation	0	0	0	0	0	0	0	0	0	0	468	0	0	0
Convergent Evolution	0	0	0	0	0	0	0	0	0	0	209	165	206	157
Adaptive Radiation	0	0	0	0	0	0	0	0	0	0	509	827	767	1,117
Relationships	241	241	340	340	340	93	0	0	0	0	805	805	805	423
Religion and Science	97	271	239	165	165	158	122	168	0	0	0	0	0	0
Defense for Study of Evolution	0	0	0	0	0	0	0	0	0	0	0	0	0	0
Special Creation	0	0	0	0	0	0	0	0	0	0	0	0	0	0
Mechanisms of Evolution:														
Mutations	0	0	676	1,017	1,017	610	208	191	257	791	537	537	537	437
Recombination	0	0	0	0	0	0	0	0	0	0	484	363	363	0
Geographic Isolation	0	0	0	0	0	218	179	179	599	599	534	611	611	447
Reproductive Isolation	0	0	187	175	175	0	0	0	0	0	241	241	241	34
Genetic Drift	0	0	0	0	0	0	0	0	0	0	175	175	175	158
Polyploidy	0	0	0	87	87	91	67	67	55	55	127	127	127	0
Natural Selection	38	38	0	0	0	359	273	702	745	745	1,584	1,553	1,557	665
Evolution of Man	135	135	0	0	0	0	0	0	0	0	216	310	513	539
Fossil Record of Man	0	0	0	686	686	42	44	0	0	675	1,342	1,178	1,372	1,096
Uniqueness of Man	834	834	587	587	587	1,054	932	292	675	408	386	373	564	602
Cultural Evolution	845	845	513	513	513	493	428	0	413	314	0	183	162	142
Race Differentation	0	0	59	59	59	0	0	0	314	342	0	52	52	61
Future of Man	0	0	0	0	0	0	0	0	342	0	97	0	52	0
Miscellaneous	0	109	36	36	36	0	0	0	0	0	308	1,079	1,079	597
Totals	4,184	4,467	5,778	7,019	7,019	7,717	5,072	4,635	8,255	8,779	13,675	18,054	18,211	13,829

TABLE III

Usage of the Term Evolution in Secondary Biology Textbooks: 1900–1977[a].

The word evolu- tion was in the	1900– 1919	1920– 1929	1930– 1939	1940– 1949	1950– 1959	1960– 1969	1970– 1977	Total
Text	5/8	7/14	8/15	7/15	6/14	16/17	10/10	59/83
Glossary	0/0*	1/10*	3/14*	4/15	7/14	6/8*	7/7*	28/69
Index	3/7*	4/14	5/14*	6/15	6/14	16/17	10/10	50/91

[a] Note: This chart should be read as follows: Five of eight books published in the period 1900–1919 had the word evolution in the text.
*Several textbooks did not have an index or glossary.

Linville (1923) asserted "so generally is the validity of the theory admitted that it is often called the law of evolution."[11] Trafton (1923) also strongly supported the validity of the concept of evolution when he wrote:

> ... gradually scientists began to accept the theory of evolution till today it is universally accepted by scientists as fact, and we rarely hear any arguments about its truth. The matter about which scientists now disagree is related to the question of how evolution takes place; but the fact of evolution no scientist doubts.[12]

After 1925, statements of this nature were infrequent until the 1960s. Also, the textbooks of Linville and Trafton, which had these strong supporting statements plus much material on evolution, were not revised and republished in the 1930s. There is evidence that textbooks by Hunter (1926) and Smallwood et al. (1929) were popular textbooks in the 1920s and early 1930s.[13] They had less discussion of evolution than most other textbooks and were revised and republished in the 1930s.

During this period many authors still accepted Lamarck's thesis concerning the inheritance of acquired characteristics. Certain authors depicted evolution as a process wherein organisms evolve into more perfect specimens. Natural selection was often dramatized as "red in tooth and claw." In discussions of eugenics, concerns often were expressed that natural selection was no longer allowed to operate on humans, and as a result, the population of defective individuals was increasing.

Definitions of evolution, fossils and other remains as evidence of evolution, cultural evolution, and the Darwinian thesis of evolution were topics discussed most frequently. Two textbooks (Gruenberg, 1925; Kinsey, 1926) had extensive coverage of natural selection.

Kinsey's (1926) *Introduction to Biology* gave more coverage to evolution than any other textbook of this period. Caldwell and Weller (1932) reported that the textbook was being used extensively in 1932 but Grabiner and Miller (1974) reported neither it nor later editions were ever used widely. Moon's (1926) *Biology For Beginners* devoted more space to evolution than most textbooks of that period and also was identified as a popular textbook by Caldwell and Weller (1932). Textbooks by Hunter (1923, 1926) and Smallwood et al. (1920, 1924, 1929) had less discussion of evolution than most other textbooks. They also continued to be revised and published in the 1930s. The textbooks by these authors also were popular. *New Biology* (Smallwood, et al., 1924) was used widely both in the 1920s and early 1930s.[14] Hunter's (1923) *New Essentials of Biology* was still used extensively in 1932 according to Caldwell and Weller (1932).

Summary: 1930–1939

Evolution received increased emphasis in the textbooks of the 1930s as certain topics appeared in a larger number of textbooks and were covered in more depth than in the previous periods. Still, the treatment of evolution tended to be brief, noncontroversial, and characterized by restraint. The word evolution appeared infrequently. Comments concerning the validity of the concept of evolution usually were avoided. The evolution of humans virtually was ignored. Questions concerning the origin of the earth and of life were avoided in all but two textbooks. Natural selection was not discussed in 5 of the 15 textbooks.

Definitions and explanations of evolution were diverse. The authors of a 1934 textbook simply stated "Change follows change through succeeding ages, until finally there may be forms very different from those now seen."[15] Hunter (1935) wrote:

> We know that in the millions of years that life has existed on the earth that there have been many changes. According to present day evidences, living things at first were very simple in structure, but as time went on more and more complex types appeared.[16]

Pieper et al. (1932) were less definite and more cautious than Hunter as they declared:

> We must not forget that changes in environment and changes in structure and modes of living of organisms were extremely gradual. It took millions of years for very small changes to take place. Just how these changes in animals can take place is still the subject of scientific investigation. Not enough facts are known to give a definite explanation.[17]

Smith (1938) was more definite about the validity of the concept of evolution than most authors of this period when she stated:

> Plants and animals have been changing and are now changing. No one acquainted with the facts doubts that evolution, or continual change in plants and animals, has taken place. No one has discovered a single fact to disprove the theory of evolution, and the facts that establish its truth are abundant.[18]

Natural selection, discussed in 9 of 15 textbooks, still was often described as a ruthless and bloody process. Concern still was expressed that human beings were interfering with the process of natural selection and eugenics needed to be practiced more intensively.

Dynamic Biology by Baker and Mills (1933) was the most widely used textbook of the 1930s according to Grabiner and Miller (1974). Evolution was discussed in the last chapter of the textbook. Seventeen topics concerned with the study of evolution were discussed using 4,125 words. This was approximately 2% of the total words in the text. Five of the 15 textbooks analyzed for this period had fewer words concerned with evolution than did *Dynamic Biology*. This textbook, as well as three subsequent editions (Baker & Mills, 1943, 1948, 1953), did not use the word evolution in the text, glossary, or index or include any material on the evolution of man, man's fossil record, or cultural evolution.

The most extensive treatment of evolution during this period was in *Exploring Biology* by Smith (1938). Twenty four topics concerned with evolution were covered in 12,240 words. This textbook was one of four textbooks containing material on human evolution. It and two other textbooks (Hunter, 1931, 1935) were open and definite in stipulating that humans had evolved.

Overall, the authors of this period claimed to have written textbooks emphasizing the principles of biology. However, they did not emphasize evolution as one of these principles.

Summary: 1940–1949

As noted in Table I there was an overall increase in emphasis on evolution in the textbooks of this period. The number of topics discussed in some individual textbooks increased. Discussions of specific topics were often longer but still lacking depth.

The biology textbooks of this period were characterized by stability. The treatment of evolution remained the same in the 1940, 1943, and 1946 editions of one textbook (Curtis et al., 1940, 1943, 1946). The coverage in these textbooks nearly coincided with that of the 1934 edition (Curtis et al., 1934). *Biology for Beginners* (Moon & Mann,

1941) was similar to earlier editions published in 1938 and 1933 (Moon & Mann, 1938, 1933).

Only 3 of the 15 textbooks used the term evolution in the index, glossary, and text. Ten of the textbooks omitted or used the term infrequently.

All 15 textbooks defined or explained evolution. Certain authors defined evolution as a progressive process. As in 1938, Smith (1943, 1949), in the two editions of *Exploring Biology* published in the 1940s, emphasized that the truth of evolution was established by several facts, whereas no facts existed that could disprove it. Smith (1949) also asserted that "Modern biologists accept evolution as proved."[19] Gruenberg and Bingham (1944) not only supported the validity of the concept of evolution but took issue with opposing views. Other authors were much more conservative in their treatment of evolution.

Natural selection no longer was depicted as a bloody and ruthless process in the 1940s. Discussions on eugenics and social Darwinism were less prevalent than during earlier periods. Evidences of evolution were stressed more than previously. Five textbooks had material on the origin of life. Some of this material was changed in revisions. Hunter (1941) stated "Later one-celled green plants must have come into existence and then one-celled animals, which feed on the green plants and bacteria."[20] In 1949, a similar statement was extended by Hunter and Hunter (1949) with the addition of the line "As you see, if you turn to the first chapter of Genesis, this is the order of Creation."[21]

Despite an overall increase in emphasis, five textbooks failed to mention human evolution, the human fossil record, or cultural evolution. There was also a deemphasis on the fossil record of humans in some textbooks during the last part of the decade. Moon and Mann (1941) discussed this topic in 700 words in 1941, whereas Moon et al. (1947) decreased the coverage to less than 50 words in 1947. Hunter (1941) utilized nearly 1,000 words on the topic in 1941, but Hunter and Hunter (1949) reduced the coverage to 235 words in 1949. Smith's (1943, 1949) textbooks followed a similar pattern from 1943 to 1949.

Despite claims by several authors that their textbooks had been organized around biological generalizations and principles, evolution was not emphasized and treated as such in the textbooks of the 1940s.

Biology and its successor, *Modern Biology,* both written by Moon and Mann (1941, 1947) were the most popular textbooks of this period according to Grabiner and Miller (1947). Details on these two textbooks are presented in Table II.

The most extensive coverage of evolution was in *Biology and Man* (Gruenberg & Bingham, 1944) and *Exploring Biology* (Smith, 1943).

They had 14,383 and 12,240 words devoted to evolution, respectively. Smith's textbooks were used widely (Grabiner et al., 1974) but there is not evidence that *Biology and Man* was popular.

Summary: 1950–1959

Fourteen textbooks were analyzed for this period. The stability that characterized the coverage of evolution in certain of the textbooks in the 1940s persisted through the 1950s. *Everyday Biology* (Curtis et al., 1953) had the same material on evolution as was used in three earlier editions (Curtis et al., 1940, 1943, 1946). Furthermore, the coverage of evolution in the 1953 edition and the 1934 edition (Curtis et al., 1934) was nearly identical.

The coverage of evolution in textbooks by Baker and Mills (1933, 1943, 1948, 1953, 1959) changed little from one edition to another. The 1959 edition had basically the same material on evolution that was in the 1953 edition, which was identical to that in the 1948 and 1943 editions. There were few differences between the 1933 and 1943 editions. The treatment of evolution in three other textbooks basically was unchanged from the 1940s to the 1950s. Revisions that did occur often resulted in less emphasis than before. Also three of the 1940s textbooks (Hunter, 1941; Gruenberg & Bingham, 1944; Hunter & Hunter, 1949) were not revised or republished in the 1950s. As a result, the available textbooks had less emphasis on evolution than those in the 1940s.

The word evolution was seen seldom in secondary school biology textbooks in the 1950s as it did not appear in 8 of 14 textbooks. Instead, expressions such as racial development, progressive development, and change were used. A 1959 textbook (Smith, 1959) used the word evolution only once in the text even though the word had been used liberally in earlier editions. All 14 textbooks of this period had the material on evolution placed in one of the final chapters. This was typical as only two of the textbooks published before 1960 had considerable emphasis on evolution in the first half of the textbook.

Statements emphasizing evolution as a valid concept were less frequent than in the 1940s. The statements by Smith (1938, 1943, 1949) attesting to the truth of evolution and quoted earlier in this article were moderated in the 1959 edition.

Direct statements about human evolution were found in 3 (Vance and Miller, 1950, 1954; Kroeber et al., 1957) out of 14 textbooks. The human fossil record was discussed in seven textbooks with the most extensive treatment in *Exploring Biology* (Smith, 1954). However, the 1959 edition of this textbook did not mention either the fossil record or evolution of humans.

Evidence of evolution received less emphasis than in the 1940s. Four textbooks advanced hypotheses regarding the origin of the earth and life. The tendency to depict natural selection as a process involving nonrandom reproduction continued. Euthenics gradually began to replace eugenics and material concerned with the influence of natural selection on eugenics and humans decreased.

Overall, there was no evidence that the concept of evolution was regarded and emphasized as a major concept in the textbooks of the 1950s.

According to Cubie (1958), textbooks adopted and/or recommended by most states during this period were *Modern Biology* (Moon et al., 1951), *Dynamic Biology Today* (Baker & Mills, 1953), *Elements of Biology* (Dodge et al., 1952), *Biology in Daily Life* (Curtis & Urban, 1953), *Biology for You* (Vance & Miller, 1954), *Exploring Biology* (Smith, 1954), and *Biology and Human Affairs* (Ritchie, 1941). Grabiner and Miller (1974) indicated that *Modern Biology* (Moon et al., 1951, 1956) dominated the market and *Exploring Biology* (Smith, 1954, 1959) was the second most popular textbook in use in the 1950s.

Summary: 1960–1969

More was written on the topics concerned with evolution in the 17 textbooks of the 1960s than in the 66 textbooks published from 1900–1959. Thirty-seven of the 44 topics received more emphasis in the 1960s than in any other time period. Twenty topics received over 50% of their total word allotment from the textbooks of the 1960s. Ninety-two percent of 33,498 total words written prior to 1970 on hypotheses concerned with the origin of life was in the textbooks of the 1960s. Sixty to 70% of the material written in the textbooks published prior to 1970 on geologic eras, cultural evolution, the fossil record of humans, and the life of Darwin was in the 17 textbooks of the period.

The unprecedented emphasis on evolution in this period clearly was the result of the utilization of evolution as a theme by the Biological Sciences Curriculum Study (BSCS) in the development of the three versions in 1961 and the revised editions of each version in 1963 and 1968. Collectively, the nine BSCS textbooks published in the 1960s contributed 261, 475 or 71% of the total 368,093 words written on the topics concerned with evolution in the 17 textbooks of this period. Also 30% of all the words written on evolution in the 83 textbooks published prior to 1970 was found in the nine BSCS textbooks. More words were written on natural selection in the nine BSCS textbooks than in all of the 52 textbooks analyzed for the period 1900–1949.

The term evolution appeared regularly in the biology textbooks for the first time since secondary school biology came into existence. Only one textbook (Moon et al., 1960) failed to include the word evolution anywhere

In previous decades, the discussion of evolution was usually restricted to one or two chapters. The BSCS textbooks of the 1960s tended to develop, define, and discuss the concept of evolution in chapters concerned with other topics as well as evolution. As a result, evolution was more integrated throughout the BSCS textbooks.

The validity and value of evolution as a biological concept was more widely acknowledged by the authors of the 1960s than in previous periods. However subsequent word changes occurred that indicated some indecision over whether evolution should be presented as a concept, principle, or theory. The 1961 BSCS textbooks tended to stress evolution as a concept or principle, but by 1968 they were inclined to stress evolution as a theory. Other changes in the BSCS textbooks indicated that words used in discussing evolution had been selected with care.

The three different BSCS versions did not cover the same topics in the same depth and breadth. Also, as the non-BSCS textbooks of this period were revised, their coverage of certain topics related to evolution exceeded the coverage in the different BSCS versions.

The BSCS versions were used widely in the 1960s. During the 1960–1961 school year, 14,000 students used the preliminary experimental BSCS materials. The three revised editions were used by 52,000 students in 1961–1962.[22] In 1963–1964 there were 300,000 students using the BSCS textbooks and it was estimated that about one-half of the new secondary school biology textbooks purchased were BSCS.[23] Of the 2,300,000 students enrolled in tenth grade biology in 1964–1965, 700,000 were using BSCS textbooks.[24] During the 1965–1966 school year, 50.16% of 1,359 biology teachers reported they used one of the three BSCS versions.[25] Despite the popularity of the BSCS textbooks, *Modern Biology* (Moon et al., 1963; Otto & Towle, 1965) was the most widely used single textbook in 1965.[26]

Overall, the coverage of evolution in the textbooks of the 1960s was the most straightforward and comprehensive of the century. Also, if the slight deemphasis of evolution in the textbooks of the 1970s is not reversed in the future, the 1960s may represent the zenith of evolution's coverage in the secondary school biology textbooks of this century.

Summary: 1970–1977

Definite changes that signalled a decline in the emphasis of evolution were made in several of the 10 textbooks analyzed for this period.

The overall coverage of the 44 topics concerned with evolution was reduced. In certain textbooks the emphasis on selected topics concerned with evolution was reduced. In certain textbooks the emphasis on selected topics concerned with evolution was drastically reduced or eliminated. Changes in wording resulted in material that was more cautious and indefinite about evolution.

The 1977 edition of *Modern Biology* (Otto & Towle, 1977) used 13,829 words to cover the topics concerned with evolution, whereas the earlier edition (Otto & Towle, 1973) utilized 18,211. The 1977 revision of *Modern Biology* had briefer explanations of both evolution and natural selection than in 1973. Many word changes in the 1973 and 1977 editions of this textbook indicated a more cautious approach.

The coverage of evolution in the 1977 edition of *Biology* (Smallwood & Green, 1977) was considerably less than in the 1974 and 1968 editions (Smallwood & Green, 1968, 1974). The 1968, 1974 and 1977 editions had 22,121, 20,344 and 11,431 words, respectively, on the 44 topics concerned with evolution. Material on the Darwinian thesis of evolution was reduced from 2,750 words in 1974 to 296 in 1977. The life of Darwin was covered in 1,373 words in 1974 and 45 in 1977. Reductions of the same magnitude were done on material related to life origin hypotheses and natural selection. Material on geologic eras, fossil formation, reproductive and geographics isolation, polyploidy, and other topics was eliminated completely in the 1977 edition of *Biology*. Significant changes in wording also occurred as the 1974 edition was revised in 1977. Theoretical became a frequent word as statements such as "This is one of the reasons birds are considered to be fairly close relatives of reptiles" (Smallwood & Green, 1974) were rewritten as "Certainly there is little doubt that the reptiles were the theoretical ancestors of the birds" (Smallwood & Green, 1977).[27] Fish were considered as "The theoretical ancestors of all other vertebrates" (Smallwood & Green, 1977), whereas Annelids were termed the "theoretical ancestors of anthropods" (Smallwood & Green, 1977).[28]

Biology (Weinberg, 1977) had approximately 3,000 fewer words devoted to explaining evolution than in an earlier edition (Weinberg, 1974). This reduction was not drastic, but still noticeable. Changes in wording were minor, but again noticeable.

The coverage of evolution in BSCS textbooks has changed from edition to edition as noted in Table IV. The 1973 editions of all three versions had fewer words devoted to the 44 topics concerned with evolution than the earlier editions. Despite the use of fewer words, the completeness of the presentations and the emphasis on evolution were still present.

TABLE IV
Number of Words Allocated to 44 Topics Concerned with Evolution in BSCS
Textbooks

Year	Green Version	Yellow Version	Blue Version
1961	22,389	34,761	23,306
1963	30,383	45,898	22,987
1968	19,732	42,621	21,717
1973	18,626	31,050	19,400

Word changes and deletions in the BSCS textbooks indicated sensitivity to pressure groups and increasing caution. The original BSCS Blue Version (1961) read:

> This fossil evidence dramatically shows that life had been gradually changing over millions of years from one form to another. There is no longer any reasonable doubt that evolution occurs.[29]

In 1963, the word "suggests" was substituted for the phrase "dramatically shows" in the aforementioned passage.[30] The second line of this quote was deleted in the 1968 and 1973 editions. The statement "Throughout this book it will be evident that the theory of evolution by natural selection is the major framework of modern biology" appeared in 1968 but not in 1973.[31] The statement "Biologists are convinced that the human species evolved from nonhuman forms of life" did not appear after 1963.[32]

These and other changes in the BSCS Blue Version editions were minor, but nevertheless indicated that objections made against the textbooks around the nation may have been effective in bringing about minor changes in the content concerned with evolution.

The later editions (BSCS, 1968, 1973) of BSCS Green Version were slightly different from earlier editions (BSCS, 1961, 1963). The decrease in word totals, as noted in Table IV, was primarily due to the decrease in coverage of geologic eras and geographic isolation as causes of variation and evolution. Word changes, deletions, and minor changes in emphasis were common as the 1968 edition of the BSCS Green Version was revised in 1973. However, despite these changes and the failure to integrate evolution throughout the textbook, the coverage of evolution in the 1973 edition of *Biological Science: An Ecological Approach* was quite comprehensive.

The 1973 BSCS Yellow Version edition of *Biological Science: An Inquiry Into Life* (BSCS, 1973) contained 11,000 and 14,000 fewer words, respectively than did the 1968 and 1963 editions (BSCS, 1968, 1963). However, the 1973 BSCS Yellow Version textbook still covered evolution in a more extensive manner than any of the other 10 text-books of the 1970s that were analyzed.

The evolution of fish, amphibians, and plants; geological data concerning the age of the earth; hypotheses explaining the origin of life; various evidences of evolution; causes of evolution and variation such as mutation, recombination, and geographic isolation; the human fossil record; and cultural evolution were among the topics that received less emphasis in the 1973 BSCS Yellow Version than in earlier editions.

Many word changes occurred in the revision of the BSCS Yellow Version in 1973. The 1968 edition referred to Darwin's theory of the origin of species, whereas in 1973 it was termed a hypothesis. In 1963, evolution was called a well-established concept. Theory was substituted for concept in 1968 and in 1973 the reference was omitted entirely. Despite these and other changes in wording, the coverage of evolution in the 1973 BSCS Yellow Version textbook was still extensive.

Overall, evolution was covered in a comprehensive manner in most textbooks of the 1970s. However, certain changes were strong evidence that critics have encountered some success in their efforts to have the study of evolution deemphasized in biology textbooks. Analysis of the 10 textbooks of the 1970s did indicate that efforts to have the Genesis account of creation included in biology textbooks had been largely unsuccessful as only three (Smallwood & Green, 1974; Weinberg, 1974, 1977) of the 10 textbooks analyzed in this study included any material on the topic.

Modern Biology (Otto & Towle, 1973, 1977) continued to be popular in the 1970s, whereas the use of the BSCS versions declined. In the early 1970s, *Modern Biology* (Otto & Towle, 1973) was used by 35–40% of the nation's biology students and one of the three BSCS versions was used by another 40% of the students.[33] However, data from 1974–1976 indicated the number of students using one of the BSCS versions had declined by 5 to 8%.[34]

CONCLUSION

If, as most educators claim, textbooks are important in dictating what, how, and when certain subject matter will be taught, the data in this study are substantial evidence that the study of evolution was a peripheral and neglected part of the biology curriculum prior to the development of the BSCS textbooks in the 1960s. The professional biologist's acceptance of evolution as an important and unifying idea of the discipline had not been powerful enough to counterbalance the rejection and suppression resulting from opinion, legislation, and pressures exerted by organized religious groups, administrative edicts, publisher's caution, threatened teachers, and numerous other forces. Some of these forces were circumvented in the 1960s, when with federal funds, professional biologists led an energetic reconstruction of the biology curriculum that resulted in new objectives, new course content, new authors, and new methods of writing curriculum materials. Evolution emerged from this reconstruction as an important theme and element of biology textbooks. This importance continues to be reflected in many contemporary biology textbooks. Other textbooks, in a retreat from the 1960s, now have deemphasized evolution. The deemphasis has not been a result of a diminishing in the power of evolution to explain and make sense out of the natural world. Instead, the deemphasis has been the result of publishers, authors, educators, and politicians responding to the strenuous efforts of antievolutionists and creationists to suppress and diminish the study of evolution. Also, citizens who do not understand evolution, the constitutional mandate that schools be neutral in matters of religious theory, doctrine, and practice, and the importance of unifying ideas to the study of discipline, stand by and allow special interest groups to influence and dictate the content of the biology curriculum in this nation. This could result in the return of evolution to the periphery of the secondary school biology curriculum. Should this happen, evolution will remain an idea within the world and work of biologists rather than diffusing outward into the minds of all educated people.

REFERENCES

1. J. C. Mayfield, *Using Modern Knowledge to Teach Evolution in the High School as Seen by the Participants in the High School Conference of the Darwin Centennial Celebration at the University of Chicago, November 24–25, 1958* (Chicago: The Graduate School of Education of the University of Chicago, 1960), p. 8.

2. Gerald Skoog, "The Topic of Evolution in Secondary School Biology Textbooks: 1900–1968" (Ph.D. diss., University of Nebraska, Lincoln, 1969; *Dissert. Abstr. Int.*, 1970, 31, 187A–188A; University Microfilms No. 70–12, 285). Full citations for all text books mentioned in this study are given in the Appendix.

3. W. Cubie, "A Comparative Analysis of the Objectives and Content of Biology Instruction in the Secondary Schools in Three Periods as Revealed by Representative Textbooks in the Field during those Periods" (Ph.D. diss. Indiana University, Bloommington, 1958).

4. O. W. Richards, "The Present Status of Biology in the Secondary Schools," *School Review* 31 (1923): 143–46.

5. J. V. Grabiner and P. D. Miller, "Effects of the Scopes Trial," *Science* 185 (1974): 832–37.

6. O. D. Frank, "Data on Textbooks in the Biological Sciences Used in the Middle West," *School Science and Mathematics* 16 (1916): 354–57.

7. See Richards, 1923.

8. G. W. Hunter, *Essentials of Biology* (Chicago: American Book, 1911), p. 8.

9. Hunter, 1911, p. 83.

10. W. M. Atwood, *Civic and Economic Biology* (Philadelphia: P. Blakiston and Sons, 1922), pp. 353–54.

11. H. Linville, *The Biology of Man and Other Organisms* (New York: Harcourt Brace, 1923), p. 465.

12. G. Trafton, *Biology of Home and Community* (New York: Macmillan, 1923), pp. 580–81.

13. O. W. Caldwell and F. Weller, "High School Biology Content as Judged by Thirty College Biologists," *School Science and Mathematics* 32 (1932): 411–24.

14. See Caldwell et al, 1932; and Grabiner et al, 1974.

15. F. D. Curtis, O. W. Caldwell, and N. H. Sherman, *Biology for Today* (Chicago: Ginn, 1934), p. 647.

16. G. W. Hunter, *Problems in Biology* (Chicago: American Book, 1935), p. 271.

17. C. J. Pieper, W. Beauchamp, and O. D. Frank, *Everyday Problems in Biology* (Chicago: Scott Foresman, 1932), p. 337.

18. E. T. Smith, *Exploring Biology* (Chicago: Harcourt Brace, 1938), p. 541.

19. Smith, 1949, p. 554.

20. G. W. Hunter, *Life Sciences: A Social Biology* (Chicago: American Book, 1941), p. 498.

21. G. W. Hunter and F. R. Hunter, *Biology in Our Lives* (Chicago: American Book, 1949), p. 372.

22. See H. Grobman, "The Rationale and Framework of the BSCS Evaluation Program," *BSCS Newsletter* 19 (1963): 6–11.

23. "The BSCS in 1963," *BSCS Newsletter* 22 (1964): 3.

24. See A. Grobman, "National Science Foundation's Role in the Development of Science Education in American Schools," *BSCS Newsletter* 27 (1965): 7–10.

25. See D. Stanke, "Implications of Teacher Preparation in the Effective Use of Teaching Aids for the New Curriculum Materials," *BSCS Newsletter* 32 (1967): 23–24.

26. See Grabiner et al, 1974.

27. W. L. Smallwood and E. R. Green, *Biology* (Morristown, NJ: Silver Burdett, 1968, 1974, 1977), pp. 267 (1974), 255 (1977).

28. Smallwood and Green, 1977, pp. 243, 232.

29. BSCS, *High School Biology, Blue Version* (Boulder: BSCS, 1961), p. 57.

30. BSCS, *Biological Science: Molecules to Man* (Boston: Houghton Mifflin, 1963), p. 29.

31. BSCS, 1968, p. 84.

32. BSCS, 1963, p. 411.

33. See S. L. Helgeson et al, *The Status of Pre-College Science, Mathematics, and Social Science Education. i. Science Education: 1955–1975* (Columbus, OH: Ohio State

University, Center for Science and Mathematics Education, 1977) (ERIC document ED 153876).

34. Helgeson et al, 1977.

APPENDIX: BIBLIOGRAPHY OF BIOLOGY TEXTBOOKS

1900–1919

Abbott, J. F. *The Elementary Principles of General Biology.* New York: Macmillan, 1914, 329 pp.
Bigelow, M., & Bigelow, A. N. *Applied Biology.* New York: Macmillan, 1911, 583 pp.
Hodge, C. F., & Dawson, J. *Civic Biology.* Chicago: Ginn, 1918, 381 pp.
Hunter, G. W. *Elements of Biology.* Chicago: American Book, 1907, 445 pp.
Hunter, G. W. *Essentials of Biology* Chicago: American Book, 1911, 448 pp.
Peabody, J. E., & Hunt, A. E. *Elementary Biology.* New York: Macmillan, 1913, 534 pp.
Peabody, J. E., & Hunt, A. E. *Elementary Biology-Animal and Human.* New York: Macmillan, 1912, 364 pp.
Smallwood, W. M., Reveley, I. L., & Bailey G. A. *Practical Biology.* Chicago: Allyn & Bacon, 1916, 421 pp.

1920–1929

Atwood, W. M. *Civic and Economic Biology.* Philadelphia: P. Blakiston Sons, 1922, 470 pp.
Clement, A. G. *Living Things.* Syracuse, NY: Iroquis, 1924; 1925, 488 pp.
Gruenberg, B. C. *Biology and Human Life.* Chicago: Ginn, 1925, 592 pp.
Hunter, G. W. *New Civic Biology.* Chicago: American Book, 1926, 448 pp.
Hunter, G. W. *New Essentials of Biology.* Chicago: American Book, 1923, 453 pp.
Kinsey, A. *An Introduction to Biology.* Chicago: J. B. Lippincott, 1926, 558 pp.
Linville, H. *The Biology of Man and Other Organisms.* New York: Harcourt, Brace, 1923, 507 pp.
Moon, T. J. *Biology for Beginners.* New York: Henry Holt, 1921, 558 pp.
Moon, T. J. *Biology for Beginners.* New York: Henry Holt, 1926, 647 pp.
Peabody, J. E., & Hunt, A. E. *Elementary Biology—Animal and Human.* New York: Macmillan, 1912, 364 pp.
Smallwood, W. M., Reveley, I. L., & Bailey, G. A. *Biology for High Schools.* Chicago: Allyn & Bacon, 1920, 590 pp.
Smallwood, W. M., Reveley, I. L., & Bailey, G. A. *New Biology.* Chicago: Allyn & Bacon, 1924, 704 pp.
Smallwood, W. M., Bailey, I. L., & Bailey, G. A. *A General Biology.* Chicago: Allyn & Bacon, 1929, 788 pp.
Trafton, G. *Biology of Home and Community.* New York: Macmillan, 1923, 614 pp.

1930–1939

Baker, A. O., & Mills, L. N. *Dynamic Biology.* Chicago: Rand McNally, 1933, 722 pp.
Curtis, F. D., Caldwell, O. W., & Sherman, N. H., *Biology for Today.* Chicago: Ginn, 1934, 692 pp.
Hunter, G. W. *New Essentials of Biology.* Chicago: American Book, 1939, 453 pp.
Hunter, G. W. *Problems in Biology.* Chicago: American Book, 1931, 706 pp.

Hunter, G. W. *Problems in Biology.* Chicago: American Book, 1935, 706 pp.

Kinsey, A. *New Introduction to Biology.* Chicago: J. B. Lippincott, 1933, 840 pp.

Kinsey, A. *New Introduction to Biology.* Chicago: J. B. Lippincott, 1938, 845 pp.

Mank, H. G. *The Living World—An Elementary Biology.* Chicago: Benjamin H. Sanborn, 1933, 673 pp.

Moon, T. J., & Mann, P. B. *Biology for Beginners.* New York: Henry Holt, 1933, 474 pp.

Moon, T. J., & Mann, P. B. *Biology.* New York: Henry Holt, 1938, 866 pp.

Peabody, J. E., & Hunt, A. E. *Biology and Human Welfare.* New York: Macmillan, 1933, 658 pp.

Pieper, C. J., Beauchamp, W., & Frank, O. D. *Everyday Problems in Biology.* Chicago: Scott Foresman, 1936; 1932, 686 pp.

Smallwood, W. M., Reveley, I. L., & Bailey, G. A. *New Biology.* Chicago: Allyn & Bacon, 1934, 604 pp.

Smallwood, W. M., Reveley, I. L., & Bailey, G. A. *New Biology.* Chicago: Allyn & Bacon, 1937, 604 pp.

Smith, E. T. *Exploring Biology.* Chicago: Harcourt, Brace, 1938, 696 pp.

1940–1949

Baker, A. O., & Mills, L. N. *Dynamic Biology Today.* Chicago: Rand McNally, 1943, 822 pp.

Baker, A. O., & Mills, L. N. *Dynamic Biology Today.* Chicago: Rand McNally, 1948, 822 pp.

Curtis, F. D., Caldwell, O. W., & Sherman, N. H. *Everyday Biology.* Chicago: Ginn, 1940, 698 pp.

Curtis, F. D., Caldwell, O. W., & Sherman, N. H. *Everyday Biology.* Chicago: Ginn, 1943, 698 pp.

Curtis, F. D., Caldwell, O. W., & Sherman, N. H. *Everyday Biology.* Chicago: Ginn, 1946, 698 pp.

Curtis, F. D., & Urban, J. *Biology in Daily Life.* Chicago: Ginn, 1949, 608 pp.

Gruenberg, B. C., & Bingham, N. E. *Biology and Man.* Chicago: Ginn, 1944, 685 pp.

Hunter, G. W., & Hunter, F. R. *Biology in Our Lives.* Chicago: American Book, 1949, 534 pp.

Hunter, G. W. *Life Sciences—A Social Biology.* Chicago: American Book, 1941, 803 pp.

Moon, T. J., & Mann, P. B. *Biology.* New York: Henry Holt, 1941, 866 pp.

Moon, T. J., Mann, P. B., & Otto, J. H. *Modern Biology.* New York: Henry Holt, 1947, 664 pp.

Smallwood, W. M., Bailey, I. L., & Bailey, G. A. *Elements of Biology.* Chicago: Allyn & Bacon, 1948, 691 pp.

Smith, E. T. *Exploring Biology.* Chicago: Harcourt, Brace, 1943, 619 pp.

Smith, E. T. *Exploring Biology.* Chicago: Harcourt, Brace, 1949, 607 pp.

Vance, B. B., & Miller, D. F. *Biology for You.* Chicago: J. B. Lippincott, 1946, 731 pp.

1950–1959

Baker, A. O. & Mills, L. N. *Dynamic Biology Today.* Chicago: Rand McNally, 1953, 822 pp.

Baker, A. O., Mills, L. N., & Tanczos, J. *New Dynamic Biology.* Chicago: Rand McNally, 1959, 616 pp.

Curtis, F. D., & Urban, J. *Biology in Daily Life.* Chicago: Ginn, 1953, 608 pp.

Curtis, F. D., & Urban, J. *Biology the Living World.* Chicago: Ginn, 1958, 705 pp.
Curtis, F. D., Caldwell, O. W., & Sherman, N. H. *Everyday Biology.* Chicago: Ginn, 1953, 698 pp.
Dodge, R., Smallwood, W. M., Reveley, I. L., & Bailey, G. A. *Elements of Biology.* Chicago: Allyn & Bacon, 1952, 746 pp.
Dodge, R., Smallwood, W. M., Reveley, I. L., & Bailey, G. A. *Elements of Biology.* Chicago: Allyn & Bacon, 1959, 740 pp.
Kroeber, E., Wolff, W., & Weaver, R. *Biology.* Boston: D. C. Heath, 1957, 608 pp.
Moon, T. J., Mann, P. B., & Otto, J. H. *Modern Biology.* New York: Henry Holt, 1951, 698 pp.
Moon, T. J., Mann, P. B., & Otto, J. H. *Modern Biology.* New York: Henry Holt, 1956, 757 pp.
Smith, E. T. *Exploring Biology.* Chicago: Harcourt, Brace, 1954, 579 pp.
Smith, E. T. *Exploring Biology.* Chicago: Harcourt, Brace, 1959, 731 pp.
Vance, B. B., & Miller, D. F. *Biology for You.* Chicago: J. B. Lippincott, 1950, 733 pp.
Vance, B. B., & Miller, D. F. *Biology for You.* Chicago: J. B. Lippincott, 1954, 652 pp.

1960–1969

BSCS. *Biological Science: An Inquiry into Life.* Chicago: Harcourt, Brace & World, 1963, 748 pp.
BSCS. *Biological Science: An Inquiry into Life.* Chicago: Harcourt, Brace & World, 1968, 840 pp.
BSCS. *Biological Science: Molecules to Man.* Boston: Houghton Mifflin, 1963, 715 pp.
BSCS. *Biological Science: Molecules to Man.* Boston: Houghton Mifflin, 1968, 823 pp.
BSCS. *BSCS Green Version High School Biology.* Chicago: Rand McNally, 1968, 823 pp.
BSCS. *High School Biology, Blue Version.* Boulder: Biological Sciences Curriculum Study, 1961, 766 pp.
BSCS. *High School Biology, BSCS Green Version.* Chicago: Rand McNally, 1963, 749 pp.
BSCS. *High School Biology, Green Version.* Boulder: Biological Sciences Curriculum Study, 1961, 650 pp.
BSCS. *High School Biology, Yellow Version.* Biological Sciences Curriculum Study, 1961, 870 pp.
Kroeber, E., Wolff, W., & Weaver, R. *Biology.* Boston: D. C. Heath, 1960, 646 pp.
Kroeber, E., Wolff, W., & Weaver, R. *Biology.* Boston: D. C. Heath, 1965, 376 pp.
Moon, T. J., Otto, J. H., & Towle, A. *Modern Biology.* New York: Henry Holt, 1960, 758 pp.
Moon, T. J., Otto, J. H., & Towle, A. *Modern Biology.* New York: Holt, Rinehart, & Winston, 1963, 758 pp.
Otto, J. H., & Towle, A. *Modern Biology.* New York: Holt, Rinehart, & Winston, 1965, 792 pp.
Smith, E. T., & Lawrence, T. G. *Exploring Biology.* Chicago: Harcourt, Brace & World, 1966, 766 pp.
Weinberg, S. *Biology.* Boston: Allyn & Bacon, 1967, 684 pp.
Vance, B. B., & Miller, D. F. *Biology for You.* Chicago: J. B. Lippincott, 1963, 660 pp.

1970–1977

BSCS. *Biological Science: An Ecological Approach.* Chicago: Rand McNally, 1973, 740 pp.
BSCS. *An Inquiry into Life.* Chicago: Harcourt Brace Jovanovich, 1973, 912 pp.

BSCS. *Molecules to Man.* Boston: Houghton Mifflin, 1973, 764 pp.

Oram, R. F. *Biology Living Systems.* Columbus: Charles E. Merrill, 1976; 1973, 706 pp.

Otto, J. H., & Towle, A. *Modern Biology.* New York: Holt, Rinehart, & Winston, 1973, 858 pp.

Otto, J. H., & Towle, A. *Modern Biology.* New York: Holt, Rinehart, & Winston, 1977, 757 pp.

Smallwood, W. L., & Green, E. R. *Biology.* Morristown, NJ: Silver Burdett, 1974, 733 pp.

Smallwood, W. L., & Green, E. R. *Biology.* Morristown, NJ: Silver Burdett, 1977, 737 pp.

Weinberg, S. *Biology.* Boston: Allyn & Bacon, 1974, 644 pp.

Weinberg, S. *Biology.* Boston: Allyn & Bacon, 1977, 565 pp.

Evolution, Creation, and Biology Teaching

by Richard D. Alexander

No laws were ever passed saying that evolution had to be taught in biology courses. The prestige of evolutionary theory has been built by its impact on the thousands of biologists who have learned its power and usefulness in the study of living things. No laws need to be passed for creationists to do the same thing.

Recently, creationists, anti-evolutionists, and others have sought to revive arguments that grave doubts should exist as to whether or not all animals and plants, and particularly the human species, are products of the slow, step-by-step, cumulative process of mutation and natural selection that biologists call evolution.[1] Persons familiar with the data supporting evolution, and others who accept the views of professional biologists without reviewing the evidence themselves, have paid little attention to the creationists' arguments, which are essentially unchanged from those prominent a half century ago.[2] A certain proportion of people who are emotionally involved, probably some on each side, are unlikely to be swayed by arguments or data. Another group, to whom this essay is principally addressed, includes those who for one reason or another remain genuinely in doubt, or unable to satisfy themselves easily and quickly on this issue, and those who seek reviews of the evidence for teaching purposes.

Creationists have concentrated their efforts on secondary and primary school biology courses where they can involve those parents for whom this may become an emotional issue, both because of apparent conflict with religious beliefs and because parents may feel some responsibility to guard their younger children against exposure to certain issues or attitudes. Success is also more likely here than at college levels, because it is easier to enact legislation affecting primary and secondary schools, and to influence classroom materials through the control of school boards. Such efforts have succeeded temporarily in states such as Tennessee, where legislation was passed and later declared unconstitutional, requiring that creation be discussed as an alternative theory whenever evolution is discussed in public schools;[3] and California, where the state school board has required that creation be included in biology textbooks and other classroom materials discussing evolution.[4]

Bills requiring discussion of creation in high school biology courses mentioning evolution are being submitted yearly to state legislatures. They are modified repeatedly to test what might eventually become acceptable to the legislature in each particular state. Recently, four such bills were presented to the Committee on Education of the Michigan Legislature: Michigan Senate Bills 66, 67; Michigan House Bills 4047, 4339, Jan. and Mar. 1973; one of these passed the House by a vote of 71–25. This is a pernicious move that calls for resistance. If evolutionists were attempting to require that evolution be taught it would be no less pernicious. When a creationist, Darwinist, Marxist, or supporter of any other theory defends his or her views publicly, he or she does everyone a service. But when anyone attempts to establish laws or rules requiring that certain theories be taught or not be taught, he or she invites us to take a step toward totalitarianism. Whether a law is to prevent the teaching of a theory or to require it is immaterial. It does not matter if equal time is being demanded or something called "reasonable" time, because there can be no reasonable time in such a law.

No teacher should be dismayed at efforts to present creation as an alternative to evolution in biology courses; indeed, at this moment creation is the only alternative to evolution. Not only is this worth mentioning, but a comparison of the two alternatives can be an excellent exercise in logic and reason. Our primary goal as educators should be to teach students to think and such a comparison, particularly because it concerns an issue in which many have special interests or are even emotionally involved, may accomplish that purpose better than most others.

The human background is a central question in the lives of thoughtful individuals who wish to understand themselves and others. Society needs nothing more, perhaps, than a thorough comprehension of human tendencies, motivations, and possibilities. These are, in large part, the issues when one is contemplating the effects of human history upon our behavior. Creation and evolution in some respects imply backgrounds about as different as one can imagine. In the sense that creation is an alternative to evolution for any specific question, a case against creation is a case for evolution and *vice versa*.

With regard to creationist theories about life, we are in a peculiar position because many people are taught from childhood that there is a Creator, who is to be revered absolutely and unquestioningly. When creation theorists strive to introduce creation into the classroom as an alternative biological theory to evolution they must recognize that they are required to give creation the status of a falsifiable idea—that is, an idea that loses any special exemption from scrutiny, that is accepted as conceivably being false, and that must be continually tested until the question is settled. A science classroom is not the place for an idea that is revered as holy. If efforts to keep creation and Creator in such status, in regard to the history of life on earth, accompany moves to incorporate them into science teaching in public schools, then such efforts would properly be viewed as efforts to introduce religion into the classroom.

The evidence supporting and detailing the facts and theory about evolution can be found in any introductory biology textbook, such as that by William Keeton.[5] The evidence, however, is complex and multifaceted. This is why evolutionary theory will always remain vulnerable to distortion by those who insist upon a quick, simple review. This essay is not intended to provide a description of the range of evidence supporting evolution. On the other hand, the evidence against creationism, as espoused by members of the Creation Research Society and others, involves relatively simple arguments and can be summarized easily. Creationist arguments are few, and they are repeated almost without change or development throughout the creationist literature of this and other decades. Their applicability to biological questions depends wholly upon a number of highly questionable or demonstrably false dichotomies. Creationist arguments can also be shown to involve significant retreats, indicative of untenable hypotheses. By treating creationism as an alternative to evolution, teachers have an excellent opportunity to demonstrate the strength and usefulness of the evolutionary model of life as a framework for biological investigation and understanding.

Comparisons between the views of creationists and evolutionary biologists are also useful because the most important change that can occur in biology is a dramatic updating of evolutionary theory and teaching. The views of evolution that I see publicized by the creationists of this decade are antique views, with little relevance to what is going on in biology today. They treat the controversy between evolution and creation as if it were static—as if nothing had happened since 1859—when, in fact, evolutionary theory has advanced steadily since Darwin. In contrast, creationist theory has inexorably retreated toward those sets of problems and ideas on which there is yet no significant evidence.

The theory of natural selection is being used today to develop and test predictive hypotheses about sex ratios, sexual dimorphism, sexual competition, sexual selection, parental investment, nepotism, social reciprocity, group-living, altruism, senescence, rates of infant mortality, and other problems to which it was not being applied significantly as recently as a decade ago. Unfortunately, high school biology teachers, who completed their formal training in biology before this new wave of evolutionary ecology and social biology had begun, are being dragged into ancient arguments and diverted from the truly exciting aspects of modern evolutionary biology.

SOME GENERAL REMARKS ABOUT EVOLUTION

The massive volume that Charles Darwin published in 1859 resulted from nearly 20 years of field observations, comparisons, experimentation, and logical thought about the nature of living organisms.[6] In it Darwin expounded his theory of evolution by natural selection. In the Galapagos Island region, he had noticed that species, believed at that time to be immutable, were in some cases more similar to one another than in other cases. Sometimes he could not tell if two populations were parts of the same species or parts of two different species. He also noticed that island species, or populations, were more similar if the islands were closer together; and that they were more similar when they occurred in different climatic regions on the same continent than if they occurred in the same climatic regions on different continents. These early observations and comparisons led Darwin to suppose that perhaps species are not immutable after all, but changeable, and that one species may sometimes give rise to two or more species. Eventually he decided that the process of change involved in this speciation, or species multiplication, must result because variants exist within every species; some variants out-reproduce others; and which ones out-reproduce in any given time or place depends upon the environment. This process of natural selection of variants, which he compared to the artificial selection that man carries out on his domestic animals and plants, would cause populations on different islands to diverge unless they had chances to interbreed; this, in turn, would cause speciation to happen whenever accidental separations lasted long enough.

From this reasonable but startling beginning, Darwin went on to even more astonishing postulates, including the following:

1. All attributes of living organisms might be the result of a cumulative process of natural selection, extending backward through time to the beginnings of life on earth.

2. The major groups of organisms alive today differ from one another because they got separated during speciation processes in the distant past.

3. The entire fossil record is a remnant from the operation of heritable changes, natural selection, and isolation in a succession of past environments. Significantly, he noted that the fossils of a given continent generally resemble the living organisms of that continent rather than the fossils of any other.

From this beginning by Darwin, we derive the three major areas of investigation in evolutionary biology: (1) speciation, or how species multiply; (2) adaptation, or precisely how natural selection works; and (3) phylogeny, or the tracing of the patterns of evolutionary change through time. For the first several decades following Darwin it was phylogeny, and the search for more fossils, that were emphasized in biology. Later, speciation became an enormously popular area of investigation. Today, the study of adaptation, or the predictive and analytical value of natural selection, is paramount.

Darwin's combinings of facts, theories, hypotheses, conjectures, speculations, and guesses made sense in 1859, and they make sense now. Darwin's arguments and his methods have been tested, retested, examined, discussed, and refined by perhaps the greatest army of diligent and skeptical investigators ever to examine any testable hypothesis in the history of man. No evidence is available to deny the evolutionary process that is accepted as the working hypothesis of probably more than 99% of the active investigators in biology today. Thus, biologists pay scant attention to the arguments of the few anti-evolutionists. What they have learned about biology and evolution leaves them convinced that evolution is the framework within which they must operate; they have no uneasiness that what they are doing will be much affected by anything that could be said in brief oral debates or dissections of the arguments of creationists.

Biologists also know that, through their journals and professional meetings, they will root out errors in their findings. On the whole, they subscribe to George Gaylord Simpson's simple definition of science as a self-correcting method of finding out about the universe.

If evolution involved only fruit flies and cabbages, creationists would not attempt to have laws passed saying which theories must or must not be mentioned in classrooms. Anti-evolutionists and creationists are concerned because ultimately the same kinds of questions and tests that evolutionary theory uses to analyze the various other organisms in the world are likely to be applied to efforts to understand ourselves. They recognize the possibility of conflicts between evolutionary theory and their particular religious or belief systems. Such conflicts may often occur when the two systems of explanation are being used to explain or reconstruct human history. No conflict exists, however, between evolution and religion (or any social, political, or economic idealogies) when the latter is concerned with plans or goals for society, or the future of human behavior. Evolution is an explanatory theory about history. Anthropologists, most of whom accept that humans have evolved, ultimately must examine tendencies toward having certain kinds of ideologies as products themselves, directly or indirectly, of the evolutionary process. They began long ago to investigate religion in that fashion. Such investigations have an unnerving aspect. But they also have an intriguing quality. Consider the paradox of an organism possessing some quality of self-awareness, trying to analyze itself, using for the analysis the very attributes that are to be analyzed, when one of the most prominent of those attributes is resistance to any such analysis. That is the most difficult challenge we are likely to extract from this universe for a long, long time.

These are the difficult problems that every thoughtful biology teacher has to consider in order to discuss organic evolution in the classroom, because evolution leads inexorably to the analysis of human beings. In fact, revolutions in our thinking about human behavior have already begun, chiefly within evolutionary biology; part of the evidence is contained in papers published by Hamilton, Williams, Trivers, Alexander, West-Eberhard, and Wilson.[7] Such revolutions can be productive, so long as they remain in the realm of open scientific debate, and so long as they never lose the quality of self-correction. But biology teachers assume an awesome responsibility when they undertake to discuss the relationship of human history to human behavior in terms of possible and probable causes, including Darwinian or natural selection.

A STATEMENT OF MODERN EVOLUTIONARY THEORY

Darwinian theory, as used by evolutionary biologists today, is simple to state, difficult to apply, and astonishing to contemplate. The evolutionary process from which it stems derives from the interaction of five basic phenomena.

1. *Inheritance:* All living organisms (phenotypes) are products of the interaction of their genetic materials (genotypes) with their developmental (ontogenetic) environments; these genetic materials can be passed from generation to generation unchanged.

2. *Mutation:* The genetic materials do change occasionally, and these changes are in turn heritable.

3. *Selection:* All genetic lines do not reproduce equally, and the causes of this variation may be consistent for long periods.

4. *Drift:* Genetic materials are sometimes lost through accidents, which are random or nonrepetitive in their effects on populations.

5. *Isolation:* Not all genetic lines are able, for various intrinsic and extrinsic reasons, to interbreed freely, and thus to continually reamalgamate their differences.

These five phenomena have all been demonstrated repeatedly, and can be demonstrated at will, as can their various interactions. No living things have been demonstrated to lack any of them, or are suspected to lack any of them. Hence, they are the factual basis of evolution.

Of the five main components of the evolutionary process, natural selection, or the differential reproduction of genetic variants, is almost universally accepted as the guiding force. The reasons for this assumption, which are not widely discussed, but which are crucial to the understanding of evolution, are: first, that altering directions of selection apparently always alters directions of change in organisms (although, because of genetic specialization or the absence of appropriate mutants, possibly in some cases only after delay); second, that the causes of mutation and the causes of selection appear to be independent; and, third, that only the causes of selection sometimes (but not always, of course) remain consistently directional for relatively long periods.

Mutations are most often caused by atmospheric radiation. Selection is caused by an updated version of what Darwin termed the "Hostile Forces of Nature": climate, weather, food shortages, predators, parasites, and diseases. This list implies competition for resources, such as food, or protection from the other hostile forces; accordingly, for all sexual species, we must include as a selective factor competition for mates, and for the best mates.

Because directions of mutation evidently remain random in regard to directions of selection (although not necessarily in any other respect), mutational changes as such are independent of adaptation, or the fine tuning that organisms exhibit in response to their physical and biotic environments. The same is true of genetic drift, for its causes are by definition without cumulative directional effects on the genetic materials. Thus, as evolution proceeds mutations must increasingly tend to become deleterious, and their rates have likely been severely selected downward. Also, directional evolutionary change has to be caused by directional selection. The only apparent exception is the concept of selection suddenly becoming absent in the environment of a complex organism with mutational changes then leading to steady reductions in complexity. Although this effect has sometimes been postulated when some particular selective pressure has evidently disappeared (e.g., reductions in size and complexity of human teeth with the advent of cooked food, or disappearance of eyes in cave animals), such cases are more appropriately explained as changes in directions of selection. In no way do they support an argument that selection itself somehow mysteriously disappeared from the organism's environment. When one direction or force of selection is removed from the environment of a species, the effect is to cause other previously opposing forces to become more powerful or effective.

These are the reasons, then, for the common tendency to refer to the theory of evolution as the theory of natural selection. They include the assumption that long-term evolutionary changes result from the effects of natural selection across long periods of time (see arguments below on this question). Refinements of evolutionary theory since Darwin have chiefly involved new understanding of adaptiveness from short-term studies of the selective process, and comparative studies of function. The results of these studies lead us to the conclusion that to apply evolutionary theory we must focus our attention on the causes and effects of differential reproduction.

MODERN CREATIONIST ARGUMENTS

Following is a list of the usual creationist arguments. All of them may be found in the controversies of the nineteenth century and the early twentieth century, as well as in the more recent references cited earlier. These arguments include: (1) Information can be divided into facts and theories. (2) Evidence can be divided into that which is conclusive and that which is only circumstantial. (3) Facts are derived only from conclusive evidence; and (4) conclusive evidence comes only from direct observations and experiments. (5) Since the essence of science is repeatability, and (6) repeatability necessarily involves experimentation, which can only be carried out through direct observation, then (7) if a conclusion does not come from directly observable phenomena, it is not scientific because the evidence is only circumstantial. Hence, (8) comparative study of the present cannot lead to facts about the past; (9) Darwin's comparative method, by which he "discovered" evolution and speciation, is neither scientific nor conclusive; and (10) we cannot study the past scientifically, especially not the distant past. (11) Questions about life can also be divided into "mechanisms" and "origins," or "means" and "ends." (12) General evolution or macro-evolution (ends) cannot be equated with natural selection, special evolution, or micro-evolution (means), for (13) natural selection deals only with mechanisms, not with origins, and (14) there is no scientific evidence about the origins of kinds of life. (15) Evolution refers to a progression from "amoeba to man" but (16) selection cannot be demonstrated to cause new organs or new species, rather it is (17) just a variation on a limited set of themes. (18) Change in living things can thus be divided into "within-kinds" change and "between-kinds" change. (19) Only "within-kinds" change can be observed directly; and (20) there are no genetic connections between major groups. (21) Mutational changes do not link major groups; nor do chromosomal rearrangements or ploidy. Therefore, (22) natural selection is different from evolution, and (23) there is no scientific evidence about "between-kinds" change. (24) The fossil record, which might be used to support evolutionists on the gradual nature of evolution, is woefully incomplete; (25) what is missing are all of the links or postulated intermediates between major groups. (26) All known dating methods are notoriously inaccurate; and (27) there is evidence both of a widespread flooding and of overlap of man with trilobites. (28) Evolution also means progressive change, but the only real source of variations upon which selection can act are mutations, and (29) all mutations are deleterious, as is witnessed by the reversion to the "wild state" by all organisms once they are released from artificial selection. Therefore, since (30) all known change in life is degenerate (because all mutations are deleterious), and (31) all known change in non-living matter under natural conditions is also from complex to simple, (32) it is doubtful whether even natural selection can be used to explain anything at all about life. (33) The scientist is like a fisherman who uses a two-inch mesh in his net: he cannot catch fish under two inches in size. (34) Creation is the superior theory because it accords with the gaps in the fossil record and can be used to explain every difficulty that confronts an evolutionary theory.

Permeating these arguments are three principal themes. The first is the idea that there are basic dichotomies in the nature of questions about the history of life, and that although support for a selective mechanism of short-range or minor change may be justified, nothing is thereby suggested about long-range or major change (see arguments 1–23, 33). The second is the argument that the fossil record is essential to evolutionary theory, yet is incomplete in ways that support creation and diminish evolution (24–27). The third is the assertion that all mutations are deleterious and all change by selection therefore degenerate unless it results from created variation as opposed to environmentally induced mutations or novelties (28–32).

REFUTING CREATIONIST DICHOTOMIES

Facts and theories are not separated by a magic line. There is no magical or profound difference in what one does with these two concepts. Scientists deal in probabilities. Arbitrarily, scientists have chosen the levels of 95% probability and 99% probability as appropriate confidence levels in statistical analyses of their data. They require that the results obtained in their tests are only 5% or 1% likely to have resulted from chance alone. They call this a positive result, even though something remains unknown about the situation that somehow accounts for that last 5% or 1%. Such a result does not mean that the problem is solved. It simply means that one can proceed to the next step in the investigation with some confidence—95% or 99%, to be "exact."

The creationists' arguments suggest that a fact is something that, once discovered, is kept forever like a coin or a preserved butterfly. Not so. Nothing is irreversibly factual. Any fact may turn out not to be a fact at all; and in scientific investigation the only useful thing one can do with a fact is to use it to build better or more complete explanations. What researchers do with facts is establish the next line of hypotheses. And if their "fact" proves vulnerable, they discard it and start over. It is a fact that 100% certainties are obvious only in useless tautologies such as: Hairless men have no hair. It is a fact that life insurance companies make money by operating on probabilities.

Conclusive evidence and circumstantial evidence are not separated by a magic line. Creationists distinguish between what they call direct or conclusive evidence and circumstantial evidence. So do courts of law. But there is a large difference. Courts admit that no magic line separates the two. Sometimes one cannot tell if the evidence is direct or merely circumstantial. Moreover, courts recognize that facts can derive from circumstantial evidence. People are still sentenced on circumstantial evidence.

We do not know who our relatives are from direct knowledge; we must rely upon what others have told us. Yet, we all consider that we know such things beyond significant doubts. In all likelihood no one ever did an experiment on whether or not the sun would rise the next day, yet we regard it as a fact that the sun rises each day. We do so because we have repeated the observation so many times as to render completely trivial the likelihood that it is accidental or random; but we have not thereby *eliminated* the *possibility* that the sun will not rise tomorrow.

There is no fundamental difference between the comparative method and the experimental method in biology. Both experiments and comparative studies attempt to discover statistically significant differences between sets of observations. The distinction is not in the amount of control or the precision of the results, but in the presence or absence of manipulation and in the usual kinds of controls employed. In experimentation, we deal with phenomena that can be manipulated, sometimes to make the comparisons easier or quicker, or more likely to yield unequivocal results. We depend upon comparisons without manipulations when we must—when, for example, we are dealing with long-term phenomena, or with variables whose effects cannot be eliminated and so must somehow be randomized.

The ideal test of the effectiveness of seat belts in reducing deleterious effects of automobile accidents would be experiments in which groups of identical automobiles driven by groups of drivers identical in weight, height, and other attributes were caused to have identical crashes. We cannot set up such experiments, but we do not simply give up on making decisions about seat belts. Instead, we search for other methods. Experiments with dummies and animal substitutes are useful. But the most important information probably has come from comparisons of unplanned accidents in which seat belts are (1) used and (2) not used. Such comparisons represent precisely the kinds of studies used by evolutionists to solve problems about long-term processes. By making appropriate comparisons we use the *natural experiments,* just as Darwin developed the theory of natural selection by comparing variously diverged populations with varying likelihoods of exchanging migrant individuals.

The problem with natural experiments is that they are not designed to answer the particular questions we want to answer. Sometimes, we can answer a question more precisely with specially designed experiments. It is not the precision of the results that represents the difference between the comparative method and the experimental method, however, but the difficulty of discovering how to make the natural experiments answer our question. This involves chiefly the manner in which we control the experiment not its precision. One controls a natural experiment not by *eliminating* the effects of irrelevant or confusing variables, as in a laboratory experiment, but by *randomizing* them.

Creationists' distinctions between origins and mechanisms depend upon all the other dichotomies. One must always ask: Origin of what? How does one tell whether he is talking about origins or mechanisms? We can sometimes demonstrate that differences between traits in organisms are due to genetic differences that derive from mutations, and some creationists do not deny this. But they distinguish between origins of *major* organs, or *major* traits, and mechanisms. Moore argues that it is scientific to require the evolutionists to reconstruct each case of speciation.[8] Unless one can tell precisely how and when and where each species formed, he suggests, to talk about speciation as a process is unscientific. Furthermore, since these questions about long-term events like formation of major organs or speciation, cannot be answered,

Moore contends that such events must be as easily attributable to creation as to evolution.

Requirements that every case of long-term change be reconstructible in detail from direct observation, however, are approximately as scientific as suggesting that life insurance companies cannot make money unless they know how and when each person insured is going to die; or that we should not fasten our seat belts until the ideal experiment, described earlier, has been carried out. Insurance companies in fact make money by knowing on average when deaths are likely. Evolutionists make progress in understanding the attributes and history of living organisms using the same kinds of information.

Erection of false dichotomies in efforts to employ creation as a theory explaining life has forced creationism to undergo significant retreats. Creationists argue as though evolutionary explanations and creationist explanations are both static, neither advancing and neither retreating. This is not true. With the adoption of an attitude demanding (and admitting) verifiable evidence both for evolution and for creation, creationists were forced to acknowledge existence of the process they came to call "micro-evolution." "Micro-evolution" is synonymous with the evolutionary process evolutionists theorize can be projected in a uniformitarian fashion to explain life in general. This left the creationists defending creation only against "macro-evolution" or long-term change, which they argue cannot be investigated scientifically. Ironically, Darwin, ignorant of both the genetic basis of life and the nature of mutational change, modelled the long-term process of speciation by comparing near and distant island species as early as 1837,[9] and may have been led only subsequently to his theory of both short- and long-term change by selection (Darwin 1859) and the slow divergence of populations in different localities with different constellations of selective forces.

Distinguishing macro- and micro-evolution forced creationists to draw the line between these phenomena. Initially, they drew this line between "within-species" and "between-species" changes, contending that these two kinds of changes were not due to the same phenomena because species were products of creation. As biologists' understanding of species developed, however, it became clear that although species ordinarily do not interbreed in nature they can often be caused to do so by altering their environments or forcing them together in the laboratory. In general, the more similar two species are, the more likely it is that they can hybridize, and that the hybrids will be fertile. Thus, no absolute genetic gap exists at the species level. It is also well known that when two different individuals in the same species are mated to produce hybrids, the hybrids are likely to be intermediate in some characteristics, like one parent in some characteristics, and like the other parent in others. The same is true when two species are hybridized.

Moreover, every biologist studying species in any group of organisms finds some populations for which there is no way of deciding whether or not they have achieved full species status, regardless of how that criterion is established. Therefore, every degree of difference, evidently down to the level of individual mutations, exists between diverging populations; and there are numerous cases in which the irreversibility of the divergence of populations is uncertain, depending upon external environmental events such as the permanence of geographic or ecological barriers, which are not entirely predictable.

Contrary to creationist arguments, all of these facts indicate that the differences between species are, like those between individuals within a species, simply accumulations of mutations. Thus, the idea that reproductive barriers between species are the result of anything alien to the basic evolutionary process as we know it, is insupportable; evidence for the opposite conclusion is abundant.

In view of this evidence, supporters of a theory of creation have retreated in two ways. First, they have centered their defense farther up the taxonomic hierarchy, sometimes referring to the genus rather than the species when speaking of the probable products of creation. Second, they have tended to become vague about the exact level at which micro- and macro-evolution became distinct from one another, often speaking of "within-kinds" and "between-kinds" change without defining kinds. In still other instances, they suggest that what was created, or what evolutionary theory cannot explain, are "major groups."

The species concept, with all its difficulties, has the real correlate of reproductive isolation under natural conditions, sometimes difficult to apply, but directly observable whenever the species involved breed at the same times and places. Genera, on the contrary, are simply groups of species placed together because of overall similarity, with generic limits a matter of opinion and convenience in classification. In fact, hybrids between species belonging to different genera are common,[10] and hybrids have even been obtained between species of fish belonging to different families.[11] Major groups are even less definite, and fewer in number. A creationist theory restricted to "major groups" is much less important than one presumed to account for lower-level groups, and too indefinite to be meaningful.

Sometimes, alterations of our views about presumed long-term trends in evolution, such as orthogenetic and "progressive" trends, the idea that ontogeny recapitulates phylogeny, or the particular phylogenetic constructions proposed for certain groups (e.g. horses), have been regarded as casting doubt on evolutionary theory in general.[12] Such arguments lack foundation because it is highly unlikely that anything as complex and poorly documented as the long-term history of life could be reconstructed without many errors and false starts; and the revisions proposed do not suggest causes other than natural selection. Moreover, every time supposed special features of long-term evolution like orthogenesis, progress, and recapitulation are diminished in importance or eliminated, the argument is strengthened that macro-evolution is nothing but micro-evolution over longer time spans.

When a theory must constantly retreat, this is evidence in favor of its alternatives. In this case, it is not only clear that there is no definite line between natural selection and evolution, but that creation must be applied at some entirely different level in this universe than that of explaining existing traits and kinds of living organisms if it is to remain a viable idea.

EVOLUTION AND THE FOSSIL RECORD

The fossil record is not really necessary to defend an evolutionary explanation of life. Nevertheless, it is extraordinarily supportive of evolution. In terms of whether or not long-term evolution by natural selection has occurred, there simply are not significant problems, just as there are no real missing links between man and proto-man. The important point is not exact dates, exact sequences, or directionality of changes. The dates themselves, or changes in dating, are not challenges to evolutionary theory, though they are often so headlined in the newspapers. The important points are two. First, dates, sequences, and directional changes, as known, generally accord with one another. Estimates of relative ages based on location in the ground roughly match the estimates of relative age based on the nature of the fossil. When isotope dating methods became possible the relative ages determined by those methods for the most part matched what had already been learned. Yet the chances of the above three complex kinds of data matching by accident, in the fashion required to support evolutionary theory, are infinitesimal.

The second important point about paleontological evidence is that due to the incompleteness of the data and the imperfection of the methods of measurement available at any given time, it is entirely predictable that slight mismatches of fossil data will occur. Moreover, increases in numbers or prominence of such cases should occur sometimes when new data or methods are acquired. Such inconsistencies do not support evolution; neither do they negate it. They always must be considered in light of the overall consistency of paleontological evidence and the apparent incompleteness of data on the particular problem involved. Most important is what happens to such cases after they have been identified. Do they tend to disappear as more knowledge is gained? Such trends cannot fail to support evolution. In the face of such trends even the persistence of a "hard core" of inconsistent cases fails to detract from evolutionary theory. Moreover, to support a creationist theory an opposite trend would be required: a growing number of cases inconsistent with evolution that fall into a definite pattern supporting a creationist explanation. Such a pattern already exists to support evolution, based upon thousands of separate cases. Hundreds of new paleontological discoveries are made each year by hundreds of paleontologists competing with one another to discover what really happened during the history of life on earth. The number of problems solved by these discoveries far exceeds the number raised.

Gaps exist in the fossil record for the following reasons:

1. Not all species are preserved.
2. The more time that has elapsed, the more chance there is for loss.
3. Earlier animals tended to be softer and small, hence less likely fossilized.
4. Evolution is sometimes more rapid, giving less opportunity for fossilizing some of its stages.

Gaps exist between major groups because:

1. We define groups as those between which gaps still exist.
2. Intermediates between major groups, as one would expect, tend to be more ancient than those between groups lower in the taxonomic hierarchy and accordingly more recent; hence they are less likely available as fossils.

We reconstruct the past just as we predict the future. Our information is incomplete in each case, and we can gain new evidence in each case to test a model or a prediction. Complaints are made about reconstructions based on sequences developed from data fragments from different places. Perhaps it would be optimal to be able to reconstruct a complete sequence from one beginning, but we really have no reason to expect animals to have been fossilized in perfect arrangements for such a purpose. To argue that the past cannot be reconstructed is even less reasonable than to argue that the future cannot be predicted.

Moore says that a major prediction of creation theory is that there will be gaps between distinct kinds of forms of living animals and plants, with different degrees of variability within known kinds of animals and plants.[13] But does such a theory predict what kinds of gaps will occur? Evolutionary theory predicts correctly that there should be more fossils of bony and shelled animals and more gaps in soft-bodied forms, more fossils of recent forms and fewer of more ancient forms, and erratic gaps because of irregular spacing and varying severity of environmental catastrophes and changing rates of evolution in different circumstances.

If as time passes, no one finds an exception to meet Darwin's challenge of universality, the theory of evolution by natural selection is further confirmed. As additional fossil discoveries continue to increase the number of attributes of organisms for which extinct intermediate forms are known—such as kinds of legs and wings, sizes and kinds of skulls—it becomes increasingly probable that the structures of organs for which no intermediates between extant forms are known were also once represented by intermediates. As the proportion of living forms unrepresented by extinct forms is steadily reduced by fossil finds, as has happened continuously since Darwin's theory was first published, the theory of gradual evolutionary change is increasingly supported. Whenever a specific gap used by creationists as evidence of creation is filled, the power of creation as an explanatory theory is further diminished.

ERRONEOUS ASPECTS OF CREATIONIST DESCRIPTIONS OF NATURAL SELECTION

Change by natural selection is not degenerative. Creationists argue that all "constructive" genetic variation was created, that all mutations are deleterious, and that all change by selection acting on mutants must be degenerative. These arguments are paradoxical for several reasons, including:

1. Selection can be shown to act upon any existing variations as well as upon demonstrably novel mutations, simply by altering the environment.

2. Some new mutations can be shown to be identical to alleles already existing (mutations are evidently recurrent).

3. What is deleterious in one environment can be shown to be advantageous in another.

Thus, a line cannot be drawn between existing variation that might have been created and that introduced by recurrent mutations, and whether a variant is advantageous or not depends entirely upon its environment and not upon whether it is a part of what appears to be the existing "natural" variation within a species or a known recent mutant.

Change by natural selection is not progressive, except in the sense of improving adaptiveness. There is no implication of progress from simple to complex, from amoeba to man, nor is there any sense of better or worse, except in relation to adaptiveness to the immediate environment. Accordingly, changes from complex to simple in modern organisms are not evidence against evolution but cases of evolution. When organisms that have been selected by man are released from that selection they are being returned to the environment where their original attributes were acquired, and through natural selection their original traits, or similar traits, once again become prominent.

Natural selection and not creationism leads to testable theories about the evolution of many aspects of life. What does it mean if such phenomena as sex ratios, amounts of sexual dimorphism, and correlation between breeding systems and parental behavior can be explained by the same theory in animals as different as primates, ungulates, and pinnipeds?[14] It means that the theory has general applicability. It also means that we have probably found out about something that has been happening gradually in each of these groups for a long time, beginning long before anyone was watching them. The only theory that has successfully made such predictions is natural selection. This indicates that natural selection can be extended into the past beyond our power to observe its action directly. Continuous ranges of variation in characters involved in phenomena like sexual dimorphism can demonstrate that sexual dimorphism evolves very slowly. So from the study of adaptation as well as the study of speciation we can successfully link short- and long-term evolutionary changes and prove that the two are not different.

Darwin (1859) specified the means for falsifying the idea that observable small changes lead to large changes which take so long that they are not directly observable: "If it could be demonstrated that any complex organ existed, which could not possibly have been formed by numerous, successive, slight modifications, my theory would absolutely break down."

Hybridization experiments showing that big differences between species are due to differences in large numbers of separately heritable genes, as well as the general relationship of genes to the development of the phenotype, indicate that Darwin's next statement, "But I can find no such case," would represent the conclusion to which modern biologists would also be drawn. Similarly, the alteration of complex organs by matings of individuals in which the organs differ slightly is a clear support for the idea that such organs have evolved through accumulations of small changes.

Evolutionary theory invokes only demonstrable mechanisms. A fundamental difference between evolution and creationism is that creationism invokes processes and mechanisms that cannot be demonstrated, and that no one has ever observed; evolutionary theory predicts on the basis of processes and mechanisms that everyone can observe and verify today. Evolutionists do not argue or require that no unobserved or unobservable, unverified or unverifiable processes and mechanisms can possibly occur. They simply build their models on the basis of the observable and verifiable, and continue to test those models. As long as predictability keeps on increasing, they keep on refining and adjusting their models and testing the new versions. No creationist has suggested an alternative testing procedure.

Natural selection is not an untestable hypothesis. A common objection to the theory of natural selection is that it is a tautology: In survival of the fittest, the fittest survive. Why do they survive? Because they are the fittest. The circularity of these statements has led people to say that natural selection explains nothing because it explains everything. Some of the same people also say that Darwin did not provide a means of falsifying his hypothesis—that he did not tell us about anything that could not be true if natural selection occurs.

We can dismiss the latter contention and introduce a compelling and provocative aspect of evolutionary theory by considering a bold challenge issued by Darwin (1859); it was:

> If it could be proved that any part of the structure of any one species had been formed for the exclusive good of another species, it would annihilate my theory, for such could not have been produced through natural selection.

Darwin thus provided, in 1859, a means by which his theory could be falsified, and he so identified it. (He said, in effect, that his theory, if correct, should explain everything *observable* but not everything *imaginable.*) Moreover, he did not say that an exception to his view of adaptation would weaken or diminish his theory, rather that it would annihilate his theory. Darwinian theory thus demands a selective background for the traits of all organisms and simultaneously rejects the possibility of certain kinds of altruism as evolved adaptations (but does not thereby exclude them from the behavioral repertoires of modern humans, who need not be bound by their evolutionary history). In other words, Darwinian evolution was, by Darwin himself, placed in a maximally vulnerable position by his clear exposition of what is required of living things if it is to be upheld. Darwin did tell us how to falsify his theory.

Although Darwin spoke only of "structure" we are obviously forced to expand the challenge to include all traits, whether morphological, physiological, or behavioral. Although he spoke only of altruism between species we cannot avoid the fact that all forms of genetic or reproductive altruism *within* species are also contrary to evolutionary theory, and should exist only as a result of accidents, or sudden environmental changes rendering an organism temporarily "maladapted." The human environment, however, includes our ability to reflect consciously and plan deliberately; we can thwart the adaptive background of our genes.

One more thing needs to be said about the supposed circularity or tautology of the phrase "survival of the fittest." If we never could *predict* differential survival or reproduction, but could only analyze it in retrospect, this criticism would be justified. (Of course, this is not so.) We can make countless accurate predictions from variations in the attributes of organisms, such as in an environment including sharp-eyed hawks and a white sand substrate, white mice will out-reproduce black mice. Thus, the concept of natural selection does not require circularity.

Darwinism is not an ideology. Darwinian natural selection may provide the core item in analyzing the causal history of the traits of living organisms, even including the general patterns of human behavior and culture. I think there is ample evidence making this an appropriate hypothesis. On the other hand, it does not follow, in any sense whatever, that Darwinism provides a basis for the construction of desirable political, economic, social, moral or ethical systems to be employed now or in the future. Darwinism's usefulness in these regards remains strictly in the realm of providing information that will assist humans in developing whatever system they may elect to strive for. It has no role in determining the nature of that system.

CONCLUSION

When one is a member of a frustrated minority, it is tempting to seek to force one's views on others. A society such as ours must constantly guard against such efforts if it is to move towards openness. Some creationists have implied repeatedly that society is already closed because editors will not publish their papers. It is easy to believe that critical referees are wrong and that one is being persecuted, and sometimes both complaints are well-founded. But there are numerous scientific publications, and scientists do not usually seek to get laws passed to protect themselves from criticism.

No laws were ever passed saying that evolution had to be taught in biology courses. The prestige of evolutionary theory has been built by its impact on the thousands of biologists who have learned its power and usefulness in the study of living things. No laws need to be passed for creationists to do the same thing. Recently creationists have reiterated that all they want is to resolve these issues on purely scientific grounds, but their behavior with regard to the law suggests otherwise. Moore in asking whether there is need for legislative intervention, implies that such legislation may be the only way to "true academic freedom" unless high school biology teachers start teaching creation.[15]

The greatest threat to society and to our children is not whether students are exposed to wrong ideas—after all, many high school biology students are legally adults with voting privileges, and all high school biology students have already been exposed to many wrong ideas. What is important is whether each has been taught how and given the freedom to test new ideas, evaluate them, and respond appropriately. The question of whether evolution or creation or both are mentioned, supported or taught in any or all of the schools is trivial by comparison. As long as biology teachers conduct their courses in the spirit of free inquiry, open debate, and self-correcting searches for predictive theories and repeatable results, no parent need fear that his or her children are being subjected to anything but the best kind of preparation for life in the technologically complex and socially demanding society in which we live.[16]

REFERENCES

1. See, for example, D. T. Gish, "A Challenge to Neo-Darwinism," *American Biology Teacher* 32 (1970): 495–97; D. T. Gish, "Creation, Evolution, and the Historical Evidence," *American Biology Teacher* 35 (1973): 132–40; Z. Levitt, *Creation: A Scientist's Choice* (Wheaton, IL: Victor Books, 1971); N. Macbeth, *Darwin Retried* (New York: Dell Publishing Co., 1971); J. N. Moore, "Paleontological Evidence and Organic Evolution. The Position of John N. Moore and Moore's Critique of Cuffey's Position," *Journal of the American Scientific Affiliation* 24 (1972): 160 ff.; J. N. Moore, "Evolution, Creation, and the Scientific Method," *American Biology Teacher* 35 (1973): 23–26, 34; J. N. Moore, "Evolution, Creation, Scientific Method, and Legislation," *Michigan Science Teachers Bulletin* 21 (1974): 6; J. N. Moore and H. S. Slusher, eds., *Biology: A Search for Order in Complexity* (Grand Rapids, MI: Zondervan Publishing House, 1970); L. M. Spears, "'Creationists' Win Right to Revise Science Texts," *National Observer* December 30, 1972; N. Wade, "Creationists and Evolutionists: Confrontation in California," *Science* 178 (1972): 724; N. Wade, "Evolution: Tennessee Picks a New Fight with Darwin," *Science* 182 (1973): 696; E. C. Lucas, et al., "Creationism and Evolutionism," *Science* 179 (1973): 953–54, 956; W. G. Peter, "Fundamentalist Scientists Oppose Darwinian Evolution," *Bioscience* 20 (1970): 1067–69.

2. See, for example, J. H. Brown, "The Teaching of Evolution," *Science* 56 (1922): 448–49; J. S. Dexter, "Anti-Evolution Propaganda in Georgia," *Science* 61 (1925): 243–50; F. D. Barker, "Evolution and the University of Nebraska," *Science* 63 (1926): 71; B. W. Wells, "Fundamentalism in North Carolina," *Science* 64 (1926): 17–18; C. Bush, "The Teaching of Evolution in Arkansas," *Science* 64 (1926): 356; E. Linton, "Note on the Scientific Method and Authority," *Science* 64 (1926): 526–27; anonymous, "The Tennessee Antievolution Law," *Science* 65 (1927): 57; anonymous, "Evolution, Creation, and Science: Review of F. L. Marsh," *Quarterly Review of Biology* 20 (1945): 267–68.

3. See Wade, 1972.

4. See C. Forbes, "Evolution and Human Behavior," *Christianity Today* 17 (1972): 35; R. A. Dodge, "Divine Creation: A Theory?" *AIBS Education Review* 2 (1973): 29–30.

5. William T. Keeton, *Biological Science,* rev. ed. (New York: W. W. Norton, 1974).

6. Charles R. Darwin, *On the Origin of Species* (1859), a facsimile of the first edition, with an introduction by Ernst Mayr (Boston: Harvard University Press, 1967).

7. W. D. Hamilton, "The Genetical Evolution of Social Behavior," *Journal of Theoretical Biology* 7 (1964): 1–52; W. D. Hamilton, "The Moulding of Senescence by Natural Selection," *Journal of Theoretical Biology* 12 (1966): 12–45; W. D. Hamilton, "Extraordinary Sex Ratios," *Science* 156 (1967): 477–88; G. C. Williams, "Pleiotropy, Natural Selection, and the Evolution of Senescence," *Evolution* 11 (1957): 398–411; G. C. Williams, *Adaption and Natural Selection* (Princeton, NJ: Princeton University Press, 1966); G. C. Williams, *Sex and Evolution* (Princeton, NJ: Princeton University Press, 1975); G. C. Williams and D. C. Williams, "Natural Selection of Individually Harmful Social Adaptions among Sibs with Special Reference to Social Insects," *Evolution* 11 (1957): 32–39; R. L. Trivers, "The Evolution of Reciprocal Altruism," *Quarterly Review of Biology* 46 (1971): 35–57; R. L. Trivers, "Parental Investment and Sexual Selection," in *Sexual Selection and the Descent of Man,* ed. by B. Campbell (Chicago: Aldine, 1972); R. L. Trivers, "Parent-Offspring Conflict," *American Zoologist* 14 (1974): 249–64; R. D. Alexander, "The Search for an Evolutionary Philosophy of Man," *Proceedings, Royal Society of Victoria* 84 (1971): 99–119; R. D. Alexander, "The Evolution of Social Behavior," *Annual Review of Ecology and Systematics* 5 (1974): 325–83; R. D. Alexander, "The Search for a General Theory of Behavior," *Behavioral Science* 20 (1975): 77–100; R. D. Alexander, "Natural Selection and the Analysis of Human Sociality," in *Changing Scenes in the Natural Sciences,* ed. by C. E. Goulden, Academy of Natural Sciences Special Publication 12 (Philadelphia: Academy of Natural Sciences, 1977); R. D. Alexander, "Evolution, Human Behavior, and Determinism," *Proceedings, Philosophy of Science Association* 2 (1977): 3; R. D. Alexander, "Natural Selection and Social Exchange," in *Social Exchange in Developing Relationships,* ed. by R. L. Burgess and T. L. Huston (New York: Academic Press, in press); M. J. West-Eberhard, "The Evolution of Social Behavior by Kin Selection," *Quarterly Review of Biology* 50 (1975): 1–34; E. O. Wilson, "The Queerness of Social Evolution," *Bulletin, Entomology Society of America* 19 (1973): 20–22; E. O. Wilson, *Sociobiology: The New Synthesis* (Cambridge, MA: Harvard University Press, 1975).

8. Moore, "Paleontological Evidence and Organic Evolution . . .", 1972; Moore, "Evolution, Creation, and the Scientific Method, 1973.

9. D. Lack, *Darwin's Finches* (New York: Harper Torchbooks, 1961).

10. A. P. Gray, "Mammalian Hybrids: A Check-list with Bibliography," *Commonwealth Bureau of Animal Breeding and Genetics, Edinburgh: Technical Communi-

cation 10 (1954): 144; A. P. Gray, "Bird Hybrids: A Check-List with Bibliography," *Commonwealth Bureau of Animal Breeding and Genetics, Edinburgh: Technical Communication* 13 (1958): 390.

11. C. Hubbs and G. E. Drewry, "Survival of F^1 Hybrids between Cyprinodot Fishes, and with a Discussion of the Correlation between Hybridization and Phylogenetic Relationship," *Bulletin, Institute of Marine Science, Austin Texas* 6 (1960): 81.

12. See Macbeth, 1971.

13. Moore, 1974.

14. See R. D. Alexander et. al., "Sexual Dimorphisms and Breeding Systems in Pinnipeds, Ungulates, Primates, and Humans," in *Evolutionary Theory and Human Social Organization* (Scituate, MA: Duxbury Press, 1978).

15. Moore, 1974.

16. Other recommended reading: D. P. Barash, *Sociobiology and Behavior* (New York: Elsevier, 1976); R. Dawkins, *The Selfish Gene* (New York: Oxford University Press, 1976); P. S. Moorehead and M. P. Kaplan, eds., *Mathematical Challenges to the Neodarwinian Concept of Evolution,* Wistar Institute Symposium Monograph 5 (Philadelphia: Wistar Institute Press, 1967); M. Sahlins, *The Use and Abuse of Biology* (Ann Arbor: University of Michigan Press, 1976).

Part 2
The Creationist Movement

Creationism

by John A. Moore

Scientists normally expect that scientific matters will be resolved by data, not emotion. They may look upon those who reject the rigors of scientific proof as somewhat benighted. But in a democracy those who dwell in the light and those who dwell in the dark share equally in deciding what shall be taught.

Except for the individuals intimately concerned, those contentious debates swirling around the teaching of evolution in the schools produce a feeling of *deja vu*. Essentially the same things have been said throughout the past century yet the problems are no more solvable today than when they were first encountered. The teacher and school administrator who is forced to deal with these matters can be assured that there is a vast and vigorous literature to which he may turn for information, inspiration, or solace. Some of this literature is truly first rate, well worth pursuing for its own sake: few can match, in prose, the reason and ridicule of grand old Thomas Henry Huxley as he thundered across the Victorian landscape.

My purpose will be to provide perspectives for the science teacher and school administrator. First, there will be a review of the attacks by creationists and others on evolutionary biology. Second, the argument will be developed that the conflict is on-going and unlikely to be resolved. Third, depending on the stance the teacher or school administrator wishes to adopt, a variety of scenarios will be suggested.

DARWIN AND HIS CRITICS

During the first half of the 19th century tremendous advances were made in geology and biology. The geologists, among whom Charles Lyell was pre-eminent, came to believe that the earth was exceedingly old—far older than suggested by Judeo-Christian tradition. Furthermore, it was recognized that the stratified rocks of the earth's crust represent material deposited long ago, usually under water. The materials were slowly changed to rock and in some instances contemporaneous remains of animals and plants were included in the deposits. Thus, it became apparent that the sedimentary rocks were, in a useful sense, a running diary of the earth's past. Careful studies, therefore, might tell much about the earth's crust and its denizens of long ago.

The biologists of that half century were concerned mainly with inventorying all species of animals and plants. By mid-century there was a general knowledge of the species of animals and plants and their geographic distribution. The incredible variety of living creatures raised in the minds of some of the more inquisitive naturalists the question of origins.

For most, however, the question had already been satisfactorily answered: the creatures of the earth were the products of a Divine Creator who had peopled the earth with the various species of plants and animals. It was believed that each species was fixed, that is, it was essentially constant in its characteristics and isolated from all others by an inability to cross-breed.

Useful references to the intellectual antecendents of the Darwinian Revolution are Gillispie (1951), Greene (1959), Eisley (1958), Adams (1938), and Geikie (1905) (see bibliography). Lyell's classic *Principles of Geology* (1830) is still a joy to read. Editions prior to 1859 will be especially useful in showing how a great mind viewed the natural and living worlds before Darwin.

Darwin's hypothesis was presented to the world in a joint article with Alfred R. Wallace in 1858 and, more definitively in 1859, in *On the Origin of Species.* Darwin agreed with numerous other scientists that evolution had occurred but, more importantly, proposed a mechanism: natural selection. Many scientists had found themselves unable to accept the notion of evolution simply because they could not suggest that species were fixed and could change only slightly from an average condition. Huxley wrote that "The *Origin* provided us with the working hypothesis we sought."[1]

It is usually assumed that the *Origin* provided a strong factual basis for evolution. It did not. Darwin looked upon the *Origin* as an abstract of a multi-volume work that was to be prepared later. The complete work would provide the full data then available. Lyell wrote, "It is a splendid cause of close reasoning, and a long sustained argument throughout so many pages; the condensation is immense, too great perhaps for the unitiated. . . ."[2]

Darwin's argument can be broken down into these five elements. First, there is considerable intraspecific variability in natural populations. Second, the rate of reproduction of all species is greater than the carrying capacity of their environment. Third, this means that there will be a struggle for insufficient resources. Fourth, in this struggle for life, presumably any individuals that were more fit would have a greater chance of surviving than the less fit. Fifth, with the passage of time, the population would come to consist of evermore fit individuals, that is, individuals better able to survive and to leave offspring. The elimination of the less fit by nature was called natural selection.

In 1859 there seemed to be no doubt that the first two elements of the argument were correct. The remaining three were not solidly based on observation or experiment but were proposed as a reasonable hypothesis of what might occur. This was the hypothesis to be tested.

And it was tested first in the forum. The debates that started in the autumn of 1859 saw a few scientists, plus even fewer churchmen and others, supporting Darwin. The majority, scientists and non-scientists alike, were against his hypothesis. Much of the opposition was on scientific grounds: it was felt that the scientific basis for evolution by natural selection was wholly inadequate. This point must be emphasized. All too often it is assumed that the opposition was based solely on the fact that Darwin's views were in conflict with those of the church. That was not the case.

The broad spectrum of opponents to the Darwinian world scheme did include, of course, the fundamentalists—those who say Genesis contains the only admissible explanation of organic diversity. The fundamentalists may have been pleased that there were scientific reasons to doubt Darwin but, even had there been none, their opposition would have been as vehement and vitriolic. In substituting a naturalistic explanation for organic diversity in place of Divine Creation, Darwin was attacking the veracity of the Bible and hence the very foundations of Western civilization.

It is instructive to outline some of the main arguments brought against Darwin in the first decade following the publication of the *Origin*. To a discouraging degree these are the same arguments advanced today by some of the creationists—long after answers satisfactory to the scientific world have been obtained.

1. There is no evidence that one species can change to another. Critics suggested that natural selection was more likely to weed out the less hardy, less prolific, and more extreme types. Thus it would tend to make species more uniform rather than lead to a slow change of one species into another. Darwin had made much of what could be done with artificial selection, especially with pigeons. Pigeons had been selected for thousands of years and many peculiar varieties had been obtained. However, no new species had been obtained as evidenced by

the fact that even the more extreme varieties can be crossed and the offspring survive. Thus, the critics saw these data refuting rather than supporting the Darwinan hypothesis.

Evolutionists then and now will agree that one does not observe one species changing into another or, as one of Darwin's critics demanded "see some tapir caught in the act of becoming a horse."[3] Data on the rate of evolution were scarce in the 1860s. Today rough estimates are possible, and it is clear that evolutionary change takes a very long time. Smith uses genetic data to estimate that it might take 300,000 generations for the evolution of a new species.[4] Paleontological data are cited by him that suggest, during the Pleistocene, 500,000 years were required for the evolution of a new species of mammals. There are more extensive data for the time it takes for genera to evolve: the numbers are in the millions of years.

With a time scale like that, the evolutionists must agree with the creationists that one does not observe tapirs changing into horses or, except in the most unusual instances, one species changing into another. These unusual instances, of which there are a few, involve hybridization of two species of plants followed by a doubling of the number of chromosomes.[5]

One must conclude, therefore, that the lack of examples of one species changing to another before our very eyes cannot be considered a useful criticism of the Darwinian evolution. Our eyes just don't last that long. Huxley dealt with this argument in his inimitable fashion: "The objection sometimes put forward, that no one yet professes to have seen one species pass into another, comes oddly from those who believe that mankind are all descended from Adam. Has any of them yet seen the production of negroes from a white stock, or *vice versa?*"[6]

2. Artificial selection has no relation to events in nature. Artificial selection, Darwin's critics pointed out, results in changes that may be very useful to human beings but are usually highly disadvantageous to the organism. Thus, domesticated plants and animals usually require careful culture and protection. They serve us well but we have molded their characteristics so much that they can rarely survive in nature. Furthermore, the variants that do arise are almost always monsters of some sort. One simply does not observe the appearance of favorable variations. Thus, artificial selection cannot be considered a model for the origin of better adapted forms since it does the reverse.

It was impossible for the evolutionists to deal with this criticism adequately until after 1900—when genetics began to tell us about genes and their mutations. Even then it was obvious that most of the new mutations *were* harmful to some degree. The mutational changes that

Morgan and his coworkers observed in Drosophila were nearly always demonstrably deleterious. How, then, could mutations lead to a better adapted individual if they were always conferred some type of disadvantage?

Eventually, this paradox was seen to have an acceptable solution. One had to add the parameter of time. The genes of a species would all be mutating, at a slow though finite rate. One could imagine that any mutation that could occur would have occurred many times before. Thus, if a rare mutation did convey some selective advantage, it would have increased in frequency long ago and become the normal allele at the gene locus. At any one time the genotype of a species would be the result of what natural selection had been doing over the ages. The chance of any beneficial mutation being encountered by an observer is, therefore, exceedingly unlikely.

Such an argument might be convincing to an evolutionist but a creationist might suggest that it is ungarnished sophistry. But adequate data are now available. If a species finds itself in a new environment— one that has not been selected for over the ages—beneficial mutations can be observed to appear. Much of these data are making news today. One reads of numerous examples of insect pests rapidly developing resistance to insecticides. The resistance is due to gene mutations that confer resistance. In an environment that does not contain the pesticides, such mutations would have been deleterious. Industrial melanism in moths and drug resistance in microorganisms are other examples.[7]

3. The fossil record does not support the Darwinian thesis. Darwin's critics were quick to point out that evolution demanded the origin of today's organisms from very different sorts of organisms living in the remote past. Thus, birds and mammals were thought to be fairly recent products of evolutionary change—possibly derived from ancient reptiles, but the links between reptile and bird and reptile and mammal were missing. In fact, all the links between major groups were missing in the 1860s. Evolutionists could offer the apology that the fossil record was inadequate but, given the fact that fossilization does occur, a necessary deduction of the evolutionary hypothesis was that fossils intermediate between major groups must exist—and, with luck, be found.

And in time they were dug up. The first major "missing link" to be discovered was Archaeopteryx, with its combination of reptilian and avian characteristics. Today there are fossils that link all the major groups of vertebrates. In some instances there are very few links (reptiles to birds) but in others there are numerous intermediate forms (reptiles to mamals). The paleontological data showing the evolutionary trends of the vertebrates are better than for any other phylum. This is correlated with the fact that this is the most recent major group to evolve and most vertebrates have hard parts (bone, teeth, and cartilage) that fossilize well. Very little is known about the evolutionary trends in those invertebrate phyla which are both very ancient and consist largely of animals with soft parts that fossilize poorly.

Good discussions of the data of paleontology are to be found in the works of Simpson, Colbert, and Romer.[8]

4. The earth is not old enough for the postulated evolutionary changes, from monad to man, to have occurred. Darwin needed lots of time, possibly as much as half a billion years from the Cambrian to the present. A generation growing up with Bishop Ussher's and Vice-chancellor Lightfoot's pinpointing of the moment of creation as 9:00 a.m. October 23, 4004 B.C. was not willing to grant Mr. Darwin all that time. Yet more and more geologists were coming to the conclusion that the earth was immensely old. Hutton had concluded his epoch-making *Theory of the Earth:* "The result, therefore, of our present enquiry is, that we find no vestige of a beginning—no prospect of an end."[9] Yet there were many physicists in the late 19th century who felt that the age of the earth was much less than the evolutionists required for their hypothesis.

Creationists today continue to believe in the earth's youth—often suggesting an age of only a few thousand years. They maintain this position long after it has become possible to date, with a fair degree of accuracy, many of the rocks of the earth's crust by means of their radioactivity. As more and more rocks are studied, with more and more methods, the times for many events in the earth's past are becoming increasingly certain. Thus, the formation of the earth's crust is put at about 4,600,000,000 years ago. The onset of the Cambrian Period, when fossils first become abundant, is about 570,000,000 years ago. The Age of Reptiles, the Mesozoic, started about 225,000,000 years ago. The onset of the Tertiary Period is placed at 65,000,000 years ago. Eicher gives a fine history of the problem of determining geologic time and provides us with the currently accepted dates.[10]

5. Complex structures and processes are so exquisitely adaptive that they cannot be imagined as a consequence of variation and selection. Thus, it is almost impossible to imagine how any organ so complex as the eye could have been formed by the selection of chance mutations. At every stage of development the structure must have been advantageous—otherwise it would not have been selected. What good is a half eye? Or as the creationist Gish expresses it:

> In the fish-to-amphibian transition, many features would have to change. It should be easy to trace the conversion of the fins of the fish into the feet and legs of the amphibian in the fossil record.

According to the evolution story the fossil record should show a fossil with 5 percent feet and legs and 95 percent fins, one with 10 percent feet and legs and 90 percent fins, one with 25 percent feet and legs and 75 percent fins, one with 50 percent feet and legs and 50 percent fins, and so forth, until almost all traces of fins have disappeared in the forms. . . . However, no one has been able to find a single transitional form showing part fins and part feet.[11]

The problem when put that way might be expected to amuse and confuse a naive audience. Is one to expect an exactly intermediate stage to have two fins and two feet? If so, one must admit that the paleontologists have not exposed such a fossil. Nevertheless, Devonian fish are known that have skeletons in their fins that are so much like the limb skeletons of the tetrapods that there seems no problem in deriving limbs from fins.[12]

Nevertheless, the details of the evolution of complex structures, such as eyes, and behavior patterns are almost always unknown. The chief difficulty is the soft structures, such as eyes, are almost never present in fossils (and behavior patterns, never). One is left, then, in the position of being able to do no more than to suggest what might have happened. In spite of this, the procedures of comparative anatomy, embryology, and more recently of comparative biochemistry generally allow one to present an acceptable hypothesis for how a complex structure may have evolved. In a similar manner, the behavior patterns of different species can often be arranged in a sequence that serves as a hypothesis for the evolution of the behavior pattern. A recent synthesis of this field has been provided by E. O. Wilson. [13]

When one turns to a consideration of structures that can be fossilized, facts replace hypotheses. Thus, the general picture of the evolution of the vertebrate skeleton is well understood. Even such remarkable evolutionary changes, such as the conversion of some bones of the jaws into the bones of the ear, have been well documented.

6. Evolution is impossible due to the constraints of the Second Law of Thermodynamics. This is a point being actively pushed by some creationists today. The argument is complex but basically it is this: the Second Law maintains that all self-contained systems gradually pass from a state of greater order to one of lesser order. Heat is lost and, with the passage of time, complex arrangements of matter become simple. Thus, the question is asked, how could evolution, which has been characterized by a slow change of organisms from simple to complex, possibly occur? This argument when advanced by a creationist and buttressed by mathematical equations is enormously effective with naive audiences. Evolution can be portrayed as not only wicked but also against the law! Most of the people who advance such an argument are

probably well aware that it is totally invalid, yet why abandon such an effective device? The correctness of the Second Law is not questioned yet one must remember that our earth is not a closed system. Energy is reaching it in large amounts from the sun and such energy can be and is used in the construction of the complex from the simple. If one will grant that the energetics of the universe are such that a fertilized ovum can develop into a complex adult, then the Second Law is not going to interdict evolution. There is no need for extra energy for evolutionary events. If there is enough for the development and life of organisms no more is required.

7. <u>Darwinism is heresy</u>. And finally we have reached the main argument. All of the others which have a scientific basis can be dealt with by scientists. This one, however, is based on mutually incompatible systems of thought and belief. The Reverend John Duns of Scotland put it well: "Mr. Darwin's work is in direct antagonism to all the findings of a natural theology, formed on legitimate inductions in the study of the works of God; and it does open violence to everything which the Creator Himself has told us in the Scriptures of truth, of the methods and results of His workings."[14] And coming to this side of the Atlantic, Francis Bowen, Alford Professor of Natural Religion, Moral Philosophy, and Civic Polity in Harvard College, puts it this way:

> After all, for the defense of the great truths of philosophy and natural theology, it is hardly necessary to spend much time in refutation of such fanciful theories of cosmogony as this by Mr. Darwin. A proper view of the nature of causation, a clear recognition of the great truth that the natural no less than the super natural, the continuance no less than the creation of existence, the origin of an individual as well as the origin of a species or a genus, can be explained only by the direct action of an intelligent creative cause—places the vital doctrine of the being and the providence of a God on ground that can never be shaken.[15]

So if one accepts the Judeo-Christian accounts of the origin and diversity of life, the findings of science must either be ignored or somehow adjusted to what is said in Genesis. The second section of this paper will deal more with this matter, but before that is done, it will be useful to mention some of the important references to Darwin and his critics. As you might suspect the literature is enormous—and still growing rapidly.

The first general statement of Darwin's views on evolution appears in a joint essay with Wallace.[16] The first edition of Darwin's *Origin* was in 1859. The sixth, and last, edition was in 1872. Peckham has provided a *variorum.* [17] The *Origin* was reviewed extensively. Important reviews that found the argument wanting are Bowen (1860), Duns (1860),

Wilberforce (1860), and Jenkins (1967).[18] Critical though favorable reviews were presented by Gray (1860) and Huxley (1859, 1860).[19] The original journals in which these reviews were published are available usually only in the larger research libraries. Fortunately there is more readily accessible material. Glick has edited a series of essays that describe the reception of Darwinism, country by country.[20] The essay on the United States, by Michele L. Aldrich, is largely bibliographic and, as such, provided an introduction to the literature. An earlier work of Kennedy provides the same service.[21] Useful anthologies of 19th century reactions to Darwin are Appleman (1970), Daniels (1968), and Hull (1973).[22]

General discussions of the impact of Darwinism are to be found in Eisley (1958), Irvine (1955), Ghiselin (1969), Hull (1973), and Russett (1976).[23]

Evolutionary biology has come a long way since Victorian times. Good introductions to current points of view are Smith (1975), Stebbins (1977), Ehrlich, Holm, Parnell (1974), Eaton (1970), and Volpe (1977).[24] Smith, Stebbins and Volpe are in paperback editions. Four classics of the modern synthesis of evolutionary theory are Dobzhansky (1937 and later editions), Mayr (1942), Simpson (1944), and Stebbins (1950).[25] Three have been extensively revised: Dobzhansky (1970), Mayr (1963, 1970), Simpson (1953). Still more recent and also important are Lewontin (1974) and Dobzhansky *et al.* (1977).[26]

I do not know of a single modern book that deals specifically with the arguments of the creationists that they claim cast doubt on the theory of evolution. Not too surprising, I guess—neither do I know of a modern book on astronomy that deals with the Ptolemaic system, a serious challenge to the Copernican system or that gives equal time to the two theories of earth shape—flat or spherical. Neither do the chemists seem too concerned whether the atomic theory is correct or not. Evolutionists simply do not consider the creationist arguments serious threats to science—though no evolutionist questions the mischief they can do to science education.

IMMISCIBLE PARADIGMS

Even the most rational human beings use a variety of thought patterns for a variety of purposes. The same person may have one paradigm for dealing with nature, another for human beings, and still another for art. Thus, a scientific description of a great painting might have little resemblance to the description of a connoisseur. We could imagine the first dealing mainly with the chemistry of pigments, the wave lengths of the reflected light, and the scientific names for any objects illustrated. The second would be more concerned with composition, balance of colors, and, more importantly, to the emotions evoked and allusions suggested. Most of us, surely, would be more interested in what the connoisseur had to say. We would look to him, more than to the scientist, to increase our pleasure and understanding.

The problem we are dealing with in this paper arises because both scientists and creationists attempt to describe nature—using conflicting paradigms. The scientist attempts to uncover the secrets of nature by observation, experiment, and reason—but reason that excludes any involvement of supernatural forces. That is, explanatory statements can only invoke materials and processes that can be demonstrated to occur in nature. These materials and processes must be discoverable by anyone with the proper training, equipment, ability, and patience. Essentially no scientific understanding becomes complete or acceptable with one discovery. First insights into a difficult problem will be incomplete and probably, to a considerable degree, erroneous. As more and more scientists study the problem, additional facts are obtained and the explanations will improve. But the statements are almost never said to be absolutely true. They are accepted as the best answer that can be obtained with the available data. This professional caution of the scientist often seems a little excessive: the belief that pure water is composed only of hydrogen and oxygen does seem rather certain.

As an illustration of how explanatory statements change, consider the question of a physical basis of inheritance. Darwin convinced himself, but few others, that there *was* a physical basis for inheritance: that is, some substance passed from parent to offspring and determined the characteristics of the latter. His particular postulated substance was shown to be inadequate. Later in the 19th century biologists came first to believe that the physical basis resided in the nucleus and later in the chromosomes. The data, however, were only suggestive. After 1900, and especially with the work of Thomas Hunt Morgan, the chromosomes and parts of them (the genes) were clearly implicated. Finally in the 1950s, a host of workers, most notably James Watson and Francis Crick, determined the chemical nature of the gene. Today we have a fairly complete understanding of the structure of genes and less, though considerable, information on how they function. Thus, we can now make some statements about the physical basis of inheritance that are true beyond a reasonable doubt. "Beyond a reasonable doubt," that is as far as a scientist should be willing to go.

Evolutionary biology has had a similar history. Darwin presented a plausible hypothesis and provided some data. The generations of scientists that followed Darwin's lead have provided a wealth of data.

The growth in understanding has been enormous. De Solla Price has estimated that scientific information doubles every 10–15 years (10 years for the total information and in 15 years for high quality information).[27] This period of doubling seems to hold generally in all sciences and has been occurring since the middle of the 17th century (when the Scientific Revolution got underway). Thus, if we use a 15-year doubling time for information in evolutionary biology, there will have been eight doublings between 1859 (date of the *Origin*) and 1979—and a 256-fold increase in our understanding of evolution. To many, this will seem like a fairly conservative figure.

The procedures now used by scientists to obtain information about the natural world did not come into general use until the 17th century. One of the founding fathers of the new science was Francis Bacon (1561–1626). For him valid knowledge was to be, first, gathering the facts and, then, seeing what statements could be induced from them. His inductive methods were at variance with the scholastic methods that had prevailed for centuries. To the Scholastics, knowledge came ultimately from reasoning or revelation. The first was primarily the domain of the philosopher, the second of the theologian. Since clear reasoners were rare and recipients of revelations excessively so, most of the Schoolmen of the Middle Ages sought wisdom from the works of the great minds in philosophy and theology. For them Agassiz's aphorism, "Study Nature, not Books," should have been rendered, "Study Books, not Nature."

There is an amusing caricature of the Scholastic mind, possibly from the pen of Francis Bacon, that does convey the flavor of the method:

In the year of our Lord 1432, there arose a grievous quarrel among the brethren over the number of teeth in the mouth of a horse. For a full 13 days the disputation raged without ceasing. All the ancient books and chronicles were fetched out, and wonderful and ponderous erudition such as was never before heard in this region was made manifest. At the beginning of the 14th day a youthful friar of goodly bearing asked his learned superiors for permission to speak, and straightway, to the wonderment of the disputants, whose deep wisdom he sore vexed, he beseeched them to unbend in a manner coarse and unheard of and to look in the open mouth of a horse and find answer to their questionings. At this, their dignity being grievously hurt, they waxed exceeding wroth; and, joining in a mighty roar, they flew upon him and smote him, hip and thigh, and cast him out forthwith. For, said they, surely Satan hath tempted this bold neophyte to declare unholy and unheard-of ways of finding truth, contrary to all the teachings of the fathers. After many days more of grievous strife, the dove of peace sat upon the assembly, and they as one man declared the problem to be an everlasting mystery because of a grievous dearth of historical and theological evidence thereof, so ordered the same writ down.[28]

A fundamentalist, to the extent that he accepts the infallibility of Genesis as a guide to the origin and history of life, follows in the footsteps of the Scholastics of the Middle Ages. Their procedures are the same: relying solely on the products of philosophy or revelation for understanding. These procedures were abandoned by many scientists in the 17th century though, in the majority of cases, this did not lead to the abandonment of a deity. But as the procedures of science began to provide acceptable explanations for events so long relegated to the sphere of the supernatural, even a deity began to seem superfluous. For many intellectuals in the 18th century Age of Enlightenment the traditional deity was abandoned or replaced by a radically different concept. In this period, reason, freedom, and humanitarianism began to gain dominance over dogmatism, repression, and intolerance. But the Enlightenment, or Age of Reason as it was also called, was of limited pervasiveness. A powerful Church, which a century earlier had been able to humble Galileo Galilei and incinerate Giordano Bruno, continued to cast its pall over the slowly-developing modern science.

There is no avoiding conflict between a system of thought based on revelation and authority and one based on observation and experiment. The current attacks of the creationists on evolution are but a skirmish in that long war. The literature on this conflict is enormous. Two classics of the 19th century are Draper (1874) and White (1896).[29] Two of the volumes of Huxley's *Collected Essays,* namely, *Science and Christian Tradition* and *Science and Hebrew Tradition* are useful and fascinating. Good bibliographies are given by Kennedy and Aldrich.[30] Some older material is also to be found in the anthologies of Appleman, R. J. Wilson, and Daniels.[31]

An especially valuable survey of the field is Barbour (1966).[32] Other useful sources are Greene (1961), Gay (1966), Dillenberger (1960).[33] Hofstadter's now classic *Anti-Intellectualism in American Life* has much to say about the fundamentalists and their attacks on evolution, humanism, and the modern world. Also of interest is his study of Social Darwinism.[34]

The Scopes Trial was one of the most dramatic confrontations of fundamentalists and evolutionists. It is dealt with by Allen, Ginger, Tompkins, and very effectively and completely by de Camp.[35] Lawrence and Lee based a play on the trial.[36]

TO TEACH OR NOT TO TEACH, THAT IS THE QUESTION

The first part of this paper sought to establish that Darwin's hypothesis of evolution, as modified by subsequent generations to become the Darwinian theory of evolution, has proved to be a powerful tool for understanding the history and diversity of life. It is the only scientifically acceptable paradigm for these phenomena. The second part of this paper emphasized that there are non-scientific, largely religious, paradigms for explaining these same phenomena. Furthermore, ardent proponents of either paradigm are unlikely to admit the correctness of the antagonistic paradigm.

We have now reached the question, "What *is* a biology teacher to do?" There is no single answer to that question. Were one to approach the question on scientific grounds alone, the answer would be simple. But often the choice cannot be made solely on scientific grounds. What if the citizens place social, religious, or political considerations above the scientific? Does the non-teaching portion of the community have the right to say what is to be taught?

What is taught will depend on how one answers these questions. I will suggest some of the more obvious answers and outline briefly how one might proceed. The scenarios will be arranged in a sequence beginning with an exclusively scientific approach and extending through those that take into account non-scientific pressures.

Science Only

This is the common pattern of teaching evolutionary biology in the schools. The Darwinian hypothesis is explained and then supported by the data of paleontology, comparative anatomy, comparative embryology, and more recent developments in population biology. Essentially all of the introductory, intermediate, and advanced books in evolutionary biology adopt this approach—as do the textbooks of general biology. All supernatural explanations are ignored. Students are assured that all scientists who have studied the matter accept fully that there has been an evolution of life, that a satisfactory general theory can accommodate the observed phenomena, and that new information is being added rapidly to the field of evolutionary biology. Some of the students wonder, no doubt, why the issue of creationism is ignored.

Many important scientists and educators believe that the "Science Only" approach is proper and sufficient. Biology is a science and biological classes should restrict their discussions to science. Nothing useful is to be gained by taking note of the creationists or their views.

Evolutionary Biology in Perspective

But do we serve the best interests of the students and of biology by pretending that there is no opposition to evolution? If one answers this question with "no," then one should deal with the opposition carefully and explicitly. This is the scenario that I would prefer—which sets me aside from most fellow evolutionists. I believe that much is to be gained for the student and for science by discussing the nature of the objections to the theory of evolution. Such an approach has a valuable carry-over to other contentious fields where society has difficulty deciding what to do.

Here the approach would be to cover evolutionary biology in the standard way (Scenario 1) and then present the main arguments of those who do not wish to accept evolution. The main emphasis here would be to contrast the procedures of evolutionists and creationists in seeking answers to questions. The scientist's answers must, in the final analysis, be based on what observation and experiment can tell us about the events in nature. The explanations can invoke only natural products and processes. All suitably trained individuals can be expected to reach essentially the same answers when they employ the same methods and can make the same observations.

The explanations of the creationists, on the other hand, are based ultimately on ancient texts, in this case *Genesis*. It is assumed that the statements in these ancient texts were passed by word of mouth for unknown generations but their origin was a revelation to one or a few individuals. These revelations are accepted as true and, therefore, the events of nature must be understood in their terms.

Once this dichotomy of approach has been established, the discussion might proceed to discuss the very different consequences of these approaches. During the century since the *Origin* was published the scientific approach has been highly productive. Our understanding of the history of life has increased enormously. We now have some knowledge of the very ancient forms of life that lived more than 700 million years ago, much data on the variety of fossil organisms that lived at later times, even more data on the interrelations of organisms that had hard structures, such as bones and teeth that fossilize well, and much data on the dynamics of the evolutionary process. Nevertheless, the student must understand that although much has been accomplished, new information is being obtained at an ever increasing rate.

Scientific Procedures Work!

On the other hand, the creationists have accumulated no new information or understanding—it's all in *Genesis*. Their efforts for the past century have been mainly in pointing out that evolution cannot explain everything—to which the evolutionist must agree.

Despite its superficial simplicity, evolutionary theory is often not easily understood by high school students. For this reason some teachers prefer to use medicine as an example of scientific and non-scientific approaches to problems. The argument might be developed as follows: Little progress was made in understanding the nature or treatment of disease so long as it was assumed that sickness was a consequence of the displeasure of some god or evil spirit. The scientific study of disease, which has become increasingly possible during the past century, has made great progress. But, once again, there is much to learn.

Thus in a comparison of the usefulness of scientific versus supernatural procedures in studying natural events (organisms, disease, etc.), most people will reach the conclusion that one works and the other does not.

Giving Equal Time

The final scenario that I will mention meets the demands of some creationists that, if the history of life is discussed at all, equal time should be given to the evolutionist's and creationist's paradigms. (There is an obvious fourth scenario—teaching only creationism—but one who selects that solution has nothing to learn from me.)

Equal time does present a problem. The basis of creationism is *Genesis I* and the first nine verses of *Genesis II,* all of which can be read in three minutes. On the other hand, even the most summary review of the evidence on which the theory of evolution is based would require at least three class hours. Nevertheless there are other ways to devote the required equal time to creationism. One is to discuss the fascinating reasons for the two very different accounts of creation in *Genesis:* one in *Genesis I* plus the first four verses of *Genesis II* and the other in *Genesis II* verses 5–9. Long ago biblical scholars concluded that *Genesis* is a compilation of several very different sources. The *Genesis II* account of creation is very old. It is thought to represent a very ancient story that was finally written down in the 10th century B.C. The *Genesis I* account of creation is much more recent. Apparently it was not formulated until after the Jews had returned from captivity in Babylonia in the 6th century. This account is remarkably similar to the creation myth of Babylon and, apparently, the priests drew heavily on that source. A good general account of the biblical scholarship that led to these conclusions is Buttrich (1952).[37]

Some creationists have insisted that creationism be treated as a scientific theory (for example, John *N.* Moore and Gish).[38] (I should emphasize the difference between the John Moores. John *N.* Moore is a creationist. I am John *A.* Moore, an evolutionist.) If this is done, one should subject the statements of *Genesis I* and *Genesis II* to the rigors of scientific procedures. Thus, the statements should be accepted as provisionally true hypotheses and deductions made from them. These deductions should then be tested to see if the hypothesis is false or possibly true. Not surprisingly, such tests give no support for the statements in *Genesis.* These are complex matters and deserve a fuller treatment than is possible here. I have dealt with this problem in much greater detail.[39]

There is another and quite serious problem associated with including creation in a science course. *Which* account of creation? There are hundreds of stories of creation, each associated with a specific race or sect. One could maintain that, in all fairness, students should be exposed to many more than the one in the Judeo-Christian tradition. But the problem here is that, since most are entirely different, all cannot be correct and there is no procedure to select a "correct" one. The point should also be made that a sect believes its own story and usually regards all others as myths. But that is what myths are: the other fellow's sacred or traditional beliefs. And one should add, none of this is science and, if it is to be treated at all, possibly it should be in anthropology or social studies courses.

Some of the problems involved in the creationist controversy are dealt with in works by Nelkin (1976a, 1976b, 1977), Chambers (1977), Mayer (1973, 1977), Cloud (1977), J. A. Moore (1974), Le Clercq (1974), Stebbins (1973), Dobzhansky (1973), Hardin (1973), J. N. Moore (1973), Gish (1973), Cory (1973), Aulie (1972), Ost (1972), and Newell (1974).[40]

POSTSCRIPT

A generation ago, when science was basking in the glory of its accomplishments in World War II, when the Space Program was beginning to conquer those other worlds, and when the gene was identified and its code cracked, few would have anticipated the antiscience climate of our times. Today we find science under attack, not because it fails to increase our understanding of nature but, more probably, because it succeeds too well. It is now clear that scientific knowledge used without restraint can lead to the advantage of the few and the disadvantage of the many, to over-exploitation today at the risk of the destruction of tomorrow. In our concern for our lives and profits we sometimes fail to remember that science is neutral, though the uses to which it can be put are not. Thus many see science as the demon when, instead, they should see themselves. The thoughtless then seek to control science when they sould seek to control human nature.

Be that as it may, some of this antiscience is reflected in attempts to have the foundation theory of the biological sciences taught as no more than probable hypothesis or tentative theory and the creation myth of the Judeo-Christian tradition promulgated as its scientific equal. A determined minority has been able to incorporate these views in some of the laws of the land.

Scientists normally expect that scientific matters will be resolved by data, not emotion. They may look upon those who reject the rigors of scientific proof as somewhat benighted. But in a democracy those who dwell in the light and those who dwell in the dark share equally in deciding what shall be taught. A generation ago we in America watched in disbelief as the Russians repealed Mendel's Laws of Inheritance. Today some of our states seem to be doing the same for even more fundamental laws of biology.

It is clear now that no amount of scientific evidence will convince creationists of the vacuousness of their position. The controversy they have started will be resolved only when the scientists convince the voting majority of our citizens that the procedures of science work when we seek to understand nature and that supernatural and transcendental procedures do not.

Biology teachers have more to teach than biology!

REFERENCES

1. T. H. Huxley, "On the Reception of the 'Origin of Species'," in *The Life and Letters of Charles Darwin,* ed. by Francis Darwin (London: John Murray, 1888), vol. 2, p. 197.

2. Francis Darwin, ed., *The Life and Letters of Charles Darwin, Including an Autobiographical Chapter.* 3 vols. (London: John Murray, 1888), vol. 2, p. 206.

3. John Duns, "On the Origin of Species. . . .," *The North British Review,* American Edition 27 (1860): 246-63.

4. John Maynard Smith, *The Theory of Evolution,* 3d ed. (Baltimore, MD: Penguin Books, 1975).

5. See Verne Grant, *Plant Speciation* (New York: Columbia University Press, 1971), chap. 13.

6. T. H. Huxley, *Science and the Hebrew Tradition* (New York: Appleton, 1895b), p. 43.

7. See T. Dobzhansky, *Genetics and the Evolutionary Process* (New York: Columbia University Press, 1970), pp. 211–15.

8. G. G. Simpson, *The Meaning of Evolution. A Study of the History of Life and Its Significance for Man,* 2d ed. (New Haven: Yale University Press, 1967); E. H.

Colbert, *Evolution of the Vertebrates: A History of the Backboned Animals through Time,* 2d ed. (New York: Wiley, 1969); A. S. Romer, *Vertebrate Paleontology,* 3d ed. (Chicago: University of Chicago Press, 1966).

9. James Hutton, "Theory of the Earth, or an Investigation of the Laws Observable in the Composition, Dissolution, and Restoration of the Land Upon the Globe," *Transactions, Royal Society of Edinburgh* 1 (1788): 304.

10. Don L. Eicher, *Geologic Time,* 2d ed. (Englewood Cliffs, NJ: Prentice Hall, 1976).

11. Duane T. Gish, *Evolution: The Fossils Say No!* (San Diego, CA: Creation-Life Publishers, 1972), p. 14.

12. See, for example, Figure 117 in Romer, 1966.

13. E. O. Wilson, *Sociobiology: The New Synthesis* (Cambridge, MA: The Belknap Press of Harvard University Press, 1975).

14. Duns, 1860, p. 260.

15. Francis Bowen, "On the Origin of Species. . . .," *North American Review* 90 (1860): 504.

16. C. Darwin and A. Wallace, "On the Tendency of Species to Form Varieties; and on the Perpetuation of Varieties and Species by Natural Means of Selection," *Journal of the Linnean Society of London—Zoology* 3 (9) (1858): 45–62.

17. Morse Peckham, *The Origin of Species by Charles Darwin. A Variorum Text* (Philadelphia: University of Pennsylvania Press, 1959).

18. Bowen and Duns, as cited above; Samuel Wilberforce, "On the Origin of Species. . . .," *Quarterly Review* 108 (1860): 225–64; Fleeming Jenkins, "The Origin of Species," *North British Review* 46 (1867): 277–318.

19. Asa Gray, "Review of Darwin's Theory on the Origin of Species," *American Journal of Science and Arts* (2d series) 29 (1860): 158–84; T. H. Huxley, "Time and Life: Mr. Darwin's 'Origin of Species'," *Macmillans Magazine* 1 (1859): 142–48; T. H. Huxley, "Darwin on the Origin of Species," *Westminster Review,* American Edition 73 (1860): 295–310.

20. Thomas F. Glick, ed., *The Comparative Reception of Darwinism* (Austin: University of Texas Press, 1972).

21. Gail Kennedy, ed., *Evolution and Religion: The Conflict Between Science and Theology in Modern America* (Boston: Heath, 1957).

22. P. Appleman, ed., *Darwin: A Norton Critical Edition* (New York: W. W. Norton, 1970); G. Daniels, ed., *Darwin Comes to America* (Waltham, MA: Blaisdell, 1968); D. Hull, *Darwin and His Critics: The Reception of Darwin's Theory of Evolution by the Scientific Community* (Cambridge, MA: Harvard University Press, 1973).

23. L. Eiseley, *Darwin's Century: Evolution and the Men Who Discovered It* (Garden City, NJ: Doubleday, 1958); W. Irvine, *Apes, Angels, and Victorians: The Story of Darwin, Huxley, and Evolution* (New York: McGraw-Hill, 1955); M. Ghiselin, *The Triumph of the Darwinian Method* (Berkeley: University of California Press, 1969); Hull, 1973 as cited above; C. E. Russett, *Darwin in America: The Intellectual Response, 1865–1912* (San Francisco: Freeman, 1976).

24. Smith, 1975, as cited above; G. L. Stebbins, *Processes of Organic Evolution,* 3d ed. (Englewood Cliffs, NJ: Prentice Hall, 1977); P. R. Ehrlich, R. W. Holm, and D. R. Parnell, *The Process of Evolution,* 2d ed. (New York: McGraw-Hill, 1974; T. H. Eaton, Jr., *Evolution* (New York: W. W. Norton, 1970); E. P. Volpe, *Understanding Evolution,* 3d ed. (Dubuque, IA: William C. Brown, 1977).

25. T. Dobzhansky, *Genetics and the Origin of Species* (New York: Columbia University Press, 1937, 1941, 1951); E. Mayr, *Systematics and the Origin of Species from the Viewpoint of a Zoologist* (New York: Columbia University Press, 1942); G. G. Simpson, *Tempo and Mode in Evolution* (New York: Columbia University Press, 1944); G. L. Stebbins, *Variation and Evolution in Plants* (New York: Columbia University Press, 1950).

26. R. C. Lewontin, *The Genetic Basis of Evolutionary Change* (New York: Columbia University Press, 1974); T. Dobzhansky, et al., *Evolution* (San Francisco: Freeman, 1977).

27. D. de Solla Price, Jr., *Little Science, Big Science* (New York: Columbia University Press, 1963).

28. cf. L.W. Taylor, *Physics: The Pioneer Science* (New York: Dover, 1941), vol. 1, p. 41.

29. J. W. Draper, *History of the Conflict Between Religion and Science* (New York: Appleton, 1874); A. D. White, *A History of the Warfare of Science with Theology in Christendom* (New York: Appleton, 1896).

30. Kennedy, 1957, as cited above, pp. 110–14; Aldrich in Glick, 1972, as cited above.

31. Appleman, 1970, as cited above; R. J. Wilson, *Darwinism and the American Intellectual: A Book of Readings* (Homewood, IL: Dorsey Press, 1967); G. Daniels, as cited above.

32. Ian Barbour, *Issues in Science and Religion* (New York: Harper and Row, 1966, 1971).

33. J. C. Greene, *Darwin and the Modern World View* (Baton Rouge, LA: Louisiana State University Press, 1961); P. Gay, *The Enlightenment: An Interpretation* (New York: Vintage Books, 1966); J. Dillenberger, *Protestant Thought and Natural Science: A Historical Interpretation* (Nashville: Abingdon Press, 1960).

34. R. Hofstadter, *Anti-Intellectualism in American Life* (New York: Vintage Books, 1963); R. Hofstadter, *Social Darwinism in American Thought* (Boston: Beacon Press, 1955).

35. L. H. Allen, *Bryan and Darrow at Dayton* (New York: Russell and Russell, 1925); R. Ginger, *Six Days or Forever? Tennessee v. John Thomas Scopes* (Boston: Beacon Press, 1958); J. R. Tompkins, ed., *D-Days at Dayton: Reflections on the Scopes Trial* (Baton Rouge, LA: Louisiana State University Press, 1965); L. S. de Camp, *The Great Monkey Trial* (Garden City, NY: Doubleday, 1968).

36. J. Lawrence and R. E. Lee, *Inherit the Wind* (New York: Random House, 1955).

37. G. A. Buttrich, *The Interpreter's Bible* (New York: Abingdon Press, 1952), vol. 1.

38. See, for example, John N. Moore, "Evolution, Creation, and the Scientific Method," *American Biology Teacher* 35 (1973): 23–26, 34; Duane T. Gish, "Creation, Evolution, and the Historical Evidence," *American Biology Teacher* 35 (1973): 132–40.

39. J. A. Moore, "On Giving Equal Time to the Teaching of Evolution and Creation," Included in this volume.

40. These and all other items in the bibliography for this article are included in the Bibliography for this volume, pp. 477–507.

"Scientific Creationism"— A New Inquisition Brewing?

by Preston Cloud

Fundamentalist creationism is not a science but a form of anti-science, whose more vocal practitioners, despite their master's and doctoral degrees in the sciences, play fast and loose with the facts of geology and biology.

THE CREATIONIST MOVEMENT

Religious bigotry is abroad again in the land. And, in addition to socialism, sex education, and birth control, it is tilting anew at an old *bête noire*—evolution. Within the larger fundamentalist movement, a small group of hard-core zealots, comprising the membership of the Creation Research Society (CRS), is riding a crest of supernaturalist fervor to battle against a basic liberating tenet of civilized peoples—the separation of church and state. The five hundred or so members and many supporters of this organization, who call themselves creationists, see the ancient Judeo-Christian creation myths of the first two chapters of Genesis as constituting a single, divinely revealed account of origins that must be restored to preeminence in public-school teaching. Toward this end, they insist, the biblical version(s) of creation must be given "equal time" with scientific accounts of the progressive development of life from simple to complex forms. It does not satisfy them that biblical creationism receives equal time with other religious accounts of origins in courses in comparative theology. They demand that creationism be presented as a "scientific" alternative to evolution in science textbooks that deal with the origin and subsequent development of life before such textbooks can be approved for use in the public schools. They have even prepared their own textbook to serve this end, *Biology: A Search for Order in Complexity* (J. N. Moore and H. S. Slusher, editors, 1974). The outlooks described are those encompassed under the terms *creationist* and *creationism* as used in the present paper.

The humanistic preference for rational thought, particularly as seen in the search for naturalistic explanations of natural phenomena, has, of course, always been unpopular among those who prefer the supernatural, whether it be benign, diabolical, or simply whimsical. But why this strong resurgence of the age-old struggle between naturalistic and mystic perceptions of the universe? How do the creationists arrive at and support their proposition?

A part of the credo to which all members of the CRS subscribe is that "all basic types of living things, including man, were made by direct creative acts of God during the creation week described in Genesis." Elaborating on this, their leading polemicist, H. M. Morris, emphasizes in his book *The Remarkable Birth of Planet Earth* (1972) that *"it is only in the Bible that we can possibly obtain any information about the methods of creation, the order of creation, the duration of creation, or any of the other details of creation."* The Bible claimed to be not only inspired, but factual, thus is seen as a scientific document, a document, moreover, that, coming from a supposedly infallible source, is not open to inquiry or interpretation. Scholarly documentation of pre-Hebraic antecedents, including the two different versions of creation given in the first two chapters of Genesis, is ignored or brushed aside, and the "days" and other terms of the King James version of the Bible are taken literally as written— *"If it really took five billion years for God to make all these things, why did He tell us it took six days?"* asks Morris.

Incredible as it may seem, such a rationale is held to comprise a "creation model," scientifically comparable to the greatly refined modern version of the theory of evolution by means of natural selection. This position is reinforced, in the creationist view, by adding that "evolutionism," like humanism, is itself a form of faith or religion anyhow. According to such an argument, it is then seen as only fair that the "creation model" be taught on an equal footing with the "evolution model." Apparently, the creationists either believe that rational judgment, in this instance, can and should be suspended, or it has not occurred to them that a balanced and critical consideration of the alternatives and their supporting sources is bound to bring out the "heathen" antecedents and internal inconsistencies of the Genesis account, its exclusive reliance on revelation for support, its predictive barrenness, and its total refutation by two centuries of geological and biological investigation and the refined measurements of modern geochronology.

Although the creationists may be irrational, they are not to be dismissed as a lunatic fringe that can best be treated by being ignored.

In California, which accounts for about 10 percent of the public-school enrollment and thus exerts great leverage on textbook publishers, they have proven themselves to be skillful tacticians, good organizers, and uncompromising adversaries. As J. A. Moore has shown in his account of the California controversy, creationists were able both to gain control of the State Board of Education from about 1963 to at least 1974, and during the same time, to get an adherent to their views elected as Superintendent of Public Instruction until 1970. This board then proceeded to revise the *Science Framework for California Public Schools,* prepared by the California State Advisory Committee on Science Education, in such a way as to distort the findings of the scientists on that committee and introduce a creationist bias: a situation that led to tough negotiation and uneasy compromise.

The position of scholarly theologians and a substantial majority of literate, practicing Christians regarding the creationism of CRS is well expressed by statements quoted by Moore of the Very Reverend C. Julian Bartlett, dean of Grace Cathedral in San Francisco, and Dr. Conrad Bonifazi, professor of philosophy and religion at the Pacific School of Religion in Berkeley. Bonifazi summarizes:

> Broadly speaking, then, the situation is thus: an extremely conservative wing of Christian sectarianism, which has little or no repute in the world of theological scholarship, adheres to a literal interpretation of the Bible, and is therefore committed to saying that evolution contradicts the biblical account of creation. Its belief in the 'infallibility' of the Bible does not even permit it to recognize that in Genesis itself there are two accounts of creation, each differing from the other in background and in content. It is also true that the major *denominations* of Protestantism and the Roman Catholic Church in the United States recognize and condone the teaching of evolution in the disciplines of natural science. These denominations represent a large majority of Christians in this country.

And Reverend Bartlett adds: "That Biblical myth-story was but one of many such which were developed by primitive religions ... it is a religious and therefore theological document and not a scientific treatise."

If the creationists are deterred by such comments from these and other religious scholars and scholarly theologians, whose judgment one might think they would respect, there is no sign of it. As Moore has observed, and with good reason: "Scientists who have dealt with fundamentalists simply do not trust them; they rather imagine that, if the fundamentalists had the power, they would happily reinstitute an inquisition." And anyone who has studied their benign manner in public debate, their tortured logic and their often scurrilous expression in books and tracts for the faithful, has little difficulty in visualizing creationist polemicists, given the opportunity, in the role of Pius V himself. Examples of creationist logic and tactics may be found in their cartoon strip "Big Daddy," in Morris's equation of evolutionism with racism and Hitlerism, in remarks about the dishonesty of geochronologists by Slusher in his "Critique of Radiometric Dating" (ICR Technical Monograph No. 2, 1973), and in many other places in these and other tracts obtainable from the Institute for Creation Research, 2716 Madison Avenue, San Diego, California 92116. Their "give 'em hell" approach is meeting much success in California and other areas with large fundamentalist followings.

Yet scientists have been negligent both in rising to the defense of rationalism against the creationist attack and in explaining briefly and clearly the available evidence for evolution. It is not enough to shrink from the creationist position; it must be exposed. It is not enough to state that evolution has occurred; at least a sense of the nature of the evidence involved must be transmitted. At risk of belaboring the obvious, therefore, I will briefly summarize the scientific and the creationist positions. Then, in the space remaining, I will attempt to deal with one and to explicate the other insofar as the bearing of fossils and rock-ages on the question of origins is concerned. Those seeking more extended discourse, but not wanting to go deeply into scientific treatises or even textbooks, will find it in two brief and readable articles by N. D. Newell (*Proceedings of the American Philosophical Society,* 1973; *Natural History,* 1974), and in longer, but equally readable, books by G. G. Simpson (*This View of Life—The World of an Evolutionist,* 1964) and W. S. Beck (*Modern Science and the Nature of Life,* 1957), among others.

THE APPROACH TO SCIENCE AND TO CREATIONISM

In trying to arrive at a balanced judgment of issues involved, the nature and methods of both science and creationism need to be understood. Science can be described as a special, *active* way of trying to understand the universe, solar system, and earth. It differs from subjects such as fundamentalist theology that seek their insights wholly from inspiration, meditation, intuition, or divine revelation, unhampered by experimental or naturalistic constraints. Inspiration, meditation, and intuition also play important parts in the mental processes of scientists, but ideas so arrived at do not become a part of science until checked against relevant evidence and found to be consistent with it.

Evidence relevant to science consists of measurements or observations that can be made or confirmed by human observers. If the evidence is experimental, the experiment must be repeatable by others, with the same results. Should the evidence be the results of natural processes, such as floods, earthquakes, climatic change, or exploding stars, the observations must be repeatable. Others observing the same results must be able to see and measure the same thing.

The rules under which science operates specify that scientists must strive for objectivity. That objectivity is difficult is a part of being human. Even the most self-disciplined are products of previous experience and social climate. Although *total* detachment is impossible, the work of the scientist is under constant scrutiny by other scientists and that promotes caution. Characteristically, the ability to do first-rate science is not fulfilled by a high level of intelligence alone. Intellectual, as well as personal, integrity, balanced and critical judgment, and independence from authority in affairs of the mind are also important. Unverifiable assumptions are not permitted. Assumptions made must be consistent with what is already known, and they must be clearly stated so that others can see, test, and challenge them.

The central assumptions of science are that there is order in the universe and that this order can be found and explained. Its twin goals are thus: (1) to search for order in the universe; and (2) when found, to attempt to explain it in terms of processes whose results can be observed and shown to be consistent with causes that do not violate the facts or laws of nature. Science may not invoke supernatural causes— not even in support of divine revelation. There can be no intellectual conflict between science and theology because they are mutually exclusive realms of thought. That involves no value judgment at all. Supernaturalism is not science, and science is not supernaturalism. It is that simple.

The approach to theory in the *scientific* sense starts, not in books, but with *data* and the formulation of *hypotheses* (or a group of related hypotheses called a model). A hypothesis is an attempt to explain the observed data. Science requires that its hypotheses be consistent with known evidence from experimentation or nature and that they have verifiable consequences—that is, they must be capable of disproof. One way of increasing objectivity is to think of as many hypotheses as possible that are consistent with the evidence and have verifiable consequences. As competing hypotheses are tested, their believability grows or shrinks as they withstand or fail opportunities for disproof. From at first being simply permissible, they either are eliminated or grow in believability until the most successful hypothesis may become a *ruling hypothesis*. If such a ruling hypothesis continues to be successful in predicting previously unsuspected facts or relationships, and withstands all opportunities for disproof, and if it has broad application, it may finally be accepted as a *theory,* often modified from the original hypothesis. These distinctions are important. Although not always made, they should be. They express different levels of probability, which is what science is about.

It is essential in science to distinguish among observations and measurements, the hypotheses and theories that integrate and propose *mechanisms* to explain the facts observed, scientific principles that specify operating procedures, and the *laws* of science. The laws of science represent the highest level of supportable generalization. In order to be accepted as a law, the generalization must have proved invariable under all of many observed circumstances; or if variations are observed, they must occur in systematic and predictable ways. The laws of science may not be broken. Angular momentum must be conserved. The entropy of a closed system may not be decreased. Water may not flow uphill without a pump. Hypotheses that run against established scientific law are not acceptable unless they can demonstrate that the law is wrong.

It is characteristic of science that it is controversial. Scientists love to explore new areas, methods, and ideas. Hypotheses, and even theories that once appeared well established, may be challenged, modified, and even overthrown as they are tested against new experimental or observational data or better measurements. As investigation continues, the explanations of science sort out at different levels of probability without ever being considered unchallengeable where new evidence suggests the possibility of other naturalistic causes. Science is thus dynamic, progressive, ever changing, never finished. It is like the expanding wave-front of a pebble flung into a sea of ignorance; its growth both widens the domain of scientific understanding and expands the surrounding circle of ignorance as new knowledge raises new questions. Moreover, previous knowledge, without necessarily being wrong, constantly needs reconsideration in the face of new knowledge or new scientific ways of looking at it. As science expands into space, Euclidean geometry yields to hyperbolic geometry and Newtonian gravity is refined by relativistic gravity. Darwinian ideas of selection evolve into more complex theory as we prove the molecule, behavioral responses, and the rocks. The stable continents of a few decades ago become moving pieces in a great new game of geologic chess because of discoveries made on the ocean floor.

Creationism, on the contrary, is seen by its adherents as fixed, immutable, divinely revealed truth—unchanged and unchanging since the writing of the original Hebrew text, or perhaps the older but similar Babylonian and Sumerian accounts. No less an authority than H. M. Morris himself, the director of the Institute for Creation Research, assures us that "the Genesis record of creation was verified by God Himself as He gave the ten commandments." Never mind the glaring discrepancies between Genesis and the evolutionary sequences of geology, of which Morris finds "at least twenty-five"—grass, herbs, and trees before the sun, for instance. To creationists, this simply demonstrates that such vegetation must have grown in the light of the divine presence itself.

In their many public debates, creationists employ a fivefold strategy: (1) Get out the vote by means of advance agents that arouse local fundamentalist groups in order to assure a strong claque of supporters in the audience. (2) Attack evolution on the grounds that, as is usual in science, some details of the sequence and mechanisms involved are not agreed upon. (3) Snow the unsophisticated with claims that evolution violates the most misunderstood of scientific generalizations, the second law of thermodynamics. (4) Deny the evidence for intermediate forms and their gradual appearance over geologically long spans of time, introducing whatever wild claims or denials appear best suited to that purpose. (5) Claim that a literal interpretation of the Bible provides the only foundation for morality in a wicked and changing age. Granted, then, that there is disagreement among evolutionists, however trivial that may be, and that the second law of thermodynamics and alleged lack of intermediate forms are seen by them as verifiable consequences of creationism, but contradictory to evolution, Genesis emerges in the eyes of the creationists as the only alternative, without need for documentation or discussion beyond the simple assertion that it is the word of God.

It is an appealing scenario to those enamored of simple, unwavering answers, but much too successful at winning over the uncritical popular mind to be brushed aside or underrated by scientists and other humanists who see reason as the quality that offers the best hope for mankind's eventual liberation from the tyranny of fear, superstition, suspicion, and hostility.

THE QUESTION OF ORIGINS

To turn now to the question of origins, creationists focus on the origin of the earth, of life, and of the diversity of life. However, since the findings of nuclear science and astrophysics on the origin of the chemical elements tell us that the stars preceded planetary formation, and that not even the chemical elements have always existed, I will start with the origin of the universe. Given a ball of neutrons at the beginning, scientists can think of naturalistic explanations of varying degrees of probability and testability for all subsequent events. There is, however, no scientific explanation for where the primordial ball of neutrons might have come from. In fact, there is no certainty that there was only one and not several balls of neutrons, or even that the universe didn't emerge from one or several black holes, or from a deity. And, of course, we have no idea what such a deity may have been like, or from where it (or She or He) may have come. That is the problem of *first causes.* Science has no answers to the problem of first causes, although it can place limits on what kinds of answers are permissible. Science does not contradict the idea of a divine origin for the embryonic universe, during which it acquired those characteristics we designate as natural laws, whose unfolding underlies all later events. It does, however, have something to say about the permissible time framework and the composition of primordial materials. It also has a lot to say about those later happenings—the ones with which the creationists occupy themselves.

Evolution implies a systematic progression of related events—a continuous or stepwise process of change from one state to another. It is hard to think of systems that do not evolve—social, political, economic, or natural. Even though they may equilibrate temporarily, change sets in sooner or later. Historical geology attempts to trace the interrelated evolutions of life, air, water, and earth's rocky crust. Its results leave *no doubt that change from simple, slightly diversified to complex and greatly diversified forms of life has taken place over billions of years of geologic time.* In charging evolutionists with dogmatism, creationists both deny that fact and confuse it with the mechanism by which changes were achieved.

That mechanism is always open to debate. Competing hypotheses have repeatedly failed to displace Charles Darwin's basic concept of progressive change brought about by selective processes acting on naturally varying systems over long periods of time in response to changing circumstances. Indeed, creationists are clever enough not to deny either natural variation or the effects of selective processes on local populations. What they do deny is time in excess of a few thousand years and the reality of the progressive changes observed. Instead, all the "basic kinds" of life are seen by them as having originated in a complete state of "perfection" during the third, fifth, and sixth days of creation, after which, giving vent to some unexplained whimsy, God decreed the second law of thermodynamics, whereby free energy decreases, order decreases, and the universe retrogresses from its initial state of perfection forever after. Things are getting worse all the time, 'tis said, and they will get still worse for those who fail to accept the gospel of H. M. Morris, D. T. Gish, (*Evolution: The Fossils Say No!,* 1937), H. S. Slusher, J. C. Whitcomb, Jr. (*The Early Earth,* 1972), and others.

In contrast to the creationist approach, the scientific way to assess evolutionary theory is to ask what it predicts, or "postdicts," about the past, about the geologic record of life. *Current* evolutionary theory, of course, is more complex than that visualized by Darwin, including a foundation of experimental evidence unknown to him, and it is still evolving. It does, however, predict the following: (1) Life either originated on earth under an essentially oxygen-free atmosphere not long after liquid water first began to accumulate, or it reached here from elsewhere in the universe. (2) The earliest forms of life were very simple cells without well-defined nuclei, which evolved in essentially oxygen-free environments until such time as their photosynthetic activities and tolerances to oxygen permitted that gas to accumulate in the atmosphere. (3) More complex, truly nucleated, and, eventually, fully sexual microorganisms evolved only after atmospheric oxygen increased to levels capable of supporting a fully oxidative metabolism. (4) Many-celled animals came later, the first of these being delicate, soft-bodied, thin-bodied forms because they depended on simple diffusion for their oxygen supply. (5) Multicellular animals acquired protective armor or external skeletons only later, as oxygen levels increased and internal oxygen-transport systems evolved. (6) There has been a general, although by no means regular, steady progression of increasing variety and complexity of life from that time until the present.

How do these predictions fit the geologic record? Not only have they been borne out by the steady growth of factual evidence, but nuclear age determinations confirm and amplify the observed sequences of geology. Such nuclear methods permit estimates in atomic years, equivalent to *present* sidereal or clock years, for about how long ago major changes occurred. Consider the prediction (postdictions) above in the order presented. (1) During the past 15 years we have learned that life and the beginnings of photosynthesis originated more than 2 billion years ago and probably more than 3.8 billion years ago. A substantial body of evidence has also accumulated in support of the chemical probability of steps leading toward the origin of life by chemical evolution from nonliving antecedents under oxygen-free conditions; that evidence, however, derives from chemically sophisticated laboratory experiments and cosmochemistry that will not be dealt with further here. (2) The oldest demonstrable organisms were very simple single-celled and filamentous forms, and the geochemical evidence indicates an absence or very low level of oxygen in the atmosphere at the time. Although some of these organisms probably were photosynthetic, oxygen did not accumulate because released oxygen was absorbed in vast sinks of reduced substances, including dissolved iron that formed our largest iron deposits during an episode of iron formation that has not been repeated on the scale earlier observed for the past 2 billion years. (3) Free oxygen first began to accumulate in the atmosphere about 2 billion years ago, as shown by the oldest records of oxidized sediments deposited on the continents of the time, while the oldest cells of a truly advanced nature so far known are younger—about 1.3 billion years old. (4) Many-celled animals are first known from rocks about 680 million years old; these delicate, soft-bodied, thin-bodied or thin-walled animals of primitive sorts are related to worms, jellyfish, and sea pens, but without shells or skeletons. (5) The first shell-bearing animals appeared about 600 million years ago; they were very simple types—trilobites and most of the main kinds of organisms did not appear until later. (6) Although early multicellular diversification was rapid, a natural consequence of the many, then unoccupied, ecologic niches and probably multiple origins, the progression was orderly. From then until now there has been an essentially continuous progression of increasing variety and complexity of multicellular animal life.

I have written above of my own field of specialization. The results I have summarized so briefly come from forty years of independent study and research on life processes in earth history.

Let me now add some words about the discontinuities in the evolutionary record, of which creationists make so much. Populations of forms, transitional from one successful form of life to another, should be small, peripheral to large populations of successful forms, and represent brief time spans. Only evolutionary successes become abundant enough to have good statistical chance of leaving a fossil record, and those numerous forms that lack hard shells or skeletons are only rarely preserved. Because land deposits tend to be weathered and eroded, while marine ones tend to be preserved, marine fossils are more common than those of land animals; and on the continents, smart animals like man rarely become fossils by accident. Nevertheless, there *are* intermediate forms, as well as gaps; and although no person was there to witness the progress of evolution in prehuman times, its results are in the rocks for all to see. Just as we do not discard Newtonian or Einsteinian gravity because we do not have measurements of the mutual attractions between every particle in the universe, so the general evolutionary progression is clear, even though fossil remains are not found for every creature that ever lived—and even though more than a few new forms appear abruptly in the geologic record either because of its incompleteness or as a result of processes not yet well understood.

Evolution as a historical phenomenon rests on as sound and extensive a factual basis as any scientific generalization we know.

The *mechanism* of evolution, as noted earlier, is a different matter. Although no one since 1859 has come up with a durable scientific alternative to the action of natural selection on varying populations or gene pools, the possibility of large jumps as a result of precocious sexual maturity or for other reasons is still debated, while the evolution of nonsexual organisms involves different patterns from that of sexual ones. The rules of science require that, if situations should be found in which natural selection is not consistent with the facts, it must be modified or abandoned. This doesn't mean, however, that it needs to be seriously reconsidered *without* the introduction of *new* evidence because a few people don't or won't understand it. Thus, among biological and biogeological scientists, natural selection in the modern sense, sometimes called "the synthetic theory of evolution," is the favored theory for the observation that organisms *have* evolved.

If evolution over a long time interval, as documented by the geologic record, is to be explained as the work of a deity, that also has some consequences, although not verifiable. The deity either set the rules by which evolution took place or personally created all of the millions of species that have ever lived, and in a generally systematic progression of increasing complexity and diversity.

Two last thoughts before turning to a consideration of particular creationist arguments against evolution. Creationists fear that without a literal acceptance of the Bible, there is no basis for morality. In contrast, it seems to me that the best testament to the basic goodness of mankind is that so many are honest and compassionate for reasons other than fear of punishment or religious conviction.

Then there is the matter of one's vision of divine cause. If one holds to the view of a supreme being, is it more elevating to think of that being as a grand architect who set the whole thing in motion with a divine plan of operation and then let it alone, or to think of her or him as the whimsical builder pictured by a literal interpretation of Genesis?

SOME OBSERVATIONS ON PARTICULAR CREATIONIST CLAIMS

Why, then, do creationists cling to an internally inconsistent six-day miracle where "creation science" is a contradiction in terms and even the word *creationism* has a different meaning to biblical scholars? On what grounds does the CRS attack evolution? Six of their central substantive arguments are briefly considered below.

1. *Creationists claim that intermediate or transitional forms predicted by evolution theory are not found in the fossil record.* I have already explained some of the reasons deficiencies in the fossil record are to be expected and, in fact, common. But real intermediate forms

are *not* lacking. The creationists are aware of this but choose to deny the evidence. Consider four examples. (a) In the case of *Archaeopteryx,* intermediate between reptiles and birds, creationist D. T. Gish insists that, since it had wings, feathers, and flew, it was clearly a bird and nothing else. It is, of course, true that among living animals, feathers are found only among birds. Contemporary with *Archaeopteryx,* however, were good reptiles that also had wings and flew. *Archaeopteryx* also had teeth, which occur in no living birds. Indeed, Yale paleontologist J. H. Ostrom, who has restudied in detail all of the few known specimens of *Archaeopteryx,* reported in the British journal *Nature* in 1973 that, were it not for the associated impressions of feathers, he would have identified these specimens unequivocally as small theropod dinosaurs with birdlike pelvises. Because it had characteristics of both reptiles and birds, therefore, *Archaeopteryx* is intermediate by definition. But our ways of classifying animals do not provide for intermediate forms. We must choose between reptile and bird or invent a new class with some features of each. As a matter of simplicity and priority, *Archaeopteryx* is classified as a bird. (b) *Ichthyostega,* a 350-million-year-old creature, also denied as transitional by creationists, has the skeleton of and is regarded as a very fishlike amphibian, yet it might equally well be considered a very amphibianlike fish. (c) As for amphibians and reptiles, the differences are so gradational that one might say that the first amphibian to lay an egg that could survive desiccation and hatch out of water (an amniotic egg) was a reptile. (d) Intermediates between reptiles and mammals are so numerous that, although current opinion favors a single main line of evolution from reptile to mammal, there could have been several ancestral reptilian lines, all evolving mammalian characteristics at the same time. The classification of intermediate forms is, in fact, a major procedural problem in modern paleontology. I do not, of course, assert that there are no gaps. They appear to exist, and they are puzzling; but if evolutionary science is to progress toward a better understanding of them, this will not be achieved by using the creationist broom to sweep the problem under the rug.

   2. In the creationist scientist-joke cartoon-strip "Have you been brainwashed?" D. T. Gish states that *"billions of highly complex animals—trilobites, brachiopods, corals, worms, jellyfish, etc.—just suddenly appear in the geological record at the base of the Cambrian."* He can be forgiven for this misstatement because part of it could be derived from careless reading of source materials, including my own writings. But it is *not* true. Since 1954, a variety of primative microorganisms have been found to occur through a long sequence of rocks dating back to more than 2 billion years ago. We now also have evidence that a

limited variety of multicellular animal life began about 680 million years ago, perhaps 80 million years before shell fossils of the Cambrian, and that higher forms appeared sequentially up to, through, and beyond the Cambrian. Also contrary to Gish, corals were never thought by people familiar with the evidence to exist in earliest Cambrian time. Moreover, all of the forms mentioned are still simple forms of life compared with those that came in *successive* waves of greater complexity and diversity over the succeeding half-billion years of geologic time. The contrast with the creationist fantasy of a six-day creation week could not be greater.

3. *Creationists assert that time has been too short for evolution.* Geochronologists and cosmochronologists, they say, are mistaken about the great age of the universe, the solar system, and the earth. Slusher's "Critique of Radiometric Dating" attacks geochronologists for "apparent intellectual dishonesty." He states: *"Most creationists . . . have viewed the evidence regarding the age of the earth as pointing to a very young age of from about 7,000 years to 10,000 years."* All the elegant and internally consistent work of a host of geochronologists the world over, using a variety of sophisticated instrumentation and self-checking systems for the last quarter-century, is rejected because it doesn't fit Genesis. One half-baked calculation by a creationist of the time required for decay of the earth's magnetic field (Slusher, 1973) gives an age that approaches the creationist preconception! *That* age is spurious because the assumptions are invalid. Earth's magnetic field, to be sure, does decay, and on a cycle of thousands of years; but it is constantly being renewed by motions in the earth's liquid core. I add only that the devoutly Christian E. A. Milne, in his 1952 deathbed treatise on "Modern Cosmology and the Christian Idea of God," found no problem with a great age for the earth or the universe. Indeed, he thought that classical clock time, based on inconstant relations, was slowing down relative to constant atomic time, so that the age of the earth in conventional clock years was probably vastly greater than its atomic age, now estimated at 4.6 to 4.7 billion years.

4. *Creationists insist that all fossils were actually deposited at the time of the Great Noachian Flood* of about one year's duration. Apart from the problem this makes with the sequence of rocks and with geochronology is that of the volume of sediments needed to fall from suspension in a bit over a year. We know continuous sequences of stratified rock as much as 20 kilometers (12 miles) thick; and if all those known in time sequence were piled in the order of their deposition, they would be many hundreds of kilometers high. A modern reservoir, say 60 meters deep, made by damming a river in a rapidly eroding area, takes about 100 years to fill with sediment, even allowing for cat-

astrophic floods. At that rate, it would take about 32,000 years for 20 kilometers of stratified sediment to accumulate, and remember that I speak of muddy river water, wet sediment (that would compact to much smaller thickness when dry), and a small reservoir. If one multiplies 32,000 by the hundreds, the years become millions, underscoring the difficulty of accounting for even a small fraction of the sedimentary rocks known by the deposits of one year's time. Indeed, the method of sedimentary rates used by early geologists to estimate the age of the earth gave numbers of around 100 to 400 million years. This, however, included little of pre-Cambrian history and left out erosion and non-deposition. We now find much longer ages, using precise, self-checking nuclear methods. I should add here, however, that the legend of a great lowland flood some thousands of years ago *is* widespread and that Cesare Emiliani, of the University of Miami, and others, in a paper in the American journal *Science,* in 1975, have suggested that such a flood may well have happened as a result of a rapid advance and melting of a very late Pleistocene ice sheet about 11,600 years ago—precisely as reported from early accounts by Pliny the Elder.

5. Similar to their claim that all fossils were formed at once during the biblical flood, *creationists assert that the use of fossils as age indicators is self-fulfilling* because when paleontologists find particular fossils they claim the assigned age. Here I would ask you to visualize the Grand Canyon, along whose walls is a succession of nearly horizontal, layered rocks that can be traced with the eye or on foot, always in the same succession with reference to one another. Geologists judge that the bottom ones were the older (deposited first) because there is no way of suspending the overlying ones above an open space while younger sediments were deposited over large areas beneath. That would indeed require a miracle! Successive layers, therefore, decrease in age before the present from the bottom upward, and everywhere the same distinctive kinds of fossils are found in the same layers, while different ones occur above and below. Similar relationships of varying time spans are found in many parts of the world. Such relationships, matched with one another like pieces of a jigsaw puzzle, allowed students of fossils over a century and a half ago to work out successions that gave *relative* ages in terms of older than, younger than, and contemporaneous with. Until measurements from nuclear-decay series became available, however, ages in years (before the present) could not be given. Now the consistency in sequence observed between atomic ages and fossil ages supports the evolutionary progression of life, the validity of nuclear geochronology, and the conclusion of both evolutionists and creationists that evolution needs lots of time—time measurable not in days but in hundreds or thousands of millions of years. The span of ages involved

in the flat-lying rocks of the Grand Canyon alone is over 300 million years. The deformed rocks beneath them extend another 1,200 million years into the past.

6. Finally, I note the curious *creationist belief that evolution violates the second law of thermodynamics.* This law states that something called entropy always increases (in a closed system). In simplest terms this says, approximately, that free or available energy will be converted to bound, and thus unavailable, energy and that disorder will increase, to the extent order is not logically restored by investments of free energy. Creationists, most notably H. M. Morris, an engineer who ought to know better, insist that life and its diversity violate the second law of thermodynamics, presenting this as evidence of supernatural intervention. This is a misconception on several counts. One defect is that the earth is not closed with respect to energy. Instead, our planet receives new energy from the sun at the average rate of 178 trillion kilowatts daily. This energy, through photosynthesis, drives all life processes in the same way a pump drives water uphill. It is, incidentally, also the source of all of our fresh water, coal, oil, hydroelectric power, and much more. The second law applies only to the universe as a whole, or to such parts of it as may exist separately as truly closed systems. Morris, in his "Entropy and Open Systems" (*ICR Impact Series,* No. 40, 1976), not unexpectedly, takes issue also with this position. He fails, however, to allow for exchange between energy and order. A simple illustration of this phenomenon is the following sequence: energy → bauxite → aluminum metal, in which energy is invested to transform disordered aluminum ore to the ordered elemental state of aluminum metal. When disorder sets in, as a result of the fabrication, use, and dispersal of beer cans from the aluminum metal, additional energy must be invested to restore order in the form of recycled aluminum ingots. Examples of this principle are seen throughout the universe. When an igneous rock crystallizes from a melt, order is created while free energy is consumed. The chemical elements themselves, the perception of whose ordered arrangement is one of the great artistic triumphs of science, are cooked in stars, novae, and supernovae as a product of the enormous temperatures found there. *Energy from the sun,* through photosynthesis, *is the driving force of life and its evolution.* Indeed, one could argue that the ever growing diversity of life is itself a kind of entropic, effect—where the maximal ordered condition might be visualized as the original population of simple unicellular organisms. In any case, through death, the molecules and elements of all living things are eventually restored to the physical system from which their substance was derived and in which they passed their lives. Entropy gets you in the end!

SUMMARY

Fundamentalist creationism is not a science but a form of anti-science, whose more vocal practitioners, despite their advanced degrees in the sciences and their bland debating postures, play fast and loose with the facts of geology and biology. Creationism has been thoroughly and repeatedly considered over the generations and rejected as being outside the realm of science by the world scientific community. It is not a *scientific* alternative to any form of evolution theory; and unlike much of the Bible, it has no bearing on morals or ethics. Like flat-earthism, which branded photographs of the earth from space as frauds, it is of interest only for its historical aspects and as a sociological aberration.

Indeed, creation "research" is a contradiction in terms, for there is no research to be done if the task is complete, perfect, and fully described in the Bible. What the research of CRS consists of, in fact, is poring through the works of evolutionists in search of trivial inconsistencies, no matter how ancient or offbeat, that can be used to reinforce their admittedly preconceived ideas.

The real issue is not whether science or divine revelation offers better insights to the truth or even whose version of divine revelation is to be presented as an alternative to evolution. There are two more important and more manageable issues. One is whether the scientifically and theologically unsophisticated student is to be confused by treating these two very different modes of thought as if they were susceptible to similar treatment in the framework of science—a distortion of both science and religion. The other is whether an extremist group of religious bigots shall be permitted to abridge the constitutionally guaranteed separation of church and state—whether fundamentalist Old Testament orthodoxy is to be granted a privileged and improper place in the public educational system. If a person wants to believe that the earth is flat or that it and everything in it was created in six days, or to reject the proofs of its great age, he or she should have every right to do so. What he or she does not have a right to under the Constitution of the United States is to have such beliefs falsely presented as *science* in the classrooms of the public-school system.

The most serious threat of creationism is that, if successful, it would stifle inquiry. If everything were already completely set forth in biblical accounts, there would be nothing more to do, apart from suppressing heretical notions like natural selection while awaiting Judgment Day. We could close down the biological and medical research laboratories of the world and those branches of the school system that deal with subjects other than fundamentalist Judeo-Christian theology, industry, and driver training. The grand ideal of the Creation Research Society would have been achieved.

An Engineer Looks at the Creationist Movement*

by John W. Patterson

There are many facets to "scientific creationism" and the movement can be discussed in any of several ways. However, it is best viewed as a loosely connected group of fundamentalist ministries led largely by scientifically incompetent engineers.

This paper is based on a presentation given in Cedar Rapids, Iowa, to the Iowa Academy of Sciences on Saturday, April 25, 1981. The observations are derived from the extensive interactions I've had with creationists and anti-creationists over the past 3 to 4 years. These interactions include written correspondence, careful evaluation of manuscripts and published papers, many conversations, attendance at hearings and debates on creationism, and participation in two creation/evolution debates. The opinions expressed are my own, not those of my university or department.

As a professor who taught thermodynamics to engineering students for many years, I first entered the creation/evolution controversy in 1978. I was motivated to combat what I then considered—and still consider—to be the promotion of grossly erroneous if not deceitful arguments concerning entropy and the second law. I viewed this as being particularly serious, not only because thermodynamics is an important engineering science (in fact, it began as an engineering analysis by Carnot) but also because I found that it was the engineers in the creationist movement who were shaping the apologetics based on the laws of thermodynamics. Indeed, I have since found that engineering educators, senior engineers, and registered professional engineers are perhaps the most prominent leaders of the creationist movement. As an

*The author wishes to acknowledge the help of Stan Weinberg whose recommendations and editorial changes substantially improved the final draft.

engineering professor and a registered engineer myself, I felt it would be professionally irresponsible to let this travesty continue without comment.

This paper attempts to expose the nature of the creationist movement, the role that professional engineers have played in its leadership, and the level of scientific incompetence (particularly in thermodynamics) that these creationist engineers have exhibited both in public speaking and in print. I would hope that similarly revealing exposes will also be forthcoming from such non-engineering perspectives as biochemistry, biology, paleontology, physics, etc., but these I will leave to those professionals whose expertise and teaching responsibilities fall in those areas.

THE NATURE OF THE SCIENTIFIC CREATIONISM MOVEMENT

There are many facets to "scientific creationism" and the movement can be discussed in any of several ways. However, it is best viewed as a loosely connected group of fundamentalist ministries led largely by scientifically incompetent engineers. It is not dedicated to the furtherance of science, education, or intellectual development, but rather to the undermining of these and to advancing the Protestant fundamentalist dogma known as Biblical inerrancy. Based on a fiercely anti-humanist, theological outlook, creationism amounts to an evangelical system of apologetics and polemics. It seeks to promote the Bible as being literally true, but does so largely by obfuscating and attacking any scientific understanding which seems to threaten their view. Though it is dressed up with scientific terminology and references to scientific journals, it is a counterfeit imitation of scientific discourse based on misrepresentation of facts. These and similar allegations may also be inferred from the writings of others, many of whom represent a fundamentalist Christian perspective on science.[1]

My own formal training overlaps significantly some of the areas which the creationists have addressed. In addition to doing research as well as graduate and undergraduate teaching in thermodynamics, I also hold a B.S. and M.S. in mining engineering which, of course, is inextricably related to the geology and the origin of sedimentary deposits. In my view, the level of confusion, obfuscation, and incompetence reflected by the foremost creationist "experts" both in thermodynamics[2] and in geological interpretation is appalling. And here again others strongly agree.[3] Of course, the creationists do not concur with my characterization of their movement. This may be inferred from the following assertions by Duane T. Gish, Associate Director and Vice-President of the San Diego based Institute for Creation Research ministry:

> . . . The creationist movement is not a fundamentalist ministry led by incompetent engineers. Rather, it is a movement led by highly competent scientists, many of whom are biologists. As a matter of fact, biologists probably constitute a higher proportion of all scientific categories within the creationist movement.[4]

Most responsible engineers will wish this were so, but I'm afraid it is not. We can understand to some extent why engineers—who are comparatively ignorant of biological processes, genetics, etc. and who are infatuated with arguments from design—might fall vulnerable to the theological arguments from design. Excuses of this sort, however, can hardly be offered on behalf of biologists, for they have long ago been apprised of the sterility of arguments from design, of teleology and so on in the realm of biology. But let us return to Gish's assertions.

First of all, there can be little doubt that the foremost creationist organizations—the Institute for Creation Research (ICR), the Creation Research Society (CRS), the Creation Science Research Center (CSRC), the Bible Science Association (BSA) and the Students for Origins Research (SOR) groups on campuses about the country—are essentially ministries. They frequently refer to themselves as ministries and as housing writing ministries, educational ministries, and so on. As an example, the section on the ICR of the catalog for Christian Heritage College, where ICR is based, describes ICR almost exclusively in terms of the various educational ministries housed within it.[5]

Are the creationist agencies connected? Here again we find in their own literature strong evidence of loose inter-connections. Much of the literature is virtually identical in message. Also, one often finds the tracts and books of different creationist groups being advertised and sold at events sponsored by others and they also share many speakers. The SOR campus ministries are particularly well stocked with slide/cassette presentations and tracts prepared by the ICR and CSRC ministries. But the most telling evidence of connectedness has to do with the overlapping memberships and especially the number of key officials—many of whom have been engineers—that ICR, CRS and CSRC have shared through the years. Henry M. Morris, a long time engineering professor, civil engineering department chairman, and professional hydraulic engineer,[6] has served as co-founder and/or president of all three of these organizations.[7] He was also co-founder and has been vice-president and president of Christian Heritage College.[8] Moreover, creationism is taught at CHC for college credit by Gish and Morris, who have held professorships in apologetics there.

Are engineers really all that prominent in the leadership of the creationist organizations? The current ICR letterhead stationery lists fourteen "prestigious" technical advisors of whom four are engineers or engineering educators. In addition to D. R. Boylan—Ph.D. and Professor of Chemical Engineering and Dean of Engineering, all at Iowa State University[9]—there is also Ed Blick, former Associate Dean of Engineering at University of Oklahoma, and now Professor of Aerospace, Mechanical and Nuclear Engineering there.[10] Also prominent on this Board of Technical Advisors is Harold R. Henry, Professor and Chairman of Civil and Mining Engineering at the University of Alabama. One of Dr. Henry's degrees is from the University of Iowa, while his Ph.D. in Fluid Mechanics is from Columbia.[11] Another technical advisor to ICR is Malcolm Cutchins, a Professor of Aerospace Engineering at Auburn, who holds a Ph.D. degree in Engineering Mechanics.[12]

A more recent ICR staff member is Henry Morris' son, John D. Morris, who holds a bachelors degree in Civil Engineering from Virginia Polytechnic Institute and a masters and Ph.D. in Geological Engineering from Oklahoma.[13] William Bauer, who holds a Ph.D. in Hydraulics from the University of Iowa, is President of his own engineering firm in the Chicago area and has been a vice-president and very active member of the Midwest Center of ICR.[14] As of this writing, the president of ICR Midwest is W. T. Brown, a retired colonel who holds a Ph.D. in Mechanical Engineering from MIT.[15] In their 1977 booklet of testimonials, *21 Scientists Who Believe in Creation,* [16] the ICR listed the credentials and backgrounds of their (then) leading "scientists." Of these 21 creationist leaders, six (more than one fourth) either were, or had been, engineers or engineering educators, all with Ph.D. degrees.

So engineers certainly *are* very prominent in the leadership of the ICR ministries.

The Creation Research Society rarely uses the word "ministries" in describing itself, its missions and its goals, yet its prominent members are by and large the same as those of ICR. To join CRS you must swear to a statement of belief in the tenets of Christian fundamentalism.[17] The statement commits the undersigner to the belief that all assertions in the Bible are scientifically true. It is only after signing this statement that one may do research on creationism under the auspices of CRS. In this organization, as in ICR, engineers again play a prominent leadership role.

Henry Morris, a past president of CRS, remains prominent on the editorial board of the CRS *Quarterly*. [18] Also on this board is one of the creationists' foremost thermodynamicist/engineers, Emmett L. Williams, who received his Ph.D. in Metallurgical Engineering from Clemson. According to the CRSQ masthead, Williams is currently vice-president of CRS. [19] The engineering representative on the CRS Board of Directors is the Dean of Engineering at ISU [Iowa State University], namely D. R. Boylan, who also serves on the Technical Advisory Board of ICR. These three engineers—Boylan, Morris, and Williams—have contributed extensively to the creationist version of thermodynamics through the CRS *Quarterly* and in a more recent book. [20]

Among other practicing engineers who populate the ranks of the creationist movement, there is General Electric engineer Luther D. Sunderland, who travels the country lobbying for creationism in schools to various state legislators. Richard G. Elmendorf of Bairdsford, Pennsylvania, a registered P.E. and a CRS member, has a standing offer of $5,000 to anyone who can prove (to his satisfaction, of course) that evolution does not contradict thermodynamics. [21] Significantly, perhaps, Richard is also something of a geocentrist, and as part of his "betting ministry" he offers $1,000 to anyone who can prove (to him) that the earth is moving, either in rotation or translation! [22] Engineers active in the creationist movement also include Stan Swinney, a self proclaimed Aerospace Engineer who markets cassettes of his anti–evolution public lectures; Ben Darlington, retired engineer who spear-headed an effort to get creation taught in Florida schools; and Bill Overn, holder of a Bachelor of Electrical Engineering from the University of Minnesota and active creationist speaker and author with the Bible Science Association of Minneapolis.

Realizing that this must only be a partial list, I once requested a membership roster from CRS to see how many members there really were, and especially how many among them were engineers. This request was denied on the ground that CRS members might be put in jeopardy. The denial leads me to suspect creationist claims about the "large number of scientists" who have gone over to their view. Of the hundreds of thousands of M.Sc. and Ph.D. scientists total, I judge that creationists can claim only a small number: perhaps a few hundred individuals, with a significant share of these being more engineers than scientists.

In summary, I don't concur with those like Gish who pretend there are more biologists, or biochemists, or members of some other professional group than there are engineers in the leadership of the creationist movement. I know of no creationist biologists, biochemists, etc. who are deans or department heads in any of the major universities, but such is not at all uncommon amongst the ICR/CRS engineers as we have just seen. Only in fundamentalist schools and Bible colleges can creationists in the life sciences gain comparable faculty prominence.

INCOMPETENCE ALLEGATIONS

The allegation of incompetence is always controversial, partly because of the seriousness of the charge and partly, too, because we are all incompetent in some areas. But being incompetent need not be regarded as a serious matter unless it can be documented in that area wherein one claims expertise or in which he or she publishes allegedly scientific papers. Even then, we should use something of a sliding scale depending on one's level of education. For example, we ought not be too harsh with an undergraduate thermodynamics student for being inept at the Ph.D. level. We *should* be harsh, however, if one flaunts himself as a Ph.D. scientist but exhibits incompetence at the undergraduate level. With creationists, interestingly enough, this is exactly what one finds. Moreover, they often exhibit very dismal command precisely in the subject areas wherein they profess to speak with authority. It is not convenient to document the many serious examples of this among creationists, but I will provide a single example from the area of engineering thermodynamics. I invite specialists in this area to check the soundness of my allegations and technical arguments.

The most error–ridden thermodynamic analysis I have seen in print is the one by creationist D. R. Boylan which appears in the December 1978 issue of *Creation Research Society Quarterly.*[23] As we discuss this paper, I want the reader to keep the following statement by Boylan in mind, for it was published the previous year (1977) as if to assure us of his scientific expertise:

> In teaching on-campus and at church, I have found that an understanding of physical laws, particularly the First and Second Laws of Thermodynamics, is essential to the defense of biblical truths. The Second Law has been particularly helpful in developing an apologetic against abiogenesis. . . .[24]

To begin with, Boylan virtually equates two of the most distinguishable introductory level concepts in engineering thermodynamics, namely *systems* and *processes.*[25] In effect, he directs his reader to "consider life processes as systems." This is like a would–be mechanic directing us to consider gas combustion (a process) as being like a tire or an engine, which are mechanical systems.

After teaching beginners the profound difference between a process and a system, the next most important issues are (a) how to define or describe the system (e.g. closed, open, isolated, etc.) to which one's analysis is to apply; (b) how to specify the system's boundaries; and (c) how to specify the nature of the processes taking place within or over these boundaries (e.g. are they reversible, irreversible, steady state, etc.). If these specifications are not done properly, the results of one's analysis can come out garbled or self-contradictory. Boylan's paper exemplifies such confusion because he fails to specify properly the system to which his analysis applies and the nature of the "life processes" of which he speaks. Only after I submitted a harsh criticism of the paper to the *Creation Research Society Quarterly* (CRSQ)—which led to a heated correspondence with editorial board members Gish and Williams[26]—were the system process specifications made clear. Williams proved to the satisfaction of both Gish and Boylan that the first and second law analysis and the derivation of the entropy change by Boylan are for an open system subjected to a special kind of steady state condition: the so-called *steady state steady flow* (SSSF) condition.[27] But this was also a blunder, since by definition of steady state there can be no change in the entropy inventory (nor of any other extensive property) for steady state systems.[28] All these properties including entropy must remain steady or fixed in value. Hence, Boylan's central result—i.e. his erroneous formula for the entropy change—should have come out to be identically zero (!) and not the non–vanishing sum whose limiting cases he discusses at great length.[29]

In other words, Boylan's analysis implies a profound and unmistakable self-contradiction. And yet it is clear from the subsequent correspondence that neither Boylan, Williams, nor Gish realized this.[30] In fact, at last contact, Gish inferred from Williams's analysis that "there are no errors at all"[31] in Boylan's paper and actually suggested that I apologize for the criticisms I had submitted which I have not done. Also, as of this date (Spring 1982) no letters questioning Boylan's analysis have appeared in the CRSQ.

Several conclusions can be drawn from all of this. First, one must conclude that Boylan, a Ph.D. and Professor in Chemical Engineering, has committed to print worse errors than those for which beginning thermodynamics students are penalized, if not failed, in their homework and examinations. Secondly, Williams, and especially Gish, are at least as devoid of thermodynamic understanding and knowledge as is Boylan. Thirdly, the same can be said for all the engineers in the

CRSQ readership who read but did not question Boylan's analysis. If there were any who did submit criticisms, I have a feeling the public will be the last to know.

Thus Boylan's paper is best viewed as a poor attempt to make a scientific case for creationism. The paper is self-contradictory, and hopelessly garbled when viewed from the perspective of science. Equally audacious attempts to rationalize the geological column in terms of fluid mechanics and hydrological sorting have also been advanced by creationist engineers, particularly by Morris;[32] here again the confusion and obfuscation betray an apologetic approach to discourse.

In other words, the so-called "scientific creationists" have done much to undermine the scientific credibility of creationism. They have inspired a vigorous counterattack from legitimate scientists who ordinarily are not easily moved to combat.

EXPLANATORY CONJECTURES

Why have engineers become so important to the young-earth, "creation science" movement? There are two major reasons: (a) the irresponsible attitude of engineers and their professional societies, and (b) the familiarity of engineers with certain difficult areas of science from which unintelligible but authoritative sounding "apologetics" can be developed.

Engineering societies seem to be uninterested in policing themselves, as regards either ethical irresponsibility or scientific incompetence. Thus engineers can publicly endorse ludicrous forms of pseudoscience without being publicly chastised by their professional societies. My experience is that examining boards simply brand the embarrassing utterance as being outside their purview even though the engineer involved may be flaunting his engineering status while proclaiming the most absurd distortions of engineering science.[33] Were biologists, geologists, or paleontologists to endorse publicly a pseudoscience such as creationism, their chances of achieving or retaining prestigious academic positions would be greatly undermined, as would their chances for high office in professional societies. Only in Bible colleges, seminaries, and creationist ministries can the latter succeed as outspoken creationists.

Hence, when creationist groups try to promote their own credibility by flaunting the professional status of selected members, they find they mainly have engineers to select from. An example of such status flaunting is the ICR practice of listing their technical advisers, with status, on their official stationery. This list contains more engineering educators who still hold respected academic positions than members of any other group, including physicists, biologists, or geologists. Other examples of creationist credential flaunting are also widely known.[34]

Another reason for engineers being so welcome to creationism derives from their backgrounds in the rather difficult subjects of thermodynamics and fluid mechanics. Creationism is so absurd scientifically that it cannot be defended by any rational arguments which are understandable to thinking laymen. Hence the need to develop confusing and yet authoritative sounding arguments which are *un*intelligible to laymen. Clearly the second law, and especially entropy, are ideally suited for this purpose, as can be inferred from a humorous anecdote due to Tribus, himself a famous engineer. According to Tribus, John Von Neumann, the renowned mathematician/physicist, was advising Claude Shannon about naming the uncertainty function he discovered in connection with modern information theory. Von Neumann confided as follows:

> You should call it entropy for two reasons. In the first place your uncertainty function has been used in statistical mechanics under that name, so it already has a name. In the second place, and *more important, "no one knows what entropy really is, so in a debate you will always have the advantage."*[35]

There is little doubt in my mind that it has been the engineers of the creation science movement—particularly Morris, Williams, and Boylan—who are responsible for fashioning entropy and the second law into one of the most effective debating tools available to the creationists. Indeed, in a 1979 article entitled "Educators against Darwin," Hatfield summed up the creationists' view of the second law argument as follows:

> . . . The famous second law of thermodynamics, which governs energy decay is even more important—indeed it is perhaps the favorite argument of creationists. In its classical form the law states the principle of entropy—that in any physical change, energy constantly decreases in utility, moving toward a final stage of complete randomness and unavailability. This descent, the creationists argue, eliminates the possibility of "a basic law of increasing organization which . . . develops existing systems into higher systems—that is evolution.[36]

It is bad enough that this "thermopolemic" against evolution is thoroughly absurd, and that the proper explanation of the apparent paradox has been known since the 1940's, when Schrödinger published it in his book, *What is Life.*[37] But the shameful irony stems from the connections with engineering both past and present. Thus thermodynamics—itself among the greatest of physical disciplines—began in 1824 with an engineering analysis by the great French engineer Sadi Carnot.[38] Yet today we have incompetent "modern engineers" corrupting these great ideas before an unwitting public. Meanwhile their irresponsible peers stand silently by, hoping sheepishly that as long as the battleground seems to be in biology, maybe no one will see the engineering connections. I hope that this paper has helped to expose the engineering incompetence and misconduct involved, and that the following conclusions and inferences aptly summarize the important issues.

CONCLUSIONS AND INFERENCES

1. The so called "scientific creationism" or "creation science" movement is best characterized as a loosely connected group of fundamentalist ministries dedicated to (a) promoting their notion of Biblical inerrancy and (b) undermining all knowledge and understanding which conflicts with their views on scriptural inerrancy.

2. The leadership of the two most active "scientific creation" ministries, namely the ICR and CRS, is dominated by professional engineers and engineering educators, many of whom hold professorships and advanced degrees from reputable universities. But the predominance of engineers is not exclusive, and many other professional groups would do well to carry out their own investigations.

3. The arguments which "creation scientists" use to counter the well-established facts and theories of science are not at all the scientific arguments they are purported to be. Instead, they are thinly disguised apologetics and polemics directed at many areas of science. Established findings refute tenets which creationists hold to be inerrant.

4. The public utterances of the top creation scientists—together with their published works, which appear in professedly authoritative "creation science" books and journals—provide unequivocal, documentable evidence that many of these authors are grossly incompetent, not only in the areas of science on which they expound without proper credentials, but also in their own professed areas of scientific and technical expertise.

5. Public schools that willfully adopt the educational materials produced by such incompetents deserve to be disaccredited, as do their irresponsible officials and staff.

6. It is the responsibility of knowledgeable scientists, of professional educators, and of their organizations, to expose the extent to which scientific incompetence and intellectual dishonesty prevail in the

"creation science" movement. Only then can school officials be held fully responsible for allowing the forced teaching of creationism as science.

REFERENCES

1. See, for example, the following works, the last ten of which (marked with *) are written by those who represent a fundamentalist Christian perspective: W. R. Overton, "Memorandum Opinion in *Rev. Bill McLean et al* vs *The Arkansas Board of Education et al* (January 5, 1982)," included in this volume; P. Cloud, "Scientific Creationism: A New Inquisition Brewing," included in this volume; W. V. Mayer, "Evolution: Yesterday, Today, and Tomorrow," *Humanist* (Jan/Feb, 1977): 16–22; P. Cloud, "Evolution Theory and Creation Mythology, *Humanist* (Nov/Dec, 1977): 53; L. R. Godfrey, "Science and Evolution in the Public Eye," *Skeptical Inquirer* 4 (1) (1979): 21–32; W. Thwaites, "Review of *Biology: A Search for Order in Complexity,*" *Creation/Evolution,* Issue I (1980): 38–40; C. G. Weber, "The Fatal Flaws of Flood Geology," *Creation/Evolution*, Issue I (1980): 24–37; F. Awbrey, "Evidence of the Quality of Creation Science Research," *Evolution/Creation*, Issue I (1980): 24–37; S. Freske, "Creationist Misunderstanding, Misrepresentation, and Misuse of the Second Law of Thermodynamics," included in this volume; J. Cole, "Misquoted Scientists Respond," *Creation/Evolution*, Issue VI (1981): 34–44; R. A. Steiner, "The Facts be Damned," *Reason* 13 (8) (1981): 27;* D. L. Willis, ed., *Origins and Change* (an evangelical perspective on science and Christian faith) (Elgin, IL: American Scientific Affiliation, 1978);* R. P. Aulie, "The Doctrine of Special Creation," in *Origins and Change*, Part II, "Catastrophism," pp. 14–18;* R. J. Cuffey, "Dialogue: Paleontological Evidence and Organic Evolution—The Position of Roger R. Cuffey," in *Origins and Change,* pp. 54–66;* J. R. Van de Fliert, "Fundamentalism and the Fundamentals of Geology," in *Origins and Change,* pp. 38–49;* D. E. Wonderly, "Non-Radiometric Data Relevant to the Question of Age," in *Origins and Change,* pp. 67–73;* P. G. Phillips, "Meteoric Influx and the Age of the Earth," in *Origins and Change,* pp. 74–76;* J. A. Cramer, "General Evolution and the Second Law of Thermodynamics," in *Origins and Change,* pp. 32–33;* R. P. Aulie, "The Post Darwinian Controversies," *Journal of the American Scientific Affiliation,* 34 (1) (1982): 24;* J. Bassi, "Review of *The Moon: Its Creation Form and Significance,*" *Journal of the American Scientific Affiliation,* 33 (4) (1981): 249;* F. Jappe, "Communication," *Journal of the American Scientific Affiliation* 33 (4) (1981): 231.

2. See J. W. Patterson, "Thermodynamics and Evolution," in *Scientists Confront Creationism,* ed. by L. Godfrey (New York: W. W. Norton, 1982).

3. See Thwaites, Aulie, Van de Fliert, Wonderly, Phillips, Cramer, and Bassi, as cited above.

4. D. T. Gish, "Creationism's Side," in Letters Column, *Cedar Rapids Gazette,* June 18, 1981.

5. *Catalogue,* Creation Heritage College, 1978–79, p. 100.

6. See *21 Scientists Who Believe in Creation* (San Diego, CA; Creation-Life Publishers, 1977), p. 25.

7. *21 Scientists Who Believe in Creation,* 1977, p. 25; see also Gish, 1981.

8. See D. H. Milne, "How to Debate with Creationists—and Win," *American Biology Teacher* 43 (5) (1981): 235–45.

9. See *21 Scientists Who Believe in Creation,* 1977, p. 7.

10. *21 Scientists Who Believe in Creation,* 1977, p. 6.

11. *21 Scientists Who Believe in Creation,* 1977, p. 15.

12. *21 Scientists Who Believe in Creation,* 1977, p. 11.

13. See *ICR Acts and Facts* 9 (11) (November 1980): 6.

14. See "The ICR Scientists," *ICR Impact,* no. 86 (August 1980).

15. See *ICR Acts and Facts* 9 (12) (December 1980): 4.

16. *21 Scientists Who Believe in Creation,* 1977, *passim.*

17. H. L. Armstrong, ed., "Creation Research Society," *Creation Research Society Quarterly* 18 (2) (September 1981): 135.

18. Armstrong, 1981; see also Masthead on inside front cover, September, 1981.

19. Armstrong, 1981.

20. E. L. Williams, ed., *Thermodynamics and the Development of Order* (Norcross, GA: Creation Research Society Books, 1982).

21. R. G. Elmendorf, "$5000 Reward and a Challenge to Evolution," *Creation/Evolution* Issue IV (1981): 1–4.

22. Elmendorf, 1981.

23. D. R. Boylan, "Process Constraints in Living Systems," *Creation Research Society Quarterly* 15 (3) (1978): 133–38.

24. *21 Scientists Who Believe in Creation,* 1977, p. 8.

25. See G. J. Van Wylen and R. E. Sonntag, *Fundamentals of Classical Thermodynamics,* 2d ed. (New York: John Wiley and Sons, 1973), pp. 17–19 (systems), pp. 21–22 (processes).

26. J. W. Patterson, E. L. Williams, D. T. Gish, and D. R. Boylan, "Letter Correspondence," 1980.

27. Patterson et al, 1980.

28. See Wylen and Sonntag, 1973, p. 235.

29. Boylan, 1978.

30. Patterson et al, 1980.

31. Patterson et al, 1980.

32. See J. C. Whitcomb and H. M. Morris, *The Genesis Flood* (Phillipsburgh, NJ: Presbyterian and Reformed Publishing Co., 1961); H. M. Morris, *The Troubled Waters of Evolution* (San Diego, CA: Creation-Life Publishers, 1974), especially pp. 93–97; H. M. Morris, *The Twilight of Evolution* (Grand Rapids, MI: Baker Book House, 1963), chs. 3,4; H. M. Morris, "Sedimentation and the Fossil Record," in *Why Not Creation,* ed. by W. E. Lammerts (Grand Rapids, MI: Baker Book House, 1970).

33. J. W. Patterson, "Correspondence with H. M. Morris and the Iowa State Board of Examiners for the Licensing of Registered Professional Engineers," 1980.

35. M. Tribus and E. C. McIrvine, "Energy and Information," *Scientific American* 225 (3) (1971): 179. (Emphasis added.)

36. L. Hatfield, "Educators against Darwin," *Science Digest,* Special Edition (Winter 1979–80): 94.

37. E. Schrödinger, *What is Life?* (New York: MacMillan, 1945).

38. See R. P. Feynman, *The Feynman Lectures on Physics* (Reading, MA: Addison-Wesley, 1964), vol. 1, ch. 44, p. 2.

Decide: Evolution or Creation?

by Frederick Edwords

**In a free country you can decide these questions for yourself. You
are free to make whatever choice you wish. But you aren't free to
have any choice you make be correct. Therefore, choose wisely.**

Intense curiosity is one of the most striking features in human beings.
From the dawn of recorded history to the present, we humans have
sought to understand the universe around us. We have probed into the
unknown, faced new frontiers, and worked to uncover the nature of
everything we see.

FUNDAMENTAL QUESTIONS

In expressing this natural curiosity, we have asked many questions
about *causes* and *origins.* Each object of our investigation has, at some
point, made us wonder, "How did this come to be?" and "Where did
it originate?"

"Did it just spontaneously arise? Was it created by some outside
agent? Did it *grow* into its present state? Or did it begin as something
better and then *decline?*"

"Perhaps it had no beginning, but was simply always here. If so,
has it forever been like this, or does it fluctuate back and forth through
time?"

Such questions have been asked, not only of the external universe,
but of our own world and the amazing variety of life forms which
surround us.

A growing group of people calling themselves "scientific creationists," however, feel that only *two* of these questions are of any impor-

tance. They believe that the great variety of human ideas about origins can be narrowed down to, "Did this gradually evolve, or was it specially created by an outside agent?"

Yet, since the days of the ancient Greeks and of the Renaissance, philosophers and scientists have considered *many* views. Then, following a century of experiment and observation, they have gathered a large body of data which is now regarded as an incredibly convincing case for evolution.

Creationists challenge this. They say the same data used to prove evolution can just as easily be used to prove creation. As a result, both theories are unfounded and basically religious in nature, they say.

THE NATURE OF SCIENCE

One of their reasons for holding that both evolution and creation are unscientific is that, quite obviously, nobody was around in prehistoric times. And, if nobody was around, there was nobody to test the two theories or observe which one was correct.

This is an invalid argument, claim scientists. They point out that our whole criminal justice system is based on the idea that, in a courtroom, a jury of 12, who were not there at the scene of the crime, can decide guilt or innocence from the *present* exhibited evidence alone. Even though it all happened in the past and they weren't there, they can still make a rational decision which will affect the freedom or confinement of another human being.

Furthermore, they argue, it is possible for an event to have *no* human witnesses and for us to still decide it took place. This is because the evidence *by itself* can often point to it, as in the case of animal tracks in the snow, a fallen tree in the forest, or a rock worn down by the pounding of the sea.

"But these are things we see happening all around us," answer the creationists. "We have seen people commit crimes, animals make tracks in the snow, trees fall, and animals get caught, etc. Yet, in all our lives, we've never seen a lower kind of animal become a higher kind."

"In a sense we have," scientists respond, and they note how they have seen the *mechanisms* at work. For example, the peppered moth shows the mechanism of natural selection. When trees in England became blackened by soot during industrial pollution, the black moths were "naturally selected" for survival because their dark color made it harder for predatory birds to see them. This is an observed fact.

BENEFICIAL MUTATIONS

Creationists, however, in looking at this, point out that natural selection is only *one* mechanism. Scientists must still prove *beneficial mutations.* They must show that mutations can first bring about positive changes before they can argue that such changes are then naturally selected. However, says creationists, this is something scientists cannot do. Laboratory experiments time and again have shown that mutations are basically *deformities.* Thus they result in the *reduced* ability to survive, and so are disadvantages.

Scientists agree that laboratory experiments often show mutations to be disadvantageous. But they also show that mutations can also be advantageous, *depending on the environment.* It is a known fact that color change can be brought about in laboratory fruit flies through induced mutations. Now, a mere color change is, by itself, neither an advantage nor a disadvantage. *It depends on the environment.* Therefore, if the color in the peppered moth is a mutation, as it certainly has been in fruit flies, then this would be a beneficial mutation in certain environments, even though it is a disadvantage in others. This is why scientists know that the *only real way* to decide if a mutation is an advantage or a disadvantage is to *look at the environment.*

In equatorial Africa, for example, the heavy fur on a Grizzly Bear would be a disadvantage. But in climates where it snows, like in the Rocky Mountains, it is a clear advantage. Taken by itself, however, it is *neutral.* Now, since mutations affecting the amount of body hair are quite common in animals, we have a clear example of possible *positive* mutations.

To this point creationists argue that even though there are changes within basic animal kinds, there are never changes *between* kinds. That is, mutated fruit flies have always remained fruit flies. They have not become some other insect. And, although various kinds of cats could evolve, or even arise out of the same gene pool, they would all be cats just the same.

This is called "horizontal evolution," the only kind of evolution creationists accept. Bobcats and mountain lions are horizontal variations of the one cat kind created by God. What evolutionists must prove is *vertical* evolution, the case of a cat evolving into something more specialized that is *not* a cat.

Now, although scientists never heard of the terms horizontal and vertical evolution until the creationists made them up, they do have a simple answer. Scientists merely point out that, without some force to *stop* these small changes from gradually going outside the limits set by creationists, so-called "vertical" evolution would indeed take place. So, given enough time, there would obviously emerge a new animal "kind."

This isn't acceptable to creationists, however. But here is where they are divided among themselves. Some creationists hold that positive mutations do not occur in nature, but that all changes we see are mere variations within the gene pool given these animals at creation. This is what keeps them from evolving out of their original created kinds. Other creationists, however, realize that animal hair changes and other mutations can indeed be positive. These changes they call "micro-evolution." But they say the small changes of microevolution are not enough to gradually cause "mega-evolution," or large scale changes. Most mutations are still degenerative, and this doesn't help the case for evolution, they argue.

Evolutionary scientists counter that, even though most mutations are degenerative, whole populations of animals do change in ways that increase their survival. Strains of veneral disease have become immune to penicillin, and some kinds of cockroaches have come to actually thrive on bug spray. These are factors which could not have always been in the gene pool—not unless God foresaw that penicillin and bug spray would be discovered or invented and wanted to make sure humanity would still be plagued by venereal disease and insects.

THE AGE OF THE EARTH

Furthermore, modern geology has shown that the earth is very old, having been in existence some five billion years. This is plenty of time for micro-evolution to become mega-evolution. In addition, the whole process is gradual, and no line can be drawn between what is micro and what is mega, what is horizontal and what is vertical.

Creationists, however, doubt the earth is that old. Most of them reject all the dating methods used by scientists to prove our planet has existed a long time. They hold that creation of the entire universe took place only 6,000 to 10,000 years ago. This, of course, is not enough time for evolution to take place.

"If the creationists are right," ask the astronomers, "then what about the supernovas?" Take the Crab Nebula for instance. It is the result of an explosion of a supernova star that was widely observed in 1054 AD. When these stars explode, clouds of material are spewed into space. Astronomers viewing other galaxies have seen such explosions a number of times. The more time that passes, the further the clouds of material spread out. It's like watching an explosion in slow motion. When astronomers measure the rates of expansion of such clouds, they can figure out how long ago the explosion occurred.

The Cygnus Loop is a good example. It is the result of an explosion which happened 50,000 years ago. Such age has been determined by measurements of the size of its spread, how fast the gas clouds are traveling, and other data. And since it is known to be this old, one can only conclude that the Cygnus Loop exploded 40,000 years *before* the supposed creation. Add to this the fact that supernovas in other galaxies are so distant it takes *millions* of years for their light to reach us, and you have a devastating case against a young universe.

Creationists answer this, though, by arguing that early man needed to see the stars to appreciate the wonderous power of the Creator; therefore God put the photons of light from the stars close enough to the earth to allow the first people to look up and see them. As for the great age of the Cygnus Loop, this is just an illusion of *apparent* age. In actual fact, say creationists, the Cygnus Loop isn't a supernova at all. And most of the other astronomical phenomena scientists talk about aren't what they seem either.

Instead, they are really special creations of the Creator. They were made in the beginning just as we see them today. They don't represent various stages of motion, but instead are various creative designs that have always looked pretty much the same. Evolutionary astronomers, then, are actually deceived.

"But if all this is true," note astronomers, "then astronomy isn't science at all, and the Creator has used deception."

"Astronomy is indeed unscientific," argue the creationists. "Astronomy is nothing but a faith. Furthermore, astronomers aren't deceived by God, but by their own presuppositions of evolution and naturalism."

A WORLDWIDE FLOOD

"This still doesn't account for the vast array of geologic formations on the earth," scientists point out. "These could not have been formed in a short time."

"The Creator can do anything," note the creationists. "For example, the Grand Canyon was formed in a single year as the result of a worldwide flood in Noah's day.

"This flood was so tremendous that, in the *same year* it formed the Grand Canyon, it also moved the continents, formed mountain ranges, laid down all the layers of strata, and deposited all the fossils. This flood happened around 6,000 B.C."

Scientists can show, however, that to hold such a view, one must ignore an awful lot of evidence. For example, modern studies of tree ring dating for the Bristle Cone pines already go back to earlier than 6,200 B.C. And this doesn't even take into account *fossilized* trees. How did these trees survive such a flood? Furthermore, sequences of glacial varves in Scandinavia go back 12,000 years.

"Again, we are dealing with *apparent* age," answer the creationists. "Neither glaciers nor fossilized trees are as old as they look. Furthermore, a flood *had* to occur—otherwise how does one account for vast fossil graveyards and the tremendous number of animals which lie buried in the sedimentary rocks and elsewhere?"

"Fossil graveyards are actually rare," scientists respond, "and they are not caused by flooding. They almost all show evidence of slow formation over a long period of time. Furthermore, if there had been a worldwide flood, there would not be the orderly fossil succession in the sedimentary layers. The geologic column would be a jumble."

Creationists respond to this by arguing that the animals were sorted by the flooding action, heavier animals falling to the bottom, lighter animals rising to the top. Also, the fleet of foot and wing made it to the mountains and were covered up last, while the slower animals were covered up in the lower strata. This is why complex animals usually wind up in the higher layers and the simpler animals are found at the bottom. The geologic column thus *supports* a worldwide flood!

Scientists, however, have found that the fossil record does not fit the creationist description. Instead, single-celled, soft-bodied, light weight, and simple animals are found at *all* levels, but trilobites only at the lower levels. Turtles, which are very *slow* runners, are yet found in the middle and *upper* layers. The great flying reptiles are found in the middle layers with the dinosaurs and marine reptiles, not up with the modern birds. Both heavier and lighter *mammals* are most abundant in the upper layers while heavier and lighter *dinosaurs* are found in the middle layers. All of this runs contrary to the predictions of the creationist flood sorting hypothesis, but is exactly what we would expect to find if evolution were true, the earth were old, and animals lived in the times scientists say they did.

Creationists are presently seeking contributions from supporters so they can have the money to build a wave tunnel that will help them prove their hydrodynamic fossil sorting theories. In the meantime, they admit they need to do more study and research to find a way out of this apparent dilemma. The Institute for Creation Research, for example, admits it is not the Institute for Creation Answers.

DINOSAURS AND HUMANS

One thing creationists believe they *can* show, however, is that dinosaurs did not live as long ago as most people think. Historical dragon legends all seem to point to these great reptiles being around in early Biblical times, contemporary with human beings. Since many of these legends tell of fire-breathing monsters, creationists suggest that the unique crests on the heads of some duck-billed dinosaurs may have been the chemical storage areas for fuel tanks for their flame throwing mechanisms.

Scientists, however, maintain that abundant evidence points to the extinction of dinosaurs long before the arrival of human beings and that dragon legends are only *one* kind of animal legend told in folklore. Other animals, which even creationists do not believe existed, are part of the stories told by primitive peoples everywhere.

Scientists argue further that the crests on the heads of many of the duck-billed dinosaurs could not possibly have been tanks for flame throwers. Not only is there no hard evidence for this outrageous assumption, but current research indicates that these crests most likely served as sound-amplifying devices.

GAPS IN THE FOSSIL RECORD

These arguments against creationism all *sound* persuasive, say creationists, but they are all based on the presupposition of evolution from the start. One way to demolish this presupposition is to look at the Cambrian rocks, the very rocks evolutionists say are the oldest.

When we look at these rocks, we see a sudden *explosion* of life. A tremendous variety of animals thrived in the so-called Cambrian seas. These animals have *absolutely no ancestors,* declare creationists. They appear suddenly and from nothing. This is startling proof that the Creator put them there in the miracle of creation. Furthermore, these animals are surprisingly complex. They aren't just simple micro-organisms, but are multi-cellular, made up of sectioned parts, and reproduce sexually.

Scientists note, however, that discoveries since the 1950s in the Precambrian show numerous examples of simple life forms. For example, tiny cell clusters only ten micrometers across have been found in sediments in central Australia dated to be 850 million years old. Other fossils found in the Precambrian include jellyfish, worms, and sponges. Some of this escaped notice before, since some of the fossils were rare and microscopic in size.

Because of these discoveries, an evolutionary tree of life can now be drawn which shows how one-celled animals evolved into multi-celled animals, and then into soft bodied metazoans. From them came the first animals with hardened bodies. Fossils resembling living cyano-bacteria and which show the jointed or sectioned structure that became so prevalent in the Cambrian era have also been found in the Precam-brian as well.

"All this is simply speculative," argue creationists. "The various trees of life used by evolutionists show obvious gaps. In fact, not only are the gaps in the fossil record widespread, they are systematic. Such gaps between basic kinds of animals are exactly what would be pre-dicted in the creation model. The Creator created each kind separately, and so gaps in the record are expected results."

Scientists are fully aware of these gaps. They realize their evidence is not a complete tree, but a series of stems. Yet, when one looks at the stems in the tree *taken as a whole,* it becomes clear that evolution, and not creation, presents itself. In fact, the very nature of these gaps, that they are *systematic,* is precisely what evolution would predict. System-atic gaps reveal a systematic cause of the missing pieces. This systematic cause is the fact that only small populations evolve, not large ones. But once a life form becomes suited to the larger environment, it thrives, multiplies, and then examples of it become fossilized for us to see. This explains why we consistently see stems with gaps and not a complete tree. Only random gaps and stems with no pattern would prove cre-ation. Systematic gaps with a pattern show evolution.

"This is arguing from lack of evidence, not evidence," answer creationists. "Evolutionists are just giving an excuse as to why they can't produce the needed data. Take for example the supposed evolu-tion of fishes into amphibians. There is a tremendous and unbreachable gap between these two life forms. They have almost no similarity. The fossil fish, even though it has bony fins which evolutionists say evolved into feet and legs, is still 100% fish."

Scientists point out, however, that the fish with the bony fins could crawl on land, much as modern catfish can do. In fact, some fish living today can even climb trees! Add to this the fact that amphibians had and still have very clumsy legs, located more where fins would be than under the body as in mammals, and you have a striking piece of evi-dence for evolution. No Creator would place legs in such an inefficient position where an animal like a salamander would have to crawl with its belly rubbing against the ground. Unless inefficient legs were a throwback to an earlier form, they wouldn't have come to be.

Dr. Gish, a creationist at the Institute for Creation Research, argues that this is all speculation based on the preconceived idea of evolution, and is an effort to get around the problem of fossil gaps. He adds that each new life form appears abruptly and without ancestor, as in the case of that great three-horned dinosaur *Triceratops.* Dr. Gish maintains that paleontologists have never been able to find ancestors for *Triceratops,* nor have they found any transitional forms having half-horns or partially developed horns.

The scientific evidence, however, seems to speak the other way. A dinosaur called *Monoclonius* appeared on the scene about fifteen million years before *Triceratops.* It had only one horn on its head and two small half-evolved horns above its eyes.

There appeared, 10 million years before *Monoclonius,* an earlier form called *Protoceratops.* This dinosaur, which lived one hundred million years ago, resembled *Monoclonius* in that it had the same kind of bony cap to protect its neck. However, there were no horns on its head and it was a much smaller animal.

This fact of *Protoceratops* being small and without horns, together with the fact that it could stand on its somewhat longer hind legs when burrowing, links it to an earlier ancestral form called *Psittacosaurus* or "parrot reptile." *Psittacosaurus* walked on its hind legs like many earlier dinosaurs, but it had the unique parrot-like beak which also marked the later *Protoceratops, Monoclonius,* and *Triceratops.*

Creationists challenge all this. They declare that what looks like an evolutionary sequence or tree of life for the ceratopsians or horned dinosaurs isn't really that at all. Instead, it is a clear case of horizontal evolution within kinds. All these dinosaurs are merely variations within the basic ceratopsian gene pool, or, in some cases, they are separately created kinds. In no way did one evolve from another. What paleontologists must produce is a true transitional form that is clearly half one kind of animal and half another.

TRANSITIONAL FORMS

This, they have failed to do, claim creationists. The usual example, Archaeopteryx, which evolutionists claim is half reptile and half bird, is actually 100% bird. It had wings with feathers, bird legs, was the size of a pigeon, and could fly.

In fact, it was designed for flight. The proof is in the feathers. Non-flying feathers are quite even, having their shaft in the center. But *flying* feathers are uneven and aerodynamically designed. Archaeopteryx had flying feathers, therefore he was a full flying bird. It was not a transitional form.

Paleontologists also note that if you compare a skeleton of Archaeopteryx with that of a little dinosaur of the same period, it's hard to tell them apart. They both reveal similar claws, leg joints, bony tails, and reptilian heads. If impressions of feathers had never appeared in the rocks, scientists would simply have classified Archaeopteryx as just another little dinosaur, for there are no birds today anything like it.

Modern birds, for example, have a skeletal structure which is less erect, have beaks instead of reptilian heads, and have large breastbones connected to wing muscles which allow them to fly efficiently and for long distances.

Not only that, they have hollow porous bones. This is something totally absent in reptiles, and yet is required for efficient flight. So, even if Archaeopteryx had flying feathers, he could probably only glide with them. This too would explain their aerodynamic structure.

THE SECOND LAW OF THERMODYNAMICS

"Evolution is still impossible," maintain creationists. "The Second Law of Thermodynamics proves it. This is the universal law of death and decay. Physicists have shown how everything in the universe is running down, leaving less and less energy available for work.

"Even in human activities we see this law in action. Things quickly deteriorate or become disorganized when left to themselves. Only fresh inputs of energy and new organization can straighten them out. This involves design and planning. Scientists, therefore, must find some 'vital force' which would make evolution work. But, since there is no such force, evolution is impossible."

Scientists clearly disagree. They point to many cases in nature where increases in order spontaneously occur, and occur in harmony with the Second Law. Storms are an example of such increases. They are created by wind systems coming together accidentally. Once this happens, the clouds begin to form into an orderly spiral, and energy is collected into a powerful working force.

By comparing storm clouds to ordinary random cloud formations one can readily see the difference in order. Almost all weather phenomena represent such upward changes. The hydrologic cycle of evaporation, cloud formation, and rain is another example. No designer, whether human or divine, is necessary to make it work.

Crystals grow without help too; and anyone who has seen the intricate structure of a snowflake can have no doubt that nature is often self-organizing.

In biology there are increases in order as well. They happen at the chromosome level and are called mutations. It has been pointed out already that beneficial mutations occur and that they are naturally selected by the environment. This all happens by natural means without the aid of any "vital force."

A CHALLENGE

It should be clear by now that scientists have been gathering a vast array of evidence in favor of their theory for over a century and a half. Through field and laboratory work they have been able to stack up an imposing case. But established theories have been overthrown before. When the evidence become strong enough for an opposing view, the former theory had to go by the wayside.

This then holds out a challenge to creationists. Can they prove an alternate model involving special creation, a worldwide flood, a young earth, and small animal populations after the flood—and prove it scientifically? This is important because, if they cannot explain all the data within the confines of this model, do they deserve equal time in public school science classes?

In a free country you can decide these questions for yourself. You are free to make whatever choice you wish. But you aren't free to have any choice you make be correct. Therefore, choose wisely.

Part 3
Stating the Creationist Position

Creation, Evolution and Public Education

by Duane T. Gish

The product, a non-theistic religion, with evolutionary philosophy as its creed under the guise of "science," is being taught in most public schools, colleges, and universities of the United States. It has become our unofficial state-sanctioned religion.

ABSTRACT

It is widely held that the special creation concept of origins can be supported by mere religious dogma while evolution theory offers not only a scientifically valid but a thoroughly substantiated explanation for the origin of the universe and the life it contains. This paper documents the fact that neither creation nor evolution has ever been observed by human witnesses, neither is subject to the experimental method, and neither is capable of falsification. It is evident, then, that neither is a valid scientific theory.

Furthermore, modern formulations of evolutionary mechanisms are vacuous and are contradictory to well-established natural laws, and, in contrast to commonly accepted views, the fossil record actually contradicts the predictions based on evolution theory. On the other hand, the major features of the fossil record conform admirably to predictions based on a creation model. When all of the scientific evidence is considered, creation provides a model for explaining origins that is superior to the evolution model.

Better science, an improved educational process, academic freedom and cessation of the indoctrination of students in a mechanistic, humanistic religious philosophy can be accomplished by presenting in textbooks and the classrooms of public schools both the creation and the evolution models for origins and all of the evidence relevant to these models, favorable and contradictory. Only then will true education be achieved.

It is commonly believed that the theory of evolution is the only scientific explanation of origins and that the theory of special creation is based solely on religious beliefs. It is further widely accepted that the theory of evolution is supported by such a vast body of scientific evidence, while encountering so few contradictions, that evolution should be accepted as an established fact. As a consequence, it is maintained by many educators that the theory of evolution should be included in science textbooks as the sole explanation for origins but that the theory of special creation, if taught at all, must be restricted to social science courses.

As a matter of fact, neither evolution nor creation qualifies as a scientific theory. Furthermore, it has become increasingly apparent that there are a number of irresolvable contradictions between evolution theory and the facts of science, and that the mechanisms postulated for the evolutionary process could account for no more than trivial changes.

It would be well at this point to define what we mean by creation and evolution. By *Creation* we are referring to the theory that the universe and all life forms came into existence by the direct creative acts of a Creator external to and independent of the natural universe. It is postulated that the basic plant and animal kinds were separately created, and that any variation or speciation that has occurred since creation has been limited within the circumscribed boundaries of these created kinds. It is further postulated that the earth has suffered at least one great world-wide catastrophic event or flood which would account for the mass death, destruction, and extinction found on such a monumental scale in geological deposits.

By *Evolution* we are referring to the General Theory of Evolution. This is the theory that all living things have arisen by naturalistic, mechanistic processes from a single primeval cell, which in turn had arisen by similar processes from a dead, inanimate world. This evolutionary process is postulated to have occurred over a period of many hundreds of millions of years. It is further postulated that all major geological formations can be explained by present processes acting essentially at present rates without resort to any world-wide catastrophe(s).

Creation has not been observed by human witnesses. Since creation would have involved unique, unrepeatable historical events, creation is not subject to the experimental method. Furthermore, creation as a theory is non-falsifiable. That is, it is impossible to conceive an experiment that could disprove the possibility of creation. Creation thus does not fulfill the criteria of a scientific theory. That does not say anything about its ultimate validity, of course. Furthermore, creation theory* can be used to correlate and explain data, particularly that available from the fossil record, and is thus subject to test in the same manner that other alleged historical events are subject to test—by comparison with historical evidence.

Evolution theory also fails to meet the criteria of a scientific theory. Evolution has never been witnessed by human observers; evolution is not subject to the experimental method; and as formulated by present-day evolutionists, it has become non-falsifiable.

It is obvious that no one has ever witnessed the type of evolution-ary changes postulated by the general theory of evolution. No one, for example, witnessed the origin of the universe or the origin of life. No one has ever seen a fish evolve into an amphibian, nor has anyone observed an ape evolve into a man. No one, as a matter of fact, has ever witnessed a significant evolutionary change of any kind.

The example of the peppered moth in England has been cited by such authorities as H. B. D. Kettlewell[1] and Sir Gavin de Beer[2] as the most striking evolutionary change ever witnessed by man. Prior to the industrial revolution in England, the peppered moth, *Biston betularia,* consisted predominatly of a light-colored variety, with a dark-colored form comprising a small minority of the population. This was so be-cause predators (birds) could more easily detect the dark-colored vari-ety as these moths rested during the day on light-colored tree trunks and lichen-covered rocks. With the on-set of the industrial revolution and resultant air pollution, the tree trunks and rocks became progres-sively darker. As a consequence, the dark-colored variety of moths became more and more difficult to detect, while the light-colored vari-ety ultimately became an easy prey. Birds, therefore, began eating more light-colored than dark-colored moths, and today over 95% of the peppered moths in the industrial areas of England are of the darker-colored variety.

Although, as noted above, this shift in populations of peppered moths has been described as the most striking example of evolution ever observed by man, it is obvious that no significant evolutionary change of any kind has occurred among these peppered moths, certainly not the type required to substantiate the general theory of evolution. For however the populations may have shifted in their proportions of the light and dark forms, all of the moths remained from beginning to end peppered moths, *Biston betularia.* It seems evident, then, that if this

---

*An idea or set of ideas may be called a theory even though it does not qualify as a *scientific* theory. The term "model" is, however, a better expression than "theory" for the concept of creation or evolution, although "theory" and "model" will be used interchangeably in this paper.

example is the most striking example of evolution witnessed by man, no real evolution of any kind has ever been observed.

The world-famous evolutionist, Theodosius Dobzhansky, while endeavoring to proclaim his faith in evolution, admitted that no real evolutionary change has ever been observed by man when he said, ". . . the occurrence of the evolution of life in the history of the earth is established about as well as events *not witnessed by human observers can be.*"[3] It can be said with certainty, then, that evolution in the present world has never been observed. It remains as far outside the pale of human observation as the origin of the universe or the origin of life. Evolution has been *postulated* but *never observed.*

Since evolution cannot be observed, it is not amenable to the methods of experimental science. This has been acknowledged by Dobzhansky when he stated, "These evolutionary happenings are unique, unrepeatable, and irreversible. It is as impossible to turn a land vertebrate into a fish as it is to effect the reverse transformation. The applicability of the experimental method to the study of such unique historical processes is severely restricted before all else by the time intervals involved, which far exceed the lifetime of any human experimenter. And yet it is just such *impossibility* that is demanded by antievolutionists when they ask for 'proofs' of evolution which they would magnanimously accept as satisfactory."[4]

Please note that Dobzhansky has said that the applicability of the experimental method to the study of evolution is an impossibility! It is obvious, then, that evolution fails to qualify as a scientific theory, for it is certain that a theory that cannot be subjected to experimental tests is not a scientific theory.

Furthermore, modern evolution theory has become so plastic, it is non-falsifiable. It can be used to prove anything and everything. Thus, Murray Eden, a professor at Massachusetts Institute of Technology and an evolutionist, has said, with reference to the falsifiability of evolution theory, "This cannot be done in evolution, taking it in its broad sense, and this is really all I meant when I called it tautologous in the first place. It can, indeed, explain anything. You may be ingenious or not in proposing a mechanism which looks plausible to human beings and mechanisms which are consistent with other mechanisms which you have discovered, but it is still an unfalsifiable theory."[5]

Paul Ehrlich and L. C. Birch, biologists at Stanford University and the University of Sidney, respectively, have said,

> Our theory of evolution has become . . . one which cannot be refuted
> by any possible observations. Every conceivable observation can be
> fitted into it. It is thus 'outside of empirical science' but not neces-

sarily false. No one can think of ways in which to test it. Ideas, either without basis or based on a few laboratory experiments carried out in extremely simplified systems, have attained currency far beyond their validity. They have become part of an evolutionary dogma accepted by most of us as part of our training.[6]

Some evolutionists have been candid enough to admit that evolution is really no more scientific than is creation. In an article in which he states his conviction that the modern neo-Darwinian theory of evolution is based on axioms, Harris says

... the axiomatic nature of the neo-Darwinian theory places the debate between evolutionists and creationists in a new perspective. Evolutionists have often challenged creationists to provide experimental proof that species have been fashioned *de novo*. Creationists have often demanded that evolutionists show how chance mutations can lead to adaptability, or to explain why natural selection has favored some species but not others with special adaptations, or why natural selection allows apparently detrimental organs to persist. We may now recognize that neither challenge is fair. If the neo-Darwinian theory is axiomatic, it is not valid for creationists to demand proof of the axioms, and it is not valid for evolutionists to dismiss special creation as unproved so long as it is stated as an axiom.[7]

In his introduction to a 1971 edition of Charles Darwin's *Origin of Species*, Matthews states,

In accepting evolution as a fact, how many biologists pause to reflect that science is built upon theories that have been proved by experiment to be correct, or remember that the theory of animal evolution has never been thus proved? . . . The fact of evolution is the backbone of biology, and biology is thus in the peculiar position of being a science founded on an unproved theory—is it then a science or a faith? Belief in the theory of evolution is thus exactly parallel to belief in creation—both are concepts which believers know to be true but neither, up to the present, has been capable of proof.[8]

It can be seen from the above discussion, taken from the scientific literature published by leading evolutionary authorities, that evolution has never been observed and is outside the limits of experimental science. Evolution theory is, therefore, no more scientific than creation theory. That does not make it necessarily false, and it can be tested in the same way that creation theory can be tested—by its ability to correlate and explain historical data, that is, the fossil record. Furthermore, since evolution is supposed to have occurred by processes still operating today, the theory must not contradict natural laws.

Evolutionists protest, of course, that these weaknesses of evolution as a theory are not necessarily due to weaknesses of the theory, per se, but are inherent in the very nature of the evolutionary process. It is claimed that the evolutionary process is so slow that it simply cannot be observed during the lifetime of a human experimenter, or, as a matter of fact, during the combined observations of all recorded human experience. Thus, as noted above, Dobzhansky is incensed that creationists should demand that evolution be subjected to the experimental method before any consideration could be given to evolution as an established process.

It must be emphasized, however, that it is for precisely this reason that evolutionists insist that creation must be excluded from science textbooks or, for that matter, from the whole realm of science, as a viable alternative to evolution. They insist that creation must be excluded from possible consideration as a scientific explanation for origins because creation theory cannot be tested by the experimental method. It is evident, however, that this is a characteristic that it shares in common with evolution theory. Thus, if creation must be excluded from science texts and discussions, then evolution must likewise be excluded.

Evolutionists insist that, in any case, the teaching of the creation model would constitute the teaching of religion because creation requires a Creator. The teaching about the creation model and the scientific evidence supporting it, however, can be done without reference to any religious literature. Furthermore, belief in evolution is as intrinsically religious as is belief in creation.

If creation must be excluded from science in general and from science textbooks and science classrooms in particular because it involves the supernatural, it is obvious that theistic evolution must be excluded for exactly the same reason. Thus the only theory that can be taught according to this reasoning, and in fact, the only theory that is being taught in almost all public schools and universities and in the texts they use, is a purely mechanistic, naturalistic, and thus atheistic, theory of evolution. But atheism, the antithesis of theism, is itself a religious belief.

The late Sir Julian Huxley, British evolutionist and biologist, has said that "Gods are peripheral phenomena produced by evolution."[9] What Huxley meant was that the idea of God merely evolved as man evolved from lower animals. Huxley desired to establish a humanistic religion based on evolution. Humanism has been defined as "the belief that man shapes his own destiny. It is a constructive philosophy, a *non-theistic religion,* a way of life." This same publication quotes Huxley as saying, "I use the word 'Humanist' to mean someone who believes that man is just as much a natural phenomenon as an animal or plant; that his body, mind, and soul were not super-naturally created but are products of *evolution,* and that he is not under the control or guidance of any supernatural being or beings, but has to rely on himself and his own powers."[10] The inseparable link between this non-theistic humanistic religion and belief in evolution is evident.

George Gaylord Simpson, Professor of Vertebrate Paleontology at Harvard University until his retirement and one of the world's best-known evolutionists, has said that the Christian faith, which he calls the "higher superstition" (in contrast to the "lower superstition" of pagan tribes of South America and Africa) is intellectually unacceptable.[11] Simpson concludes his book, *Life of the Past,* with what Sir Julian Huxley has called "a splendid assertion of evolutionist view of man."[12] "Man," Simpson writes, "stands alone in the universe, a unique produce of a long, unconscious, impersonal, material process with unique understanding and potentialities. These he owes to no one but himself, and it is to himself that he is responsible. He is not the creature of uncontrollable and undeterminable forces, but his own master. He can and must decide and manage his own destiny."[13]

Thus, according to Simpson, man is alone in the Universe (there is no God), he is the result of an impersonal, unconscious process (no one directed his origin or creation), and he is his own master and must manage his own destiny (there is no God to determine man's destiny). That, according to Simpson and Huxley, is the evolutionist's view of man. That this is the philosophy held by most biologists has been recently emphasized by Dobzhansky. In his review of Monod's book, *Chance and Necessity,* Dobzhansky said, "He has stated with admirable clarity, and eloquence often verging on pathos, the mechanistic materialist philosophy shared by most of the present 'establishment' in the biological sciences."[14]

No doubt a majority of the scientific community embraces the mechanistic materialistic philosophy of Simpson, Huxley, and Monod. Many of these men are highly intelligent, and they have woven the fabric of evolution theory in an ingenious fashion. They have then combined this evolution theory with humanistic philosophy and have clothed the whole with the term "science." The product, a non-theistic religion, with evolutionary philosophy as its creed under the guise of "science," is being taught in most public schools, colleges and universities of the United States. It has become our unofficial state-sanctioned religion.

Furthermore, a growing number of scientists are becoming convinced that there are basic contradictions between evolution theory and empirical scientific data as well as known scientific laws. On the other hand, these scientists believe special creation provides an excellent model for explaining and correlating data related to origins which is free of such contradictions. Even some evolutionists are beginning to realize that the formulations of modern evolution theory are really incapable of explaining anything and that an adequate scientific theory of evolution, if ever attainable, must await the discovery of as yet unknown natural laws.

The core of modern evolution theory, known as the neo-Darwinian theory of evolution, or the modern synthetic theory, is the hypothesis that the evolutionary process has occurred through natural selection of random mutational changes in the genetic material, selection being in accordance with alterations in the environment. Natural selection, it-self, is not a chance process, but the material it must act on, mutant genes, is produced by random, chance processes.

It is an astounding fact that while at the time Darwin popularized it, the concept of natural selection seemed to explain so much, today there is a growing realization that the presently accepted concept of natural selection really explains nothing. It is a mere tautology, that is, it involves circular reasoning.

In modern theory, natural selection is defined in terms of differential reproduction. In fact, according to Lewontin, differential reproduction *is* natural selection.[15] When it is asked, what survives, the answer is, the fittest. But when it is asked, what are the fittest, the answer is, those that survive! Natural selection thus collapses into a tautology, devoid of explanatory value. It is not possible to explain *why* some varieties live to reproduce more offspring—it is only known that they do.

In discussing Richard Levins' concept of fitness set analysis, Hamilton stated, "This criticism amounts to restating what I think is the admission of most evolutionists, that we do not yet know what natural selection maximizes."[16] Now if evolutionists do not know what natural selection maximizes, they do not know what natural selection selects.

In a review of the thinking in French scientific circles, it was stated, "Even if they do not publicly take a definite stand, almost all French specialists hold today strong mental reservations as to the validity of natural selection."[17] Creationists maintain that indeed natural selection could not result in increased complexity or convert a plant or animal into another basic kind. It can only act to eliminate the unfit.

Macbeth has recently published an especially incisive criticism of evolution theory and of the concept of natural selection as used by evolutionists.[18] He points out that although evolutionists have abandoned classical Darwinism, the modern synthetic theory they have proposed as a substitute is equally inadequate to explain progressive change as the result of natural selection, and, as a substitute is equally inadequate to explain progressive change as the result of natural selection, and, as a matter of fact, they cannot even define natural selection in non-tautological terms. Inadequacies of the present theory and failure of the fossil record to substantiate predictions based on the theory leave macro-evolution, and even micro-evolution, intractable mysteries according to Macbeth. Macbeth suggests that no theory at all may be preferable to the present theory of evolution.

Using Macbeth's work as the starting point for his own investigation of modern evolution theory, Bethell, a graduate of Oxford with a major in philosophy, has expressed his complete dissatisfaction with the present formulations of evolution theory and natural selection from the viewpoint of the philosophy of science.[19] Both Macbeth and Bethell present excellent reviews of the thinking of leading evolutionists concerning the relationship of natural selection to evolution theory. While both are highly critical, neither profess to be creationists.

According to modern evolutionary theory, ultimately all of evolution is due to mutations.[20] Mutations are random changes in the genes or chromosomes which are highly ordered structures. Any process that occurs by random chance events is subject to the laws of probability.

It is possible to estimate mutation rates. It is also possible to estimate how many favorable mutations would be required to bring about certain evolutionary changes. Assuming that these mutations are produced in a random, chance manner, as is true in the neo-Darwinian interpretation of evolution, it is possible to calculate how long such an evolutionary process would have required to convert an amoeba into a man. When this is done, according to a group of mathematicians, all of whom are evolutionists, the answer turns out to be billions of times longer than the assumed five billion years of earth history.[21]

One of these mathematicians, Murray Eden, stated, "It is our contention that if 'random' is given a serious and crucial interpretation from a probabilistic point of view, the randomness postulate is highly implausible and that an *adequate scientific theory of evolution must await the discovery and elucidation of new natural laws—physical, physico-chemical, and biological.*"[22] What Eden and these mathematicians are saying is that the modern neo-Darwinian theory of evolution is totally inadequate to explain more than trivial change and thus we simply have no basis at present for attempting to explain how evolution may have occurred. As a matter of fact, based on the assumption that the evolutionary process was dependent upon random chance processes, we can simply state that evolution would have been impossible.

Furthermore, evolution theory contradicts one of the most firmly established laws known to science, the Second Law of Thermodynamics. The obvious contradiction between evolution and the Second Law of Thermodynamics becomes evident when we compare the definition of this Law and its consequences by several scientists (all of whom, as far as we know, accept evolutionary philosophy) with the definition of evolution by Sir Julian Huxley, biologist and one of the best-known spokesmen for evolution theory.

> There is a general natural tendency of all observed systems to go from order to disorder, reflecting dissipation of energy available for future transformations—the law of increasing entropy.[23]

> All real processes go with an increase of entropy. The entropy also measures the randomness, or lack of orderliness of the system: the greater the randomness, the greater the entropy.[24]

> Another way of stating the second law then is: 'The universe is constantly getting more disorderly!' Viewed that way, we can see the second law all about us. We have to work hard to straighten a room, but left to itself it becomes a mess again very quickly and very easily. Even if we never enter it, it becomes dusty and musty. How difficult to maintain houses, and machinery, and our own bodies in perfect working order: how easy to let them deteriorate. In fact, all we have to do is nothing, and everything deteriorates, collapses, breaks down, wears out, all by itself—and that is what the second law is all about.[25]

Now compare these definitions or consequences of the Second Law of Thermodynamics to the theory of evolution as defined by Huxley:

> Evolution in the extended sense can be defined as a directional and essentially irreversible process occurring in time, which in its course gives rise to an increase of variety and an increasingly high level of organization in its products. Our present knowledge indeed forces us to the view that the whole of reality is evolution—a single process of self-transformation.[26]

There is a natural tendency, then, for all observed natural systems to go from order to disorder, towards increasing randomness. This is true throughout the entire known universe, both at the micro and macro levels. This tendency is so invariant that it has never been observed to fail. It is a natural law—the Second Law of Thermodynamics.

On the other hand, according to the general theory of evolution, as defined by Huxley, there is a general tendency of natural systems to go from disorder to order, towards an ever higher and higher level of complexity. This tendency supposedly operates in every corner of the universe, both at the micro and macro levels. As a consequence, it is believed, particles have evolved into people.

It is difficult to understand how a discerning person could fail to see the basic contradiction between these two processes. It seems apparent that both cannot be true, but no modern scientist would dare challenge the validity of the Second Law of Thermodynamics.

The usual, but exceedingly naive, answer given by evolutionists to this dilemma is that the Second Law of Thermodynamics applies only to closed systems. If the system is open to an external source of energy, it is asserted, complexity can be generated and maintained within this system at the expense of the energy supplied to it from the outside.

Thus, our solar system is an open system, and energy is supplied to the earth from the sun. The decrease in entropy, or increase in order, on the earth during the evolutionary process, it is said, has been more than compensated by the increase in entropy, or decrease in order, on the sun. The overall result has been a net decrease in order, so the Second Law of Thermodynamics has not been violated, we are told.

An open system and an adequate external source of energy are necessary *but not sufficient* conditions, however, for order to be generated and maintained, since raw undirected, uncontrolled energy is destructive, not constructive. For example, without the protective layer of ozone in the upper atmosphere which absorbs most of the ultraviolet light coming from the sun, life on earth would be impossible. Bacterial cells exposed to such radiation die within seconds. This is because ultraviolet light, or irradiation of any kind, breaks chemical bonds and thus randomizes and destroys the highly complex structures found in biologically active macromolecules, such as proteins and DNA. Biological activity of these vitally important molecules is destroyed and death rapidly follows.

That much more than merely an external energy source is required to form complex molecules and systems from simpler ones is evident from the following statement by Simpson and Beck: "... the simple expenditure of energy is not sufficient to develop and maintain order. A bull in a china shop performs work, but he neither creates nor maintains organization. The work needed is *particular* work; it must follow specifications; it requires information on how to proceed."[27]

Thus a green plant, utilizing the highly complex photosynthetic system it possesses, can trap light energy from the sun and convert this light energy into chemical energy. A series of other complex systems within the green plant allows the utilization of this energy to build up complex molecules and systems from simple starting material. Of equal importance is the fact that the green plant possesses a system for directing, maintaining, and replicating these complex energy conversion mechanisms—an incredibly complex genetic system. Without the genetic system, no specifications on how to proceed would exist, chaos would result, and life would be impossible.

For complexity to be generated within a system, then, four conditions must be met: 1. The system must be an open system. 2. An adequate external energy source must be available. 3. The system must possess energy conversion mechanisms. 4. A control mechanism must exist within the system for directing, maintaining, and replicating these energy conversion mechanisms.

The seemingly irresolvable dilemma, from an evolutionary point of view, is how such complex energy conversion mechanisms and genetic systems arose in the *absence* of such systems, when there is a general natural tendency to go from order to disorder, a tendency so universal it can be stated as a natural law, the Second Law of Thermodynamics. Simply stated, machines are required to build machines, and something or somebody must operate the machinery.

The creationist thus opposes the wholly unscientific evolutionary hypothesis that the natural universe with all of its incredible complexity was capable of generating itself, and maintains that there must exist, external to the natural universe, a Creator, or supernatural Agent, who was responsible for introducing, or creating, the high degree of order found within this natural universe. While creationism is extra-scientific, it is not anti-scientific, as is the evolutionary hypothesis which contradicts one of the most well-established laws of science.

Finally, but of utmost significance, is the fact that the fossil record is actually hostile to the evolution model, but conforms remarkably well to predictions based on the creation model.[28] Complex forms of life appear abruptly in the fossil record in the so-called Cambrian sedimentary deposits or rocks. Although these animals, which include such highly complex and diverse forms of life as brachiopods, trilobites, worms, jellyfish, sponges, sea urchins, and sea cucumbers, as well as other crustaceans and molluscs, supposedly required about two to three billion years to evolve, not a single ancestor for any of these animals can be found anywhere on the face of the earth.[29] George Gaylord Simpson has characterized the absence of Precambrian fossils as "the major mystery of the history of life."[30] This fact of the fossil record, incomprehensible in the light of the evolution model, is exactly as predicted on the basis of the creation model.

The remainder of the fossil record reveals a remarkable absence of the many transitional forms demanded by the theory of evolution.[31] Gaps between all higher categories of plants and animals, which creationists believe constituted the created kinds, are systematic. For example, Simpson has admitted that "Gaps among known orders, classes, and phyla are systematic and almost always large."[32] Richard B.

Goldschmidt, well-known geneticist and a rabid evolutionist, acknowledged that "practically all orders or families known appear suddenly and without any apparent transitions."[33] E. J. H. Corner, Cambridge University botanist and an evolutionist, stated, ". . . I still think, to the unprejudiced, the fossil record of plants is in favor of special creation."[34]

Recently, the well-known evolutionary paleontologist, David B. Kitts, stated, "Despite the bright promise that paleontology provides a means of 'seeing' evolution, it has presented some nasty difficulties for evolutionists, the most notorious of which is the presence of 'gaps' in the fossil record. *Evolution requires intermediate forms between species and paleontology does not provide them. . . .*"[35]

Lord Solly Zuckerman, for many years the head of the Department of Anatomy at the University of Birmingham, was first knighted and then later raised to the peerage as recognition of his distinguished career as a research scientist. After over 15 years of research on the subject, with a team that rarely included less than four scientists, Lord Zuckerman concluded that *Australopithecus* did not walk upright, he was not intermediate between ape and man, but that he was merely an anthropoid ape. *Australopithecus* (Louis Leakey's "Nutcracker Man," and Donald Johanson's "Lucy") is an extinct ape-like creature that almost all evolutionists believe walked erect and showed many characteristics intermediate between ape and man. Lord Zuckerman, although not a creationist, believes there is very little, if any, science in the search for man's fossil ancestry. Lord Zuckerman states his conviction, based on a life-time of investigation, that if man has evolved from an ape-like creature he did so without leaving any trace of the transformation in the fossil record.[36] This directly contradicts the popular idea that paleontologists have found numerous evidences of ape-like ancestors for man, but rather suggests they have found none at all.

The explosive appearance of highly complex forms of life in Cambrian and other rocks with the absence of required ancestors, and the abrupt appearance of each major plant and animal kind without apparent transitional forms are the facts of greatest importance derivable from a study of the fossil record. These facts are highly contradictory to predictions based on the evolution model, but are just as predicted on the basis of the creation model of origins.

The facts described above are some of the reasons why creationists maintain that, on the basis of available scientific evidence, the creation model is not only a viable alternative to the evolution model, but is actually a far superior model. Furthermore, after more than a century of effort to establish Darwinian evolution, even some evolutionists are beginning to express doubts. This is evidently true, for example, of

Pierre P. Grassé, one of the most distinguished of French scientists. In his review of Grassé's book, *L'Evolution du Vivant,*[37] Dobzhansky states,

> The book of Pierre P. Grassé is a frontal attack on all kinds of 'Darwinism.' Its purpose is 'to destroy the myth of evolution as a simple, understood, and explained phenomenon,' and to show that evolution is a mystery about which little is, and perhaps can be, known. Now, one can disagree with Grassé but not ignore him. He is the most distinguished of French zoologists, the editor of the 28 volumes of 'Traite de Zoologie,' author of numerous original investigations, and ex-president of the Academie des Sciences. His knowledge of the living world is encyclopedic . . .[38]

In the closing sentence of his review, Dobzhansky says, "The sentence with which Grassé ends his book is disturbing: 'It is possible that in this domain biology, impotent, yields the floor to metaphysics.' "[39] Grassé thus closes his book with the statement that biology is powerless to explain the origin of living things and that it may possibly have to yield to metaphysics (supernatural creation of some kind).

In his Presidential Address to the Linnaean Society of London, "A Little on Lung-Fishes," Errol White said, "But whatever ideas authorities may have on the subject, the lung-fishes, like every other major group of fishes I know, have their origin firmly based on *nothing* . . ." He then said, "I have often said how little I should like to have to prove organic evolution in a court of law." He closed his address by saying, "We still do not know the mechanics of evolution in spite of the over-confident claims in some quarters, nor are we likely to make further progress in this by the classical methods of paleontology or biology; and we shall certainly not advance matters by jumping up and down shrilling 'Darwin is God and I, So-and-so, am his prophet'—the recent researches of workers like Dean and Hinshelwood (1964) already suggest the possibility of incipient cracks in the seemingly monolithic walls of the Neo-Darwinian Jericho."[40] White thus seems to be suggesting that the modern neo-Darwinian theory of evolution is in danger of crashing down just as did the walls of Jericho!

Thus, today we have a most astounding situation. Evolution has never been observed by human witnesses. Evolution cannot be subjected to the experimental method. The most sacred tenet of Darwinism —natural selection—in modern formulation is incapable of explaining anything. Furthermore, even some evolutionists are conceding that the mechanism of evolution proposed by evolutionary biologists could account for no more than trivial change in the time believed to have been available, and that an adequate scientific theory of evolution, based on present knowledge, seems impossible. Finally, the major features of the fossil record accord in an amazing fashion with the predictions based on special creation, but contradict the most fundamental predictions generated by the theory of evolution. And yet the demand is unceasing that evolution theory be accepted as the only scientific explanation for origins, even as an established fact, while excluding creation as a mere religious concept!

This rigid indoctrination in evolutionary dogma, with the exclusion of the competing concept of special creation, results in young people being indoctrinated in a non-theistic, naturalistic, humanistic religious philosophy in the guise of science. Science is perverted, academic freedom is denied, the educational process suffers, and constitutional guarantees of religious freedom are violated.

This unhealthy situation could be corrected by presenting students with the two competing models for origins, the creation model and the evolution model, with all supporting evidence for each model. This would permit an evaluation of the students of the strengths and weaknesses of each model. This is the course true education should pursue rather than following the present process of brainwashing students in evolutionary philosophy.

REFERENCES

1. H. B. D. Kettlewell, "Darwin's Missing Evidence," *Scientific American* 200 (3) (1959): 48–53.

2. Gavin de Beer, "Review of Everett C. Olson, *Evolution of Life,* 1965," *Nature* 206 (1965): 331–32.

3. T. Dobzhansky, "Evolution at Work," *Science* 127 (1958): 1091.

4. T. Dobzhansky, "On Methods of Evolutionary Biology and Anthropology: Part 1, Biology," *American Scientist* 45 (1957): 388.

5. M. Eden, "Inadequacies of Neo-Darwinian Evolution as a Scientific Theory," in *Mathematical Challenges to the Neo-Darwinian Interpretation of Evolution,* ed. by P. S. Moorehead and M. P. Kaplan (Philadelphia: Wistar Institute Press, 1967), p. 71.

6. L. C. Birch and P. R. Ehrlich, "Evolutionary History and Population Biology," *Nature* 214 (1967): 352.

7. C. Leon Harris, "An Axiomatic Interpretation of the Neo-Darwinian Theory of Evolution," *Perspectives in Biology and Medicine* 18 (1975): 179–84.

8. L. Harrison Matthews, "Introduction" to *The Origin of Species* (London: J. M. Dent and Sons, Ltd., 1971), pp. x, xi.

9. J. Huxley, *The Observer,* July 17, 1960, p. 17.

10. *What is Humanism?* The Humanist Community of San Jose, San Jose, CA 95106.

11. G. G. Simpson, "The World into which Darwin Led Us," *Science* 131 (1960): 966.

12. J. Huxley, "Review of G. G. Simpson, *Life of the Past,*" *Scientific American* 189 (1953): 90.

13. G. G. Simpson, *Life of the Past* (New Haven: Yale University Press, 1953).

14. T. Dobzhansky, "A Biologist's World View: Review of Jacques Monod, *Chance and Necessity: An Essay on the Natural Philosophy of Modern Biology,*" *Science* 175 (1972): 49.

15. R. C. Lewontin, "Selection in and of Populations," in *Ideas in Modern Biology* (Garden City, NY: Natural History Press, 1965), p. 304.

16. W. D. Hamilton, "Ordering the Phenomena of Ecology: Review of Richard Levins, *Evolution in Changing Environments,* 1968," *Science* 167 (1970): 1478.

17. Z. Litynski, "Should We Burn Darwin?" *Science Digest* 50 (1961): 61.

18. N. Macbeth, *Darwin Retried* (Boston: Gambit, Inc., 1971).

19. T. Bethell, "Darwin's Mistake," *Harper's Magazine* 252 (February, 1976): 70–72, 74–75.

20. See, for example, G. L. Stebbins, *Processes of Organic Evolution* (Englewood-Cliffs, NJ: Prentice-Hall, 1971), p. 3; E. Mayr, *Animal Species and Evolution* (Cambridge, MA: Harvard University Press, 1966), p. 176.

21. R. Bernhard, "Heresy in the Halls of Biology: Mathematicians Question Darwin," *Scientific Research* 2 (November, 1967): 59–66; P. S. Moorehead and M. P. Kaplan, eds., *Mathematical Challenges to the Neo-Darwinian Interpretation of Evolution* (Philadelphia: Wistar Institute Press, 1967).

22. Eden, 1967, p. 109.

23. R. B. Lindsay, "Physics: To What Extent Is It Deterministic," *American Scientist* 56 (1968): 100.

24. H. Blum, "Perspectives in Evolution," *American Scientist* 43 (1955): 595.

25. I. Asimov, *Smithsonian Institute Journal* (June 1970): 6.

26. J. Huxley, "Evolution and Genetics" in *What is Science,* ed. by J. R. Newman (New York: Simon and Schuster, 1955), p. 272.

27. G. G. Simpson and W. S. Beck, *Life: An Introduction to Biology,* 2d ed. (New York: Harcourt, Brace and World, 1965), p. 466.

28. D. T. Gish, *Evolution: The Fossils Say No!* (San Diego: Creation-Life Publishers, 1973).

29. Gish, 1973; see also Preston Cloud, "Pseudofossils: A Plea for Caution," *Geology* 1, 3 (1973): 123–27; D. Axelrod, "Early Cambrian Marine Fauna," *Science* 128 (1958): 7–9; T. N. George, "Fossils in Evolutionary Perspective," *Science Progress* 48 (1960): 1–5.

30. G. G. Simpson, *The Meaning of Evolution* (New Haven: Yale University Press, 1953), p. 18.

31. See Gish, 1973.

32. G. G. Simpson in *The Evolution of Life,* ed. by Sol Tax (Chicago: University of Chicago Press, 1960), p. 149.

33. R. B. Goldschmidt, "Evolution as Viewed by One Geneticist," *American Scientist* 40 (1952): 97.

34. E. J. H. Corner in *Contemporary Botanical Thought,* ed. by A. M. Macleod and L. S. Cobley (Chicago: Quadrangle Books, 1961), p. 97

35. David B. Kitts, "Paleontology and Evolutionary Theory," *Evolution* 28 (1974): 467.

36. S. Zuckerman, *Beyond the Ivory Tower* (New York: Taplinger Publishing Co., 1970), p. 64; see also *Journal of the Royal College of Surgeons of Edinburgh* 11 (1966): 87–115.

37. Editions Albin Michel, Paris, 1973.

38. T. Dobzhansky, "Darwinian or 'Oriented' Evolution: Review of P. P. Grassé, *L'Evolution du Vivant,*" *Evolution* 29 (1975): 376.

39. Dobzhansky, 1975, p. 378.

40. E. White, "Presidential Address: A Little on Lung-Fishes," *Proceedings, Linnean Society of London* 177 (1966): 8.

A Two-Model Approach to Origins: A Curriculum Imperative

by Richard B. Bliss

It is unconscionable for any educator to knowingly instill a *dogmatic world view* into his or her students and thus attempt to make them carbon copies, coached in their own bias. The data must be presented; the decision must be their own. What an empty mind it must be that can only regurgitate someone else's ideas.

Within the last decade, the whole concept of science and social science instruction has centered around the inquiry approach. Researchers have gathered data that show many satisfying spinoffs in reading, language arts, and logical thought within science and social studies. The inquiry approach centers around a presentation of data by the teacher, through reading and via experimentation. This data is then acted upon by the students from a logical framework. This, of course, means that the role of the professional teacher now takes the form of a person skilled in questioning techniques rather than a disseminator of all knowledge. Inquiry in its best context never reflects teacher bias, but rather is objective and develops logical thought and decision-making skills among students. The objectives of good instruction and inquiry will be centered around process skills, such as:

1. Observation (sight, hearing, smell, etc.)
2. Classification (placing information in categories)
3. Inferring (general assumptions about data)
4. Predicting (always done from considerable data)
5. Measuring (using standards for measurement)
6. Communicating (oral, written charts, graphs, etc.)
7. Interpreting Data (from all aspects of the investigation: classifying, communicating, inferring, etc.)

8. Making Operational Definitions (definitions that are clear and expressive of the process)

9. Formulating Questions and Hypothesis (a sophisticated stage, including all he presently knows)

10. Experimentation (one of the best ways of making judgments relative to a problem. Empirical knowledge)

11. Formulating Models (models are temporary structures that quickly change when new information becomes available)

These skills teach young people to make critical observations when dealing with data. Are the data complete? Are they significant to the problem? Teachers cannot justify instruction that does not develop these skills as the ramifications have broad social, as well as scientific implications. Young people are not just naturally operational to these skills; they must be taught. Instruction, then, must come from educators who are willing to coach the student in these skills without developing a dependence upon the teacher's bias.

Some considerations brought out by the National Education Association in the publication "The Spirit of Science," were values in *affective domain*. These values are what educators in private and public education, K–12, as well as college, should be more concerned about, for it reflects the final product of an education system. Reflecting on these values, they are reported as follows:

1. Respect for Logic
2. Search for Data and Its Meaning
3. Long for Knowledge and Understanding
4. Consideration of Consequences
5. Consideration of Premises
6. Demand for Verification
7. Question All Things

Values, to be sure, are not easy to evaluate or test in a paper and pencil way, but are, nevertheless, values that can become operational to young minds if they are trained through objective, unbiased inquiry. This will, in fact, develop the logical thought patterns, decision-making ability, and critical thinking that is so often missing among our high school and university graduates. We cannot escape the fact that this young person is going to be the decision-maker of the future, and we have the role of guiding him toward opportunities that will enable him to be a wise decision-maker. It is unconscionable for any educator to knowingly instill a *dogmatic world view* into his or her students and thus attempt to make them carbon copies, coached in their own bias. The data must be presented; the decision must be their own. What an empty mind it must be that can only regurgitate someone else's ideas.

Science should be a search for truth; however, too often we hear comments such as those stated to this writer by two of his graduate students. "I will decide what is good science for my class to consider and what is not"; "I don't care how many faults there are to *evolution*. I will insist that my students modify the evolution model. I will in no instance present *creation* as an alternative to evolution"; "Evolution is science, creation is religion, we cannot have religion in the classroom." All too often this is the rule when it comes to the manner in which teachers perceive their role in the instruction of origins in the classroom. Fortunately, this type of thinking does not prevail in the majority of cases.

Those who have been exposed to a Two-Model Approach in college methods classes are less likely to take the view expressed above, for they cherish the freedom they had to examine all data and the freedom to come to their own conclusions. Those teachers who have been exposed to severe dogmatism in their high school or college training are not likely to have such an open mind on the subject, nor are they likely to tolerate others that oppose them.

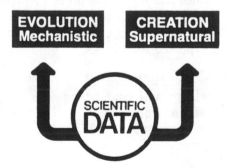

A Two-Model Approach to Origins should not include sectarian religion for the public schools; the approach should base its emphasis on the interpretation of scientific data presently available. It is conceivable, even desirable, that sectarian schools will embellish the scientific limits of the model by making open reference to biblical history. A Two-Model Approach, in essence, is significant only when students have had an opportunity to hear, see, or read all pertinent data on topics relating to origins. When students have had opportunity to make these observations and evaluate data from *their level* of understanding, then a teacher might ask them to determine which model they feel the data fits best. Educators must somehow recognize how crucial it is to train their students in objective inquiry and that non-testable dogma is, in fact, a religion of its own. The tests that we must give to determine whether any model can fit into the context of a valid scientific theory, which neither evolution nor creation can, must at least follow these criteria:

1. Can it explain what has been observed?
2. Can it predict what has not yet been observed?
3. Can it be modified as new data emerge?
4. Can it be subjected to a test of falsification?

Should we not ask our students to question theories from these criteria? It is evident that theories often become gods in the minds of many, and as such, cannot be tampered with. Evolutionary theory seems to be developing in just such a way. Students should be cautioned to hold their theories with a light hand and thus spare themselves the agony of seeing these theories fall apart before their eyes. A current example of this lies within the evolution model.

For the past 100 years or more, Darwin's ideas have become a *not-to-be-challenged by word of science.* Interestingly enough, there are growing amounts of data and a significant number of scientists that are beginning to feel free about challenging these ideas. There are many that question interpretations as they review this emerging data. They are now ready to look at these data in a new light and from a competing model. Questions are arising as to whether the evolution model will be able to hold the attention of the thoughtful mind for the not-too-distant future. Evolutionary dogma is facing the most difficult challenge of its life. We must, as educators, allow our young people the option of this challenge and to present them with truly a better scientific view based upon inquiry. In this respect, they can observe data and its meaning without fear of demeaning reprisal. There is no question in my mind concerning teachers that are dedicated to a student product that is capable of making rational judgments on its own, for they will welcome this view; after all, isn't this what education is all about?

The following statements have been made by biology students after completing a unit on Origins from A Two-Model Approach (conducted in the Racine, Wisconsin Public System):

> I was on the creationist side of our debate, but I am 'in the middle of the road,' so to speak. In the modules, the lectures you gave, and in the debate, so many good arguments were given by both sides. I thought the debate was a very good way to teach and to make us want to dig up more information to prove our point. I thought it was a good way to make us think about both sides. I also thought both sides did very well in supporting their model and I think it was one of the best things we have ever done to learn about subjects. Thank you for coming to our class and helping us research our sides and learn about

both sides and for not letting your own personal views interfere on the subject. You gave us a chance to think for ourselves.

I think the best arguments for creation are the Laws of Thermodynamics, the complexity of natural laws and systems. Also there is the evidence of sediments and other rocks, fossilization, continental drift, volcanoes, pluviation, and mountain building. A good argument for creation is the explanation of many pieces of data such as different races, languages, and adaptation (which in no sense of the word is evolution). In fact, one could say that the earth and its interacting laws and systems and structures is the only needed evidence, since it would be impossible to argue that this huge universe of complexity could have come from complete and absolute randomness which came from an anti-everything VOID.

I have learned very much from this debate. Both sides put up good arguments. I still haven't decided what I am, a creationist or evolutionist. The creationists put up a good argument about the missing links. It would be possible to say that one missing link is Bigfoot. There is only one problem—no evidence there is a Bigfoot. Another one is that Sidney Fox thought of a good theory that shows living things all started from one cell and that they could have started from what is supposedly the earth's first atmosphere.

I thought the best evolution argument was the way in which organisms evolve from simple to complex was stated. They used this argument to support the theory of all living things evolving from one cell. Mainly I was interested in the way autotrophic organisms stretch from little few-celled organisms to more complex organisms like trees, etc.

These unedited comments from biology students are typical responses made in biology classes that were given two models from which to choose. There can be no question in the minds of objective teachers that students will appreciate the opportunity to make their own decisions based upon logical thought processes, as opposed to dogmatic views that are fast being shown to be wrong. I could think of nothing more embarrassing than to have a student come back to me after he has taken my class to say: "You denied me the opportunity to make my own decision; you denied me the opportunity to see, in a fair context, what scientific data was available for me to study on this subject."

It has been said that "it is no mean pedagogical feat to teach a child the facts of science, but it is a pedagogical triumph to teach these facts in the context of scientific inquiry." All teachers can view the outcome of this student product with elation when the instructor realizes that he has taught the child to think and make his own decisions. True, these decisions may not be in keeping with the instructor's view, but nevertheless, the student will have been given the opportunity to make these decisions for himself.

The world is flat! Does this sound familar? Well, for many years the historical view of the world was just so. Only the inquiring mind of man and the freedom to hear and collect all data allowed him to find that another model would fit the data better.

Consider the following general format in the context of a biology class studying the idea of origins:

First Phase: Discussion

(Teacher directed)

1. Introduction by the teacher: motivational
2. Origins: where and how did it all begin: (a) very general: some ideas about the first cell; (b) some ideas about the Geologic Column; (c) some ideas about Flood Geology; (d) some ideas about time and earth history; (e) some ideas about genetics and natural selection; (f) other topics

Second Phase: Team Organization and Research

(Divide class into groups of 4 or 5)

1. Research topics in evolution and creation (teacher allows group to choose topics)
2. Materials: (a) *Origins: A Two-Model Approach* (general); (b) *Time and Earth History* (module); (c) Geology module (general); (d) Creation filmstrips on topic; (e) Evolution filmstrips on topic, etc.

Third Phase:

1. Teacher brings new ideas (factual data only) to students without biased comment
2. Students work in groups on research project
3. Students are encouraged to continue research after school

Fourth Phase:

1. Teacher introduces an evolution-slanted film to class for critical examination
2. Students work in groups

Fifth Phase:

1. Teacher introduces creation-slanted (design) film to class
2. Students work in groups

Sixth Phase:

Somewhere in this phase of the lesson, the students prepare to react on a discussion panel with another group.

The foregoing can be modified to the teacher's needs, and certainly needs additions to complete the topics for a lesson plan; however, this general approach has been used in the Racine Unified District with very encouraging results.

Presently, the Institute for Creation Research is developing supportive materials that will enable the teacher to instill variety and excitement to a stimulating, interdisciplinary topic, and what is even more significant, this is being done in a way that will meet the very important skill and value objectives in inquiry based instruction.

Summary of Scientific Evidence for Creation*

by Duane T. Gish, Ph.D., Scientist
Richard B. Bliss, Ed.D., Science Educator
Reviewed by Wendell R. Bird, J. D., Attorney

The scientific *model of creation,* in summary, includes the scientific evidence for a sudden creation of complex and diversified kinds of life, with systematic gaps persisting between different kinds and with genetic variation occurring within each kind since that time. The scientific *model of evolution,* in summary, includes the scientific evidence for a gradual emergence of complex and diversified kinds of life from simpler kinds and ultimately from nonliving matter.

INTRODUCTION

Public schools in many localities are teaching two scientific models— the creation model and the evolution model—of the origin of the universe, of life, and of man. There is apparent scientific evidence for creation, which is summarized in this pamphlet, just as there is apparent scientific evidence for evolution. The purpose of this pamphlet is to summarize the evidence that shows that:

> "The creation model is at least as scientific as the evolution model, and is at least as nonreligious as the evolution model."

---

*This Impact pamphlet was written by a scientist, and a science educator, and reviewed by an attorney, to provide a brief summary of the scientific evidence supporting creation. The text materials and references listed at the end together give a more thorough discussion of this scientific evidence. Other staff scientists at ICR who helped prepare this summary include Dr. Henry M. Morris (Ph.D., University of Minnesota, Hydraulics); Dr. Kenneth B. Cumming (Ph.D., Harvard University, Biology); Dr. Gary E. Parker (Ed.D., Ball State University, Biology); Dr. Theodore W. Rybka (Ph.D., University of Oklahoma, Physics); and Dr. Harold S. Slusher (M.S., University of Oklahoma, Geophysics).

This scientific evidence for both models can be taught in public schools without any mention of religious doctrine, whether the Bible or the Humanist Manifesto. There are text materials and teacher handbooks that have been prepared for a fair presentation of both models, creation and evolution. There are also seminars and audiovisuals for training teachers to offer both models of origins.

> "This scientific evidence both for creation and for evolution can and must be taught without any religious doctrine, whether the Bible or the Humanist Manifesto."
>
> "Creation-science proponents want public schools to teach all the scientific data, censoring none, but do not want any religious doctrine to be brought into science classrooms."

DEFINITIONS OF THE CREATION MODEL AND THE EVOLUTION MODEL

The scientific *model of creation,* in summary, includes the scientific evidence for a sudden creation of complex and diversified kinds of life, with systematic gaps persisting between different kinds and with genetic variation occurring within each kind since that time. The scientific *model of evolution,* in summary, includes the scientific evidence for a gradual emergence of present life kinds over aeons of time, with emergence of complex and diversified kinds of life from simpler kinds and ultimately from nonliving matter. The creation model questions vertical evolution, which is the emergence of complex from simple and change between kinds but it does not challenge what is often called horizontal evolution or microevolution, which creationists call genetic variation or species or subspecies formation within created kinds. The following chart lists seven aspects of the scientific model of creation and of the scientific model of evolution:

The creation model includes the scientific evidence and the related inferences suggesting that:	The evolution model includes the scientific evidence and the related inferences suggesting that:
I. The universe and the solar system were suddenly created.	I. The universe and the solar system emerged by naturalistic processes.
II. Life was suddenly created.	II. Life emerged from nonlife by naturalistic processes.
III. All present living kinds of animals and plants have remained fixed since creation, other than extinc-	III. All present kinds emerged from simpler earlier kinds, so that single-celled organisms evolved

The creation model . . .	The evolution model . . .
tions, and genetic variation in originally created kinds has only occured within narrow limits.	into invertebrates, then vertebrates, then amphibians, then reptiles, then mammals, then primates, including man.
IV. Mutation and natural selection are insufficient to have brought about any emergence of present living kinds from a simple primordial organism.	IV. Mutation and natural selection have brought about the emergence of present complex kinds from a simple primordial organism.
V. Man and apes have a separate ancestry.	V. Man and apes emerged from a common ancestor.
VI. The earth's geologic features appear to have been fashioned largely by rapid, catastrophic processes that affected the earth on a global and regional scale (catastrophism).	VI. The earth's geologic features were fashioned largely by slow, gradual processes, with infrequent catastrophic events restricted to a local scale (uniformitarianism).
VII. The inception of the earth and of living kinds may have been relatively recent.	VII. The inception of the earth and then of life must have occurred several billion years ago.

I. The Universe and the Solar System Were Suddenly Created.

The First Law of Thermodynamics states that the total quantity of matter and energy in the universe is constant. The Second Law of Thermodynamics states that matter and energy always tend to change from complex and ordered states to disordered states. Therefore the universe could not have created itself, but could not have existed forever, or it would have run down long ago. Thus the universe, including matter and energy, apparently must have been created.

The "big-bang" theory of the origin of the universe contradicts much physical evidence and seemingly can only be accepted by faith.[1] This was also the case with the past cosmogonic theories of evolutionists that have been discarded, such as Hoyle's steady-state theory.

The universe has "obvious manifestations of an ordered, structured plan or design." Similarly, the "electron is materially inconceivable and yet it is so perfectly known through its effects," yet a "strange rationale makes some physicists accept the inconceivable electrons as real while refusing to accept the reality of a Designer." "The inconceivability of some ultimate issue (which will always lie outside scientific resolution) should not be allowed to rule out any theory that explains the interrelationship of observed data and is useful for prediction," in the words of Dr. Wernher von Braun, the renowned late physicist in the NASA space program.

II. Life Was Suddenly Created.

Life appears abruptly and in complex forms in the fossil record,[2] and gaps appear systematically in the fossil record between various living kinds.[3] These facts indicate that basic kinds of plants and animals were created.

The Second Law of Thermodynamics states that things tend to go from order to disorder (entropy tends to increase) unless added energy is directed by a conversion mechanism (such as photosynthesis), whether a system is open or closed. Thus simple molecules and complex protein, DNA, and RNA molecules seemingly could not have evolved spontaneously and naturalistically into a living cell;[4] such cells apparently were created.

The laboratory experiments related to theories on the origin of life have not even remotely approached the synthesis of life from nonlife, and the extremely limited results have depended on laboratory conditions that are artifically imposed and extremely improbable.[5] The extreme improbability of these conditions and the relatively insignificant results apparently show that life did not emerge by the process that evolutionists postulate.

> "One example of the scientific evidence for creation is the sudden appearance of complex fossilized life in the fossil record, and the systematic gaps between fossilized kinds in that record. The most rational inference from this evidence seemingly is that life was created and did not evolve."

III. All Present Living Kinds of Animals and Plants Have Remained Fixed Since Creation, Other than Extinctions, and Genetic Variation in Originally Created Kinds Has Only Occurred within Narrow Limits.

Systematic gaps occur between kinds in the fossil record.[6] None of the intermediate fossils that would be expected on the basis of the evolution model have been found between single-celled organisms and invertebrates, between invertebrates and vertebrates, between fish and amphibians, between amphibians and reptiles, between reptiles and birds or mammals, or between "lower" mammals and primates.[7] While evolutionists might assume that these intermediate forms existed at one time, none of the hundreds of millions of fossils found so far provide the missing links. The few suggested links such as *Archaeopteryx* and the horse series have been rendered questionable by more detailed data. Fossils and living organisms are readily subjected to the same criteria of classification. Thus present kinds of animals and plants apparently were created, as shown by the systematic fossil gaps and by the similarity of fossil forms to living forms.

A kind may be defined as a generally interfertile group of organisms that possesses variant genes for a common set of traits but that does not interbreed with other groups of organisms under normal circumstances. Any evolutionary change between kinds (necessary for the emergence of complex from simple organisms) would require addition of entirely new traits to the common set and enormous expansion of the gene pool over time, and could not occur from mere ecologically adaptive variations of a given trait set (which the creation model recognizes).

IV. Mutation and Natural Selection Are Insufficient to Have Brought About Any Emergence of Present Living Kinds from a Simple Primordial Organism.

The mathematical probability that random mutation and natural selection ultimately produced complex living kinds from a simpler kind is infinitesimally small even after many billions of years.[8] Thus mutation and natural selection apparently could not have brought about evolution of present living kinds from a simple first organism.

Mutations are always harmful or at least nearly always harmful in an organism's natural environment.[9] Thus the multation process apparently could not have provided the postulated millions of beneficial mutations required for progressive evolution in the supposed five billion years from the origin of the earth until now, and in fact would have produced an overwhelming genetic load over hundreds of millions of years that would have caused degeneration and extinction.

Natural selection is a tautologous concept (circular reasoning), because it simply requires the fittest organisms to leave the most offspring and at the same time it identifies the fittest organisms as those that leave the most offspring. Thus natural selection seemingly does not provide a testable explanation of how mutations would produce more fit organisms.[10]

V. Man and Apes Have a Separate Ancestry.

Although highly imaginative "transitional forms" between man and apelike creatures have been constructed by evolutionists based on very fragmentary evidence, the fossil record actually documents the separate origin of primates in general,[11] monkeys,[12] apes,[13] and men. In fact, Lord Zuckerman (not a creationist) states that there are no "fossil traces" of a transformation from an ape-like creature to man.[14]

The fossils of Neanderthal Man were once considered to represent a primitive sub-human *(Homo neanderthalensis)*, but these "primitive" features are now known to have resulted from nutritional deficiencies and pathological conditions; he is now classified as fully human.[15] *Ramapithecus* was once considered to be partially man-like, but is now known to be fully ape-like.[16] *Australopithecus,* in the view of some leading evolutionists, was not intermediate between ape and man and did not walk upright.[17]

The strong bias of many evolutionists in seeking a link between apes and men is shown by the near-universal acceptance of two "missing links" that were later proved to be a fraud in the case of Piltdown Man *(Eoanthropus)* and a pig's tooth in the case of Nebraska Man *(Hesperopithecus).*[18]

VI. The Earth's Geologic Features Were Fashioned Largely by Rapid, Catastrophic Processes that Affected the Earth on a Global and Regional Scale (Catastrophism).

Catastrophic events have characterized the earth's history. Huge floods, massive asteroid collisions, large volcanic eruptions, devastating landslides, and intense earthquakes have left their marks on the earth. Catastrophic events appear to explain the formation of mountain ranges, deposition of thick sequences of sedimentary rocks with fossils, initiation of the glacial age, and extinction of dinosaurs and other animals. Catastrophism (catastrophic changes), rather than uniformitarianism (gradual changes), appears to be the best interpretation of a major protion of the earth's geology.

Geologic data reflect catastrophic flooding. Evidences of rapid catastrophic water deposition include fossilized tree trunks that penetrate numerous sedimentary layers (such as at Joggins, Nova Scotia), widespread pebble and boulder layers (such as the Shinarump Conglomerate of the southwestern United States), fossilized logs in a single layer covering extensive areas (such as Petrified Forest National Park), and whole closed clams that were buried alive in mass graveyards in extensive sedimentary layers (such as at Glen Rose, Texas).

Uniform processes such as normal river sedimentation, small volcanoes, slow erosion, and small earthquakes appear insufficient to explain large portions of the geologic record. Even the conventional uniformitarian geologists are beginning to yield to evidences of rapid and catastrophic processes.[19]

VII. The Inception of the Earth and of Living Kinds May Have Been Relatively Recent.

Radiometric dating methods (such as the uranium-lead and potassium-argon methods) depend on three assumptions: (a) that no decay product (lead or argon) was present initially or that the initial quantities can be accurately estimated, (b) that the decay system was closed through the years (so that radioactive material or product did not move in or out of the rock), and (c) that the decay rate was constant over time.[20] Each of these assumptions may be questionable: (a) some nonradiogenic lead or argon was perhaps present initially;[21] (b) the radioactive isotope (uranium or potassium isotopes) can perhaps migrate out of, and the decay product (lead or argon) can migrate into, many rocks over the years;[22] and (c) the decay rate can perhaps change by neutrino bombardment and other causes.[23] Numerous radiometric estimates have been hundreds of millions of years in excess of the true age. Thus ages estimated by the radiometric dating methods may very well be grossly in error.

Alternate dating methods suggest much younger ages for the earth and life. Estimating by the rate of addition of helium to the atmosphere from radioactive decay, the age of the earth appears to be about 10,000 years, even allowing for moderate helium escape. Based on the present rate of the earth's cooling, the time required for the earth to have reached its present thermal structure seems to be only several tens of millions of years, even assuming that the earth was initially molten.[24] Extrapolating the observed rate of apparently exponential decay of the earth's magnetic field, the age of the earth or life seemingly could not exceed 20,000 years.[25] Thus the inception of the earth and the inception of life may have been relatively recent when all the evidence is considered.[26]

"There is scientific evidence for creation from cosmology, thermodynamics, paleontology, biology, mathematical probability, geology, and other sciences."

"There are many scientists in each field who conclude that the scientific data best support the creation model, not the evolution model."

REFERENCES

1. Slusher, Harold S., *The Origin of the Universe,* San Diego: Institute for Creation Research (ICR), 1978.

2. See, for example, Kay, Marshall and Colbert, Edwin H., *Stratigraphy and Life History,* New York: John Wiley & Sons, 1965, p. 102; Simpson, George G., *The Major Features of Evolution,* New York: Columbia University Press, 1953, p. 360: [Paleontologists recognize] that most new species, genera and families, and that nearly all categories above the level of families, appear in the record suddenly and are not led up to by known gradual, completely continuous transitional sequences.

3. Note 6 *infra.*

4. See, for example, Smith, Charles J. "Problems with Entropy in Biology," *Biosystems,* V. 7, 1975, pp. 259, 264. "The earth, moon, and sun constitute an essentially closed thermodynamic system. . . ." Simpson, George G., "Uniformitarianism," in Hecht, Max A. and Steeres, William C., eds., *Essays in Evolution and Genetics,* New York: Appleton-Century-Crofts, 1970, p. 43.

5. Gish, Duane T., *Speculations and Experiments Related to the Origin of Life (A Critique),* San Diego: ICR, 1972.

6. See, for example, Simpson, George G., "The History of Life," in Tax, Sol, ed., *Evolution after Darwin: The Evolution of Life,* Chicago:. Univ. of Chicago Press, 1960, pp. 117, 149: Gaps among known orders, classes, and phyla are systematic and almost always large.

7. See, for example, Kitts, David B., "Paleontology and Evolutionary Theory," *Evolution,* V. 28. 1974. pp. 458, 467: (See Impact No. 95 for accompanying text.) Evolution requires intermediate forms between species and paleontology does not provide them. For examples of the lack of transitional fossils, Ommaney, F. D., *Vertebrate Paleontology,* Chicago: Univ. of Chicago Press, 3d ed., 1966, p. 36 (vertebrate fish to amphibians); Swinton, W. E., *Biology and Comparative Physiology of Birds,* Marshall, A. J., ed., New York: Academic Press, V. 1, 1960, p. 1 (reptiles to birds); Simpson, George G. *Tempo and Mode in Evolution,* New York: Columbia Univ. Press, 1944, p. 105 (reptiles to mammals); Simons, E. L., *Annals N.Y. Acad. Science,* V. 167, 1969, p. 319 (mammals to primates).

8. See, for example, Eden, Murray, "Inadequacies of Neo-Darwinian Evolution as a Scientific Theory," in Moorhead, Paul S. and Kaplan, Martin M., eds., *Mathematical Challenges to the Neo-Darwinian Interpretation of Evolution,* Philadelphia: Wistar Inst. Press, 1967, p. 109: It is our contention that if "random" is given a serious and crucial interpretation from a probabilistic point of view, the randomness postulate is highly implausible and that an adequate scientific theory of evolution must await the discovery and elucidation of new natural laws. . . .

9. See, for example, Martin, C. P., "A Non-Geneticist Looks at Evolution," *American Scientist,* V. 41, 1954, p. 100.

10. See, for example, Popper, Karl, *Objective Knowledge,* Oxford: Clarendon Press, 1975, p. 242.

11. See, for example, Kelso, A. J., *Physical Anthropology,* 2nd ed., Philadelphia: J. B. Lippincot, 1974, p. 142.

12. See, for example, *Ibid.,* pp. 150, 151.

13. See, for example, Simons, E. L., *Annals N.Y. Acad. Science,* V. 102, 1962, p. 293; Simons, E. L., "The Early Relatives of Man," *Scientific American,* V. 211, July 1964, p. 50.

14. See, for example, Zuckerman, Sir Solly, *Beyond the Ivory Tower,* New York: Taplinger Pub. Co., 1970, p. 64.

15. See, for example, Ivanhoe, Francis, "Was Virchow Right about Neander-t[h]al?", *Nature,* V. 227, 1970, p. 577.

16. See, for example, Zuckerman, pp. 75–94; Eckhardt, Robert B., "Population Genetics and Human Origins," *Scientific American,* V. 226, 1972, pp. 94, 101.

17. See, for example, Oxnard, Charles E., "Human Fossils: New Views of Old Bones," *American Biology Teacher,* V. 41, 1979, p. 264.

18. See, for example, Straus, William L., "The Great Piltdown Hoax," *Science,* V. 119, 1954, p. 265 (Piltdown Man); Gregory, William K., "Hesperopithecus Appar-ently Not an Ape Nor a Man," *Science,* V. 66, 1927, p. 579 (Nebraska Man).

19. See, for example, Bhattacharyya, A., Sarkar, S. & Chanda, S. K., "Storm Deposits in the Late Proterozoic Lower Bhander Sandstone of . . . India," *Journal of Sedimentary Petrology.* V 50, 1980, p. 1327: Until recently noncatastrophic unifor-mitarianism had dominated sedimentologic thought reflecting that sediment formation and dispersal owe their genesis chiefly to the operation of day-to-day geologic events. As a result, catastrophic events, e.g., storms, earthquakes, etc. have been denied their rightful place in ancient and recent sedimentary records. Of late, however, there has been a welcome rejuvenation of [the] concept of catastrophism in geologic thought. J. Harlen Bretz recently stated, on receiving the Penrose Medal (the highest geology award in America), "Perhaps, I can be credited with reviving and demystifying legend-ary Catastrophism and challenging a too rigorous Uniformitarianism." Geological Society of America. "GSA Medals and Awards," *GSA News & Information,* V. 2, 1980, p. 40.

20. See, for example, Stansfield, William D., *The Science of Evolution,* New York: Macmillan Publishing Co., 1977, pp. 83–84; Faul, Henry, *Ages of Rocks, Planets and Stars,* New York: McGraw Hill Co., 1966, pp. 19–20, 41–49. See generally Slusher, Harold S., *Critique of Radiometric Dating.* San Diego: ICR, 1973.

21. See, for example, Kerkut, G. A., *Implications of Evolution,* New York: Perga-mon Press, 1960, pp. 138, 139.

22. See, for example, Faul, p. 61.

23. See, for example, Jueneman, Frederick, "Scientific Speculation," *Industrial Research,* Sept 1972, p. 15.

24. Slusher, Harold S. and Gamwell, Thomas P., *The Age of the Earth,* San Diego ICR, 1978.

25. Barnes, Thomas G., *Origin and Destiny of the Earth's Magnetic Field,* San Diego: ICR, 1973.

26. Slusher, Harold S., *Age of the Cosmos,* San Diego: ICR, 1980: Slusher, Harold S. and Duursma, Stephen J., *The Age of the Solar System,* San Diego: ICR, 1978.

The Scientific Case for Creation: 108 Categories of Evidence

by Walter T. Brown

All arguments for evolution are outdated, illogical, or wishful thinking.

It will be shown that *THE SCIENTIFIC EVIDENCE CONCERNING ORIGINS SUPPORTS THE CREATION MODEL AND OPPOSES THE EVOLUTION MODEL.* This overall conclusion, which will be referred to as the Level I conclusion, will be supported by examining 108 categories of scientific evidence in each of the three basic areas of science: the life sciences, the astronomical sciences, and the earth sciences. (See the diagram on the following page.)

The evidences in each of these major areas of science lead to three Level II conclusions, and these in turn are each supported by three Level III conclusions—or a total of nine conclusions at the third level. The nine Level III conclusions are each supported by about a dozen categories of evidence at Level IV, making a total of 108 categories of evidence in all.

THE THEORY OF ORGANIC EVOLUTION IS INVALID.

Evolution* Has Never Been Observed.

1. Spontaneous generation (the emergence of life from inorganic material) has never been observed.

---

*By *evolution* we mean a naturally occurring, beneficial change that produces *increasing complexity*. When referring to the evolution of life, this increasing complexity would be shown if the offspring of one form of life had a different, improved, and reproducible set of vital organs that its ancestors did not have. This is sometimes called organic evolution, the molecules-to-man theory, or *macro*evolution. *Micro*evolution, on the other hand, involves only such changes as different shapes, colors, sizes, or minor chemical alterations—changes that both creationists and evolutionists agree are relatively trivial and easily observed. It is macroevolution, then, that is being so hotly contested today, and this is what we will mean by the term evolution.

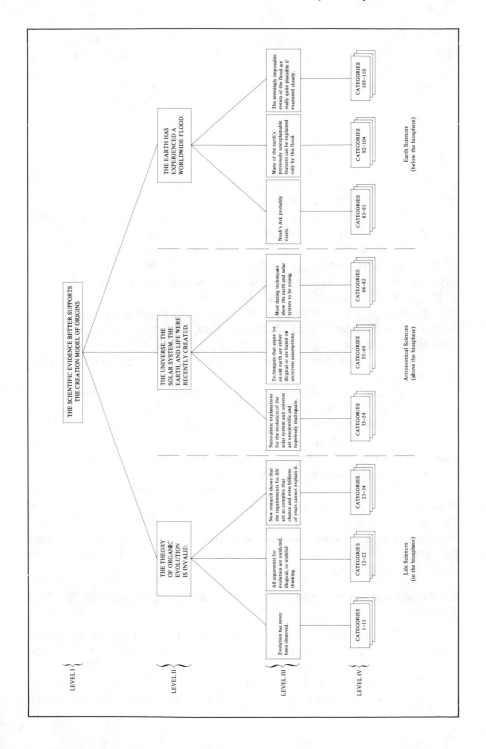

2. Mendel's laws of genetics explain almost all of the physical variations that are observed within life categories such as the dog family. A logical consequence of these laws and their modern day refinements is that there are *limits* to such variation.[a] Breeding experiments have also confirmed that these boundaries exist.[b]

3. Acquired characteristics cannot be inherited.

4. Natural selection cannot produce *new* genes; it only *selects* among preexisting characteristics.

5. Mutations are the only proposed mechanism by which new genetic material becomes available for evolution.[a] However, almost all (perhaps all) observable mutations are harmful; many are lethal.

6. No known mutation has ever produced a form of life having both greater complexity and greater viability than its ancestors.[a]

7. Over seventy years[a] of fruit-fly experiments, equivalent to 2700 consecutive human generations, give no basis for believing that any natural or artificial process can cause an increase in complexity and viability. No clear genetic improvement has ever been observed despite the many unnatural efforts to increase mutation rates.

8. There is no reason to believe that mutations could ever produce any new organs such as the eye, the ear, or the brain.

9. No verified form of extraterrestrial life of any kind has ever been observed.

10. If languages evolved, the earliest languages should be the simplest. On the contrary, language studies reveal that the more ancient the language (for example Latin, 200 B.C.; Greek, 800 B.C.; and Vedic Sanskrit, 1500 B.C.), the more complex it is with respect to syntax, cases, genders, moods, voices, tenses, and verb forms. The best evidence indicates that languages *de*volve.[a-c]

11. Studies of the thirty-six documented cases of children who were raised without contact with other humans (feral children) show that human speech appears to be learned only from other humans. Humans apparently have no inborn ability to speak. Therefore, the first humans must have been endowed with a speaking ability; there is no evidence that language has evolved.[a]

All Arguments for Evolution Are Outdated, Illogical, or Wishful Thinking.

12. The many similarities between different species do not necessarily imply a genealogical relationship; they may imply a common designer.

13. The existence of human organs whose function is unknown does not imply that they are vestiges of organs from our evolutionary

ancestors. In fact, as medical knowledge has increased, the functions of almost all of these organs have been discovered.

14. As an embryo develops, it does not pass through the adult stages of its alleged evolutionary ancestors. Embryologists no longer consider the superficial similarity that exists between a few embryos and the adult forms of simpler animals as evidence of evolution.[a-f]

15. There are many single cell forms of life, but there are no forms of life with 2, 3, . . ., or even 20 cells. If organic evolution happened, these forms of life should exist in great abundance. None do. The evolutionary tree has no trunk.

16. Stories claiming that primitive, ape-like men have been found are overstated.[a-c] Piltdown man was an acknowledged hoax.[d] The fragmentary evidence that constituted Nebraska man was a pig's tooth. The discoverer of Java man later acknowledged that it was a large gibbon[e] and that he had withheld evidence to that effect.[f, g] The fossil evidence concerning Peking man has disappeared. Ramapithecus consists merely of a handful of teeth and jaw fragments. It is now known that these fragments were pieced together incorrectly so as to resemble portions of the human jaw.[h] Detailed computer   of the Australopithecines have conclusively shown that they are not intermediate between man and apes.[i] The Australopithecines, which were made famous by Louis and Mary Leakey, are actually quite distinct from both man and apes. For about 100 years the world was led to believe that Neanderthal man was stooped and ape-like. Recent studies show that this was based upon some Neanderthal men who were crippled with arthritis and rickets.[j-l] Neanderthal man, Heidelberg man, and Cro-Magnon man were completely human. Artists depictions, especially of the fleshy portions of the body, are quite imaginative and are not supported by evidence.[m] Furthermore, the dating techniques are questionable.

17. Many of the world's fossils show, by the details of their soft fleshy portions, that they were buried before they could decay. This, together with the occurrence of polystrate fossils (fossils that traverse two or more strata of sedimentary rock) in Carboniferous, Mesozoic, and Cenozoic formations, is strong evidence that this sedimentary material was deposited rapidly—not over hundreds of millions of years.

18. Bones of many modern-looking humans have been found deep in rock formations that are supposedly many millions of years older than evolutionary theory would predict. Examples include: the Calaveras Skull,[a-d] the Castenedola Skull,[e,f] Reck's skeleton,[g] and many others.[h-k] These remains are ignored by evolutionists.

19. The vertical sequencing of fossils is frequently not in the assumed evolutionary order.[a]

20. If evolution had occurred, the fossil record should show continuous and gradual changes from the bottom to the top layers and between all forms of life. Just the opposite is found. Complex species, such as jelly fish, worms, corals, trilobites, and brachiopods, appear suddenly in the lowest layers (Cambrian).[a] Furthermore, many gaps and discontinuities appear throughout.[b-e] No fossil links have been found between protozoa and invertebrates, or invertebrates and vertebrates, or vertebrate fish and amphibians, or amphibians and reptiles, or reptiles and mammals, or primates and other mammals. So many fossils have been found that it is safe to conclude that these gaps are real; they will never be filled.[f-h] The fossil record does not support evolution.

21. The vast majority of the sediments, which encase practically all fossils, were laid down through water.

22. The worldwide fossil record is evidence of the rapid death and burial of animal and plant life by a flood; it is not evidence of slow change.

New Research Shows that the Requirements for Life Are So Complex that Chance and Even Billions of Years Cannot Explain It.

23. If the earth, early in its alleged evolution, *had oxygen* in its atmosphere, the chemicals needed for life would have been removed by oxidation. But if there had been *no oxygen,* then there would have been no ozone, and without ozone all life would be quickly destroyed by the sun's ultraviolet radiation.[a,b]

24. There have been many imaginative but unsuccessful attempts to explain how just one single protein could form from any of the assumed atmospheres of the early earth. The necessary chemical reactions all tend to move in the opposite direction from that required by evolution. Furthermore, each possible energy source, whether the earth's heat, electrical discharges, or the sun's radiation, would destroy the protein products millions of times faster than they could be formed.[a,b]

25. If, despite the virtually impossible odds, proteins arose by chance processes, there is not the remotest reason to believe that they could ever form a self-reproducing, membrane-encased, living cell. There is no evidence that there are any stable states between the assumed naturalistic formation of proteins and the formation of the first living cells. No scientist has ever advanced a testable procedure whereby this fantastic jump in complexity could have occurred—even if the universe were completely filled with proteins.

26. If life is ultimately the result of random chance, then so is thought. Your thoughts—such as what you are now thinking—would in the final analysis be a consequence of accident only and therefore would have no validity.

27. Computer-generated comparisons have been made of the sequences of amino acids that comprise a protein that is common to 47 forms of animal and plant life. The results of this study seriously contradict the predictions of the theory of evolution.[a-c]

28. The genetic information contained in *each cell* of the human body is roughly equivalent to a library of 4000[a] volumes. For chance mutations and natural selection to produce this amount of information, assuming that matter and life somehow got started, is analogous to continuing the following procedure until 4000 volumes have been produced:[b,c]

(a) Start with a meaningful phrase.
(b) Retype the phrase but make some errors and insert some additional letters.
(c) Examine the new phrase to see if it is meaningful.
(d) If it is, replace the original phrase with it.
(e) Return to step (b).

To accumulate 4000 volumes that are meaningful, this procedure would have to produce the equivalent of far more than 10^{3000} animal offspring. (To just begin to understand how large 10^{3000} is, realize that the visible universe has less than 10^{80} *atoms* in it.)

29. DNA can only be produced with the help of certain enzymes. But these enzymes can only be produced at the direction of DNA.[a] Since each requires the other, a satisfactory explanation for the origin of one must simultaneously explain the origin of the other. No evidence exists for any such naturalistic explanation.[b]

30. Amino acids, when found in nonliving matter, come in two forms that are chemically equivalent: about half can be described as "right-handed" and the other half as "left-handed" (a structural description—one is the mirror image of the other). However, the protein molecules found in all forms of life, including plants, animals, bacteria, molds, and even viruses, have only the left-handed variety. The mathematical probability that chance processes could produce *just one* tiny protein molecule with only left-handed amino acids is virtually zero.[a]

31. The simplest form of life consists of 600 different protein molecules. The mathematical probability that *just one* molecule could form by the chance arrangement of the proper amino acids is far less than 1 in 10^{527}.[a] (The magnitude of the number 10^{527} can begin to be

appreciated by realizing that the visible universe is about 10^{28} inches in diameter.)

32. There are many instances where quite different forms of life are completely dependent upon each other. Examples include: fig trees and the fig gall wasp,[a,b] the yucca plant and the pronuba moth,[c] many parasites and their hosts, and pollen-bearing plants and the honey-bee family consisting of the queen, workers, and drones. If one member of each interdependent group evolved first (such as the plant before the animal), the other members could not have survived. Since all members of the group obviously have survived, they must have come into existence at essentially the same time.

33. Detailed studies of various animals have revealed certain physical equipment and capabilities that cannot be duplicated by the world's best designers using the most sophisticated technologies. A few examples include: the miniature and reliable sonar systems of the dolphins, porpoises, and whales; the frequency-modulated radar and discrimination system of the bat; the efficiency and aerodynamic capabilities of the hummingbird; the control systems, internal ballistics, and combustion chambers of the bombardier beetle;[a,b] and the precise and redundant navigational systems of many birds and fish. The many components of these complex systems could not have evolved in stages without placing a selective disadvantage on the animal. All evidence points to a designer.

34. If sexual reproduction in plants, animals, and humans is a result of evolution, an absolutely unbelievable series of chance events would have had to occur. First, the amazingly complex and completely different reproductive systems of the male must have *completely* and *independently* evolved at about the *same time and place* as those of the female. A slight incompleteness in just one of the two would make both systems useless, and natural selection would oppose their survival. Second, the physical and emotional systems of the male and female would also need to be compatible. Third, the complex products of the male reproductive system (pollen or sperm) would have to have an affinity for and a mechanical and chemical compatibility with the eggs from the female reproductive system. Fourth, the intricate and numerous processes occurring at the molecular level inside the fertilized egg would have to work with fantastic precision—processes that scientists can only describe in an aggregate sense. And finally, the environment of this fertilized egg, from conception until it also reproduced with another sexually capable "brother or sister" that was also "accidentally" produced, would have to be controlled to an unbelievable degree. Either this series of incredible events occurred by random processes or else an intelligent designer created sexual reproduction.

THE UNIVERSE, THE SOLAR SYSTEM, THE EARTH, AND LIFE WERE RECENTLY CREATED.

Naturalistic Explanations for the Evolution of the Solar System and Universe Are Unscientific and Hopelessly Inadequate.

According to all theories of the evolution of the solar system:

35. The planets should all rotate on their axes in the same direction, but Venus and Uranus rotate backwards.[a,b]

36. All 49 moons of the various planets should revolve in the same direction, but at least 11 revolve backwards.[a,b]

37. The orbits of these 49 moons should all lie in the equatorial plane of the planet they orbit, but many, including the earth's moon, are high inclined.[a]

38. The material of the earth (and Mars, Venus, and Mercury) should almost all be hydrogen and helium—similar to that of the sun and the rest of the visible universe; actually much less than 1% of the earth's mass is hydrogen or helium.[a]

39. The sun should have 700 times more angular momentum than the planets; in fact, the planets have 200 times more angular momentum than the sun.[a,b]

40. Detailed analyses indicate that stars could not have formed from interstellar gas clouds. To do so, either by first forming dust particles[a] or by direct gravitational collapse of the gas, would require vastly more time than the alleged age of the universe. The only alternative is that stars must have been created.

41. The sun's tidal forces are so strong that dust clouds or gas clouds lying within the orbit of Jupiter could never condense to form planets.[a]

42. Saturn's rings could not have formed from the disintegration of a former satellite or from the capture of external material; the particles in these rings are too small and too evenly distributed throughout orbits that are too circular. Therefore, the rings appear to be remnants of Saturn's creation.

43. The moon was not torn from the earth, nor did it congeal from the same material as the earth since its orbital plane is too highly inclined and the relative abundances of its elements are too dissimilar from those of the earth. If the moon formed from particles orbiting the earth, other particles should be easily visible inside the moon's orbit; none are. The moon's circular orbit is also strong evidence that it was never torn from or captured by the earth.[a,b] If the moon was not pulled from the earth, was not built up from smaller particles near its present orbit, and was not captured from outside its present orbit, only one

possibility remains. The moon must have been created in its present orbit.

44. No scientific theory exists to explain the origin of matter, space, or time. Since each is intimately related to or even defined in terms of the other, a satisfactory explanation for the origin of one must also explain the origin of the others.[a] Naturalistic explanations have completely failed.

45. The First Law of Thermodynamics states that the energy of our universe is constant, or conserved. Countless experiments have shown that regardless of the energy conversion process, the total amount of energy (or its mass equivalent) remains constant. A corollary of the First Law is that natural processes cannot create energy. Since the universe obviously has energy, that energy must have been created in the past when The First Law was not operating. Since the energy of the universe could not have created itself, something external must have created it.

46. If the entire universe is considered to be a closed system, then according to the Second Law of Thermodynamics, the energy in the universe that is available for useful work has always been decreasing. But as one goes back further in time, the amount of energy available for useful work would eventually exceed the total energy in the universe that, according to the First Law of Thermodynamics, remains constant. This is an impossible condition. It therefore implies that the universe had a beginning.

47. Heat always flows from hot bodies to cold bodies. If the universe were infinitely old, the temperature throughout the universe should be uniform. Since the temperature of the universe is not uniform, the universe is not infinitely old. Therefore, the universe had a beginning.[a]

48. A further consequence of the Second Law is that when the universe began, it was in a more organized state than it is today—not in a highly disorganized state as assumed by evolutionists and proponents of the Big Bang Theory.

49. The cosmic background radiation, that is considered by many to be the major evidence supporting the Big Bang Theory, can be explained as radiation from dust particles scattered throughout space.[a,b] Furthermore, recent measurements of this radiation above the earth's atmosphere indicate that it is not consistent with the Big Bang hypothesis.[c-e] If the Big Bang occurred, then the universe should not contain either lumpy[f,g] or rotating bodies. Since both types of bodies are seen,[h] it is doubtful that the Big Bang occurred.

50. Computer simulations of the motions of spiral galaxies show them to be highly unstable; they should completely change their shape

in only a small fraction of the assumed age of the universe.[a] The simplest explanation for why so many spiral galaxies exist, including our own Milky Way Galaxy, is that they and the universe are much younger than has been assumed.

51. Theory[a] and experiment[b] indicate that nuclear reactions are not the predominant energy source for the sun. If nonnuclear energy sources are producing most of the sun's energy, then the sun should deplete this energy supply in much less than ten million years. Our star, the sun, must therefore be young. If the sun is young, then so is the earth.

52. If stars evolve, we should see about as many star births as star deaths. The deaths of stars are bright and sudden events called "novas" and "supernovas." Similarly, the birth of a star should be accompanied by sudden and easily detectable emissions from the new star. We have *never* seen a star born, but we have seen thousands of stars die. There is no evidence that stars evolve.

53. Stellar evolution is assumed in estimating the age of stars. These age estimates are then used to establish a framework for stellar evolution. This is circular reasoning.[a]

54. There is no evidence that galaxies evolve.[a,b]

Techniques that Argue for an Old Earth Are Either Illogical or Are Based on Unreasonable Assumptions.

55. Any estimated date prior to the beginning of written records must necessarily assume that the dating clock has operated at a known rate, that the initial setting of the clock is known, and that the clock has not been disturbed. These assumptions are almost always unstated or overlooked.

56. A major assumption that underlies all radioactive dating techniques is that the rates of decay, which have been essentially constant over the past 70 years, have also been constant over the past 4,600,000,000 years. This bold, critical, and untestable assumption is made even though no one knows what causes radioactive decay.

57. The public has been greatly misled concerning the consistency, reliability, and trustworthiness of radiometric dating techniques (the Potassium-Argon method, the Rubidium-Strontium method, and the Uranium-Thorium-Lead method). Many of the published dates can be checked by comparisons with the assumed ages for the fossils that sometimes bracket radiometrically[b] dated rock. In over 400 of these published checks, the radiometrically determined ages were at least one geologic age in error—indicating major errors in methodology. An

unanswered question is, "How many other dating checks were *not published* because they too were in error?"[a,b]

58. Radiohalos, tiny spheres of discoloration produced by the radioactive decay of particles that are encased in various crystals, show that the earth's crust was never in a molten state.[a] Based upon the specific patterns seen in many of these rocks, one can only conclude that these rocks had to have come into existence almost instantaneously. Furthermore, some halos suggest that the rate of radioactive decay has not always been constant but, in fact, varied by many orders of magnitude from that observed today.[b]

59. Geological formations are almost always dated by their fossil content, especially by certain *index fossils* of extinct animals. The age of the fossil is derived from the assumed evolutionary sequence, but the evolutionary sequence is based on the fossil record. This reasoning is circular. Furthermore, this procedure has produced many contradictory results.[a,b]

60. Practically nowhere on the earth can one find the so-called "geologic column." In fact, at most locations on the earth over half of the "geologic periods" are missing, and 15–20% of the earth's surface has less than one-third of these periods appearing in the "correct" order.[a] Even at the Grand Canyon, only a fraction of this column is found.[b,c] Using the assumed geologic column to date fossils and rocks is fallacious.

61. Since 1908, human footprints have been found alongside dinosaur footprints in the rock formations of the Paluxy riverbed in Texas.[a-c] Recently television cameras have recorded the discovery of new human, dinosaur, and saber-toothed tiger footprints, as well as a human hand print, underneath slabs of *undisturbed* limestone.[d] This indicates that man and dinosaurs lived at the same time *and* the same place. But evolutionists claim that dinosaurs became extinct about 65 million years before man supposedly began to evolve. Something is wrong.

62. Many different people have found, at different times and places, man-made artifacts encased in coal. Examples include an 8-carat gold chain,[a-c] a spoon,[b] a thimble, an iron pot, a bell, and other objects of obvious human manufacture. Many other "out-of-place artifacts" such as a metallic vase, a screw, nails,[a] a strange coin,[c] a doll,[c, d] and others[e] have been found buried deeply in solid rock. By evolutionary dating techniques, these objects would be hundreds of millions of years old; but man supposedly did not begin to evolve until 2–4 million years ago. Again, something is wrong.

63. In rock formations in Utah,[a] Pennsylvania,[b] Missouri,[b] and Kentucky,[b] human *footprints* that are supposedly 150–600 million

years old have been found and examined by many different authorities. Obviously, there is a major error in chronology.

64. Since there is no worldwide unconformity in the earth's sedimentary strata, the entire geologic record must have been deposited rapidly. (An *unconformity* is an erosional surface between two adjacent rock formations representing a time break of unknown duration.) *Conformities* imply a continuous and rapid deposition. Since one can always trace a continuous path from the bottom to the top of the geologic record that avoids these unconformities, the sediments along that path must have been deposited continuously.[a]

65. Radiocarbon dating, which has been accurately calibrated by counting the rings of living trees that are up to 3,500 years old, is unable to extend this accuracy to date organic remains that are more ancient. A few people have claimed that ancient wood exists which will permit this calibration to be extended even further back in time, but these people have not let outside scientists examine their data. On the other hand, measurements made at hundreds of sites worldwide[a,b] indicate that the concentration of radiocarbon in the atmosphere rose quite rapidly at some time prior to 3,500 years ago. If this happened, the maximum possible radiocarbon age obtainable with the standard techniques (approximately 50,000 years) could easily correspond to an actual age of 5,000 years.

Most Dating Techniques Show the Earth and Solar System to Be Young.

Evolution requires an old earth and an old solar system. Without billions of years, virtually all informed evolutionists will admit that their theory is dead. But by hiding the "origins question" behind the veil of vast periods of time, the unsolvable problems of evolution become difficult for scientists to see and laymen to imagine. Our media and textbooks have implied for over a century that this almost unimaginable age is correct, but practically never do they or the professors examine the shaky assumptions and growing body of contrary evidence. Therefore, most people instinctively believe that things are old, and it is disturbing (at least initially) to hear evidence that our origins are relatively recent.

Actually most dating techniques show that the earth and solar system are young—possibly less than 10,000 years old. Listed below are just a few of these evidences.

66. Direct measurements of the earth's magnetic field over the past 140 years show a steady and rapid decline in its strength. This

decay pattern is consistent with the theoretical view that there is an electrical current inside the earth which produces the magnetic field. If this view is correct, then 25,000 years ago the electrical current would have been so vast that the earth's structure could not have survived the heat produced. This would imply that the earth could not be older than 25,000 years.[a]

67. As tidal friction decreases the spin rate of the earth, the laws of physics require the moon to recede from the earth. But at just the present spreading rate, the moon would have moved from the earth to its present distance in much less than the assumed age of the moon *or* earth. Actually gravitational forces would have pulled the moon apart if it had *ever* been within 10,000 miles of the earth. Furthermore, the tidal friction and therefore the spreading rate would have been much greater in the past if the moon were closer to the earth. Consequently, the earth-moon system must be vastly younger than evolutionists assume.

68. Over twenty-seven billion tons of sediments, primarily from our rivers, are entering the oceans each year. Obviously, this rate of sediment transport has not been constant and has probably been decreasing as the looser top soil has been removed. But even if it has been constant, the sediments that are now on the ocean floor would have accumulated in only 30 million years. Therefore, the continents and oceans cannot be one billion years old.[a]

69. The atmosphere has less than 40,000 years worth of helium, based on just the production of helium from the decay of uranium and thorium. There is no known means by which large amounts of helium can escape from the atmosphere. The atmosphere appears to be young.[a,b]

70. The rate at which elements such as copper, gold, tin, lead, silicon, mercury, uranium, and nickel are entering the oceans is very rapid when compared with the small quantities of these elements already in the oceans. Therefore, the oceans must be very much younger than a million years.

71. Evolutionists believe that the continents have existed for at least one billion years. However, the continents are being eroded at a rate that would level them in much less than twenty-five million years.[a,b]

72. The occurrence of abnormally high gas and oil pressures within relatively permeable rock implies that these fluids were formed or encassed less than 10,000 years ago. If these hydrocarbons had been trapped *over* 10,000 years ago, leakage would have dropped the pressure to a level far below what it is today.[a]

73. Meteorites are only found in surface rocks.[a,b] If the sediments, which have an average depth of one mile, were laid down over hundreds

of millions of years, many of these steadily falling meteorites should have been discovered. Therefore, the sediments appear to have been deposited rapidly. Furthermore, since no meteorites are found immediately above the basement rocks on which these sediments rest, these basement rocks could not have been exposed to meteoritic bombardment for any great length of time.

74. The rate at which meteoritic dust is accumulating on the earth is such that after five billion years, the equivalent of 182 feet of this dust should have accumulated. Because this dust is high in nickel, there should be an abundance of nickel in the crustal rocks of the earth. No such concentration has been found—on land or in the oceans. Consequently, the earth appears to be young.[a-c]

75. If the moon were billions of years old, it should have accumulated extensive layers of space dust—possibly a mile in thickness. Before instruments were placed on the moon, NASA was very concerned that our astronauts would sink into a sea of dust. This did not happen; there is very little space dust on the moon. Conclusion: the moon is young.

76. The sun's radiation applies an outward force on small particles orbiting the sun. Particles less than 100,000th of a centimeter in diameter should have been "blown out" of the solar system if the solar system were billions of years old. These particles are still orbiting the sun. Conclusion: the solar system is young.

77. Since 1836, over one hundred different observers at the Royal Greenwich Observatory and the U.S. Naval Observatory have made *direct* visual measurements which show that the diameter of the sun is shrinking at a rate of about .1% each century or about five feet per hour! Furthermore, records of solar eclipses indicate that this rapid shrinking has been going on for at least the past 400 years.[a] Several *indirect* techniques also confirm this gravitational collapse, although these inferred collapse rates are only about 1/7th as much.[b] Using the most conservative data, one must conclude that had the sun existed a million years ago, it would have been so large that it would have heated the earth so much that life could not have survived. Yet, evolutionists say that a million years ago all the present forms of life were essentially as they are now, having completed their evolution that began a *thousand* million years ago.

78. Short period comets "boil off" some of their mass each time they pass the sun. Nothing should remain of these comets after about 10,000 years. There are no known sources for replenishing comets. If comets came into existence at the same time as the solar system, the solar system must be less than 10,000 years old.[a-d]

79. Jupiter and Saturn are each radiating more than twice the energy they receive from the sun.[a-c] Calculations show that it is very unlikely that this energy comes from nuclear fusion,[d] radioactive decay, or gravitational contraction. The only other conceivable explanation is that these planets have not existed long enough to cool off.[e]

80. The sun's gravitational field acts as a giant vacuum cleaner which sweeps up about 100,000 tons of micrometeroids per day. If the solar system were older than 10,000 years, no micrometeroids should remain near the center of the solar system since there is no significant source of replenishment. A large disk-shaped cloud of these particles is orbiting the sun. Conclusion: the solar system is less than 10,000 years old.[a]

81. Stars that are moving in the same direction at significantly different speeds, frequently travel in closely-spaced clusters.[a] This would not be the case if they had been traveling for billions of years because even the slightest difference in their velocities would cause their dispersal after such great periods of time. Similar observations have been made of galaxy and of galaxy-quasar combinations that apparently have vastly different velocities but which appear to be connected.[b-d]

82. Galaxies are usually found in tight clusters which often contain hundreds of galaxies. The apparent velocities of individual galaxies within these clusters are so high in comparison to the calculated mass of the entire cluster that these clusters should be flying apart. But since the galaxies within clusters are so close together, they could not have been flying apart for very long. A 10–20 billion year old universe is completely inconsistent with what we see.[a-c]

All dating techniques, to include the *few* that suggest an old earth and an old universe, lean heavily on the assumption that a process observed today has always proceeded at a known rate. In many cases this assumption may be grossly inaccurate. But in the case of the many dating "clocks" that show a young earth, a much better understanding usually exists for the mechanism that drives the clock. Furthermore, the extrapolation process is over a much shorter time and is therefore more likely to be correct.

THE EARTH HAS EXPERIENCED A WORLDWIDE FLOOD.

Archaeological Evidence Indicates that Noah's Ark Probably Exists.[a-e]

83. Ancient historians such as Josephus, the Jewish-Roman historian, and Berosus of the Chaldeans mentioned in their writings that

the Ark existed. Marco Polo also stated that the Ark was reported to be on a mountain in Greater Armenia.

84. In about 1856, a team of three skeptical British scientists and two Armenian guides climbed to Ararat to demonstrate that the Ark did not exist. The Ark was supposedly found, but the British scientists threatened death to the guides if they reported it. Years later one of the Armenians (then living in the United States) and one of the scientists independently reported that they had actually located the ark.

85. Sir James Bryce, a noted British scholar and traveler of the mid-nineteenth century, conducted extensive library research concerning the Ark. He became convinced that the Ark was preserved on Mount Ararat. Finally, in 1876, he ascended to the summit of the mountain and found, at the 13,000 level (2,000 feet above the timber line), a large piece of hand-tooled wood that he believed was from the Ark.

86. In 1883, a series of newspaper articles reported that a team of Turkish commissioners, while investigating avalanche conditions on Mount Ararat, unexpectedly came upon the Ark projecting out of the melting ice at the end of an unusually warm summer. They claimed that they entered a portion of the Ark, but other press reports maintained only an attitude of scoffing at the account.

87. In the unusually warm summer of 1905, an Armenian boy, Georgie Hagopian, and his uncle climbed to the Ark that was sticking out of an ice pack. The boy climbed over the Ark and was able to describe it in great detail. A tape recording of his detailed and very credible testimony made shortly before his death in 1972 has undergone voice analyzer tests that showed no indication of lying.

88. A Russian pilot, flying over Ararat in World War I (1915), thought he saw the Ark. The news of his discovery reached the Czar, who dispatched a large expedition to the site. The soldiers were able to locate and explore the boat, but before they could report back to the capitol, the Russian Revolution of 1917 had occurred. The report disappeared, and the soldiers were scattered. Some of them eventually reached the United States. Various relatives and friends have since confirmed this story.

89. At about the time of the Russian sighting, five Turkish soldiers, crossing Mount Ararat, claim to have accidentally encountered the Ark; however, they did not report their story until 30 years later when they offered to guide an American expedition to the site. The expedition did not materialize, and their services were never sought until after their deaths.

90. During World War II, a group of Russian flyers on at least two occasions took aerial photographs that showed the Ark protruding out

of the ice. In Berlin after the war, these photos were shown to an American doctor who subsequently disclosed this story.

91. An oil geologist, George Greene, in 1953 took a number of photographs of the Ark from a helicopter. After returning to the United States, Greene showed his photographs to many people, but was unable to raise financial backing for a ground-based expedition. Finally, he went to South America, where he was killed. Although the pictures have not been located, over 30 people have given sworn written testimony that they saw these photographs that clearly showed the Ark protruding from the melting ice field at the edge of a precipice.

There are many other stories in which people claim to have seen the Ark. Some of these are of questionable validity, and others are inconsistent with many of the known details. Only the most credible are cited above.

Many of the Earth's Previously Unexplainable Features Can Be Explained Only by this Flood.[a]

The origin of each of the following features of the earth is a subject of controversy within the earth sciences. Each typically involves numerous hypotheses and unexplainable aspects. Yet all of these features can be viewed as direct consequences of a singular and unrepeatable event—a flood whose waters burst forth from worldwide, subterranean, and interconnected chambers with an energy release in excess of one trillion megatons of TNT. The cause and effect sequence of the events involved phenomena that are either well understood or are observable in modern times.

92. coal formations
93. mountains
94. ocean trenches
95. submarine canyons
96. mid-oceanic ridge
97. continental drift
98. magnetic patterns of the ocean floor
99. strata
100. glaciers and the ice age
101. continental shelves and slopes
102. submarine volcanoes and guyots
103. salt domes
104. metamorphic rock

The Seemingly Impossible Events of a Worldwide Flood Are Really Quite Plausible if Examined Closely.

105. Every major mountain range on the earth contains fossils of sea life.

106. Practically every culture on earth has legends telling of a traumatic flood in which only a few humans survived in a large boat.[a]

107. The majority of the earth's mountains were formed after most of the sediments were deposited. If these mountains were again flattened out (while the ocean basins were allowed to rise in compensation for this downward flow of mass), the oceans would flood the entire earth. Therefore, there is enough water on the earth to cover the smaller mountains that existed prior to the flood.

108. Seeds can still germinate after soaking for a year in salt water.[a]

Author's Note: Dealing with a subject as broad as this subject of origins has continually impressed me with the vast gap between what we know and what we would like to know. Any person attempting to bridge this gap must continually look for new data, be willing to have his old interpretations challenged, and attempt to minimize the preconceptions and biases that we all bring to such a foundational issue as origins. I would appreciate, especially from scientists having evolutionary presuppositions, being informed of any significant data or interpretations of data which I have omitted or which are either supportive of or in opposition to the very brief summary that is contained in this paper.

W. T. B.
1319 Brush Hill Circle
Naperville, IL 60540

REFERENCES

2. (a) Monroe W. Strickberger, *Genetics,* 2nd Edition (New York: Macmillan Publishing Co., 1976), p. 812.
   (b) Francis Hitching, *The Neck of the Giraffe: Where Darwin Went Wrong* (New Haven, Connecticut: Ticknor and Fields, 1982), p. 55.
5. (a) Theodosius Dobzhansky, "On Methods of Evolutionary Biology and Anthropology," *American Scientist,* Winter, December 1957, p. 385.
6. (a) Pierre-Paul Grassé, *Evolution of Living Organisms* (New York: Academic Press, 1977), p. 88.
7. (a) Strickberger, p. 44.

10. (a) David C. C. Watson, *The Great Brain Robbery* (Chicago: Moody Press, 1976), pp. 83–89.
    (b) Henry M. Morris, "Language, Creation, and the Inner Man," *ICR Impact Series,* No. 28 (San Diego: Institute for Creation Research).
    (c) Les Bruce, Jr., "On the Origin of Language," *ICR Impact Series,* No. 44 (San Diego: Institute for Creation Research).
11. (a) Arthur Custance, *Genesis and Early Man* (Grand Rapids: Zondervan Publishing House, 1975), pp. 250–271.
14. (a) George Gaylord Simpson and William Beck, *Life: An Introduction to Biology* (New York: Harcourt, Brace and World, 1965), p. 241.
    (b) Hitching, pp. 202–205.
    (c) *"The biogenic law has become so deeply rooted in biological thought that it cannot be weeded out in spite of its having been demonstrated to be wrong by numerous subsequent scholars."*(Walter J. Bock, Department of Biological Sciences, Columbia University, "Evolution by Orderly Law," *Science,* Vol. 164, May 9, 1969, pp. 684–685.)
    (d) *"We no longer believe we can simply read in the embryonic development of a species its exact evolutionary history."* (H. Frings and M. Frings, *Concepts of Zoology,* p. 267.)
    (e) *"This law has been so seriously questioned and is so obviously inapplicable in many instances that as a law it is now of historical interest only."* (W. R. Breneman, *Animal Form and Function,* p. 521.)
    (f) *"The Biogenetic law as a proof of evolution is valueless."* (W. R. Thompson, Introduction, *Origin of the Species.*)
16. (a) M. Bowden, *Ape-Men: Fact or Fallacy?* (Great Britain: Sovereign Publications, 1977).
    (b) Duane T. Gish, "Multivariate Analysis: Man . . . Apes . . . Australopithecines . . . ," *Battle for Creation* (San Diego: Creation Life Publishers, 1976), pp. 298–305.
    (c) Duane T. Gish, "Richard Leakey's Skull," *Battle for Creation* (San Diego, Creation Life Publishers, 1976), pp. 193–200.
    (d) Stephen J. Gould, "The Piltdown Conspiracy," *Natural History,* Vol. 89, No. 8, August 1980, pp. 8–28.
    (e) Herbert Wendt, *In Search of Adam* (Westport, Connecticut: Greenwood Press, 1955), p. 299.
    (f) Bowden, pp. 125–129.
    (g) Hitching, pp. 208–209.
    (h) A. Zihlman and J. Lowenstein, "False Start of the Human Parade," *Natural History,* Aug/Sept. 1979, pp. 86–91.
    (i) Charles E. Oxnard, "The Place of the Australopithecines in Human Evolution: Grounds for Doubt?", *Nature,* Vol. 258, December 4, 1975, pp. 389–395.
    (j) Bowden, pp. 157–159.
    (k) Francis Ivanhoe, "Was Virchow Right About Neanderthal?," *Nature,* Vol. 227, August 8, 1970, pp. 577–578.
    (l) William L. Straus, Jr., and A. J. E. Cave, "Pathology and the Posture of Neanderthal Man," *The Quarterly Review of Biology,* December 1957, pp. 348–363.
    (m) Boyce Rensberger, "Facing the Past," *Science 81,* October 1981, p. 49.
18. (a) Bowden, pp. 64–66.
    (b) Frank W. Cousins, *Fossil Man,* (A. E. Norris & Sons Ltd., Emsworth), pp. 50–52, 82, 83.

(c) William H. Holmes, "Review of the Evidence Relating to Auriferous Gravel Man in California," *Smithsonian Institutional Annual Report,* 1899, pp. 419–472.

(d) W. H. B., "Alleged Discovery of An Ancient Human Skull in California," *American Journal of Science,* Vol. 2, 1866, p. 424.

(e) Bowden, pp. 66–67.

(f) Cousins, pp. 48–50, 81.

(g) Bowden, pp. 169–179.

(h) Bowden, pp. 66–67.

(i) William L. Strauss, Jr., "A New Oreopithecus Skeleton," *Science,* Vol. 128, September 5, 1958, p. 523.

(j) William L. Straus, Jr., "Oreopithecus bambolii," *Science,* Vol. 126, August 23, 1957, pp. 345–346.

(k) F. A. Barnes, "The Case of the Bones in Stone," *Desert Magazine,* Vol. 38, February 1975, pp. 36–39.

19. (a) "Mesozoic Milk Teeth," *Scientific American,* December 1981, pp. 78, 80.

20. (a) *"Not a single, indisputable, multicellular fossil has been found anywhere in the world in rocks supposedly older than Cambrian rocks."* Preston Cloud "Pseudo-fossils: A Plea for Caution," *Geology,* Vol. 1, November 1973, p. 123.

(b) Francisco J. Ayala and James W. Valentine, *Evolving, The Theory and Processes of Organic Evolution,* pp. 258, 266–267, 1979, as reported by John N. Moore in his article "Genes, Genetics and Creationism in the Public Schools," on p. 129 contained in the proceedings of the 8th Annual Bible-Science Convention, 1980.

(c) Stephen J. Gould, "The Return of Hopeful Monsters," *Natural History,* Vol. 86, pp. 22–30.

(d) David M. Raup, "Conflicts Between Darwin and Paleontology," *Field Museum of Natural History Bulletin,* January 1979, pp. 23, 25.

(e) *"But whatever ideas authorities may have on the subject, lungfishes, like every other major group of fishes I know, have their origins firmly based in Nothing."* (Errol White, "A Little on Lungfishes," *Proceedings of the Linnaean Society of London,* Vol. 177, Presidential Address, 1966, p. 8.)

(f) *"The extreme rarity of transitional forms in the fossil record persists as the trade secret of paleontology."* (Stephen Jay Gould, "Evolution's Erratic Pace," *Natural History,* Vol. 5, May 1977, p. 14.)

(g) *"The fossil material is now so complete that the lack of transitional series cannot be explained by the scarcity of the material. The deficiencies are real, they will never be filled."* (N. Heribert-Nilsson of Lund University, Sweden, as quoted by Francis Hitching in *The Neck of the Giraffe: Where Darvin Went Wrong,* p. 22.)

(h) *"Despite the bright promise that paleontology provides a means of 'seeing' evolution, it has presented some nasty difficulties for evolutionists the most notorious of which is the presence of 'gaps' in the fossil record. Evolution requires intermediate forms between species and paleontology does not provide them."* (David B. Kitts, School of Geology and Geophysics, University of Oklahoma, "Paleontology and Evolutionary Theory," *Evolution,* Vol. 28, September 1974, p. 467.)

23. (a) Duane T. Gish, *Speculations and Experiments Related to Theories on the Origin of Life,* ICR Technical Monograph No. 1 (San Diego: Institute for Creation Research, 1972).

(b) Hitching, p. 65.

24. (a) Gish, *Speculations and Experiments Related to Theories on the Origin of Life.*
    (b) Duane T. Gish, "Gish Debates Russell Doolittle at Iowa State," *Acts and Facts,* Vol. 9, No. 12, December 1980, p. 2.

27. (a) Personal communication from Robert Bayne Brown.
    (b) Ginny Gray, "Student Project 'Rattles' Science Fair Judges," *Issues and Answers,* December 1980, p. 3.
    (c) *Abstracts: 31st International Science and Engineering Fair* (Washington D.C.: Science Service, 1980), p. 113.

28. (a) Carl Sagan, *The Dragons of Eden* (New York: Random House, 1977), p. 25.
    (b) Murray Eden, as reported in "Heresy in the Halls of Biology: Mathematicians Question Darwinism," *Scientific Research,* November 1967, p. 64.
    (c) Paul S. Moorhead and Martin M. Kaplan, editors, *Mathematical Challenges to the Neo-Darwinian Interpretation of Evolution,* Proceedings of a symposium held at the Wistar Institute of Anatomy and Biology, April 25 and 26, 1966 (Philadelphia: The Wistar Institute Press, 1967).

29. (a) Richard E. Dickerson, "Chemical Evolution and the Origin of Life," *Scientific American,* Vol. 239, September 1958, p. 73.
    (b) Hitching, p. 66.

30. (a) James F. Coppedge, *Evolution: Possible or Impossible?* (Grand Rapids: Zondervan Publishing House, 1973), pp. 71–79.

31. (a) Coppedge, pp. 71–72.

32. (a) Oscar L. Brauer, "The Smyrna Fig Requires God for Its Production," *Creation Research Society Quarterly,* Vol. 9, No. 2, September 1972, pp. 129–131.
    (b) Bob DeVine, *Mr. Baggy-Skin Lizard* (Chicago: Moody Press, 1977) pp. 29–32.
    (c) DeVine, pp. 17–20.

33. (a) Robert E. Kofahl and Kelly L. Segraves, *The Creation Explanation* (Wheaton, Illinois: Harold Shaw Publishers, 1975), pp. 2–9.
    (b) Thomas Eisner and Daniel J. Aneshansley, "Spray Aiming in Bombardier Beetles: Jet Deflection by the Coanda Effect," *Science,* Vol. 215, January 1, 1982, pp. 83–85.

35. (a) Donald H. Menzel, *Astronomy* (New York: Random House, 1970), pp. 178, 198–199.
    (b) John C. Whitcomb, Jr., *The Origin of the Solar System* (New Jersey: Presbyterian and Reformed Publishing Co., 1977), p. 16.

36. (a) Laurence A. Soderblom and Torrence V. Johnson, "The Moons of Saturn," *Scientific American,* January 1982, p. 101.
    (b) Whitcomb, p. 6.

37. (a) Whitcomb, p. 15.

38. (a) *Van Nostrand's Scientific Encyclopedia* (Van Nostrand Reinhold Co., fifth edition, 1976), pp. 493–494.

39. (a) R. A. Lyttleton, *Mysteries of the Solar System* (Oxford: Clarendon Press, 1968), p. 16.
    (b) Fred Hoyle, *The Cosmology of the Solar System* (Enslow Publishers, 1979), p. 11.

40. (a) Harwit, *Astrophysical Concepts* (New York: John C. Wiley, 1973), p. 394.

41. (a) Paul M. Steidl, *The Earth, the Stars, and the Bible* (Grand Rapids: Baker Book House, 1979), p. 106.

43. (a) Steidl, pp. 2–79.
    (b) M. Mitchell Waldrop, "The Origin of the Moon," *Science,* Vol. 216, May 7, 1982, pp. 606–607.
44. (a) Nathan R. Wood, *The Secret of the Universe* (Grand Rapids: Eerdman's Publishing Co., 1936, 10th edition).
47. (a) Sir Isaac Newton, source unknown.
49. (a) Russell Akridge, Thomas Barnes, and Harold S. Slusher, "A Recent Creation: Explanation of the 3° K Background Black Body Radiation," *Creation Research Society Quarterly,* Vol. 18, No. 3, December 1981, pp. 159–162.
    (b) Hannes Alfven and Mendis Asoka, "Interpretation of Observed Cosmic Background Radiation," *Nature,* Vol. 266, April 21, 1977, pp. 698–699.
    (c) H. P. Gush, "Rocket Measurement of the Cosmic Background Submillimeter Spectrum," *Physical Review Letters,* Vol. 47, No. 10, September 7, 1981, pp. 745–748.
    (d) Kandiah Shivanandan, James R. Houck, and Martin O. Harwit, "Preliminary Observations of the Far-Infrared Night-Sky Background Radiation," *Physical Review Letters,* November 11, 1968, pp. 1460–1462.
    (e) "Freak Result Verified," *Nature,* Vol. 223, 1969, pp. 779–780.
    (f) G. Burbidge, "Was There Really a Big Bang?", *Nature,* Vol. 233, 1971, pp. 36–40.
    (g) Ben Patrusky, "Why Is the Cosmos 'Lumpy'?", *Science 81,* June 1981, p. 96.
    (h) Stephen A. Gregory and Laird A. Thompson, "Superclusters and Voids in the Distribution of Galaxies," *Scientific American,* March 1982, pp. 106–114.
50. (a) David Fleischer, "The Galaxy Maker," *Science Digest,* October 1981, pp. 12, 116.
51. (a) A. B. Severny, et. al., *Nature,* Vol. 259, 1976, p. 87.
    (b) Steidl, pp. 92–94.
53. (a) Steidl, pp. 134–136.
54. (a) Harold S. Slusher, "Clues Regarding the Age of the Universe," *ICR Impact Series,* No. 19 (San Diego: Institute for Creation Research), pp. 2–3.
    (b) Steidl, pp. 143–145.
57. (a) John Woodmorappe, "Radiometric Geochronology Reappraised," *Creation Research Society Quarterly,* Vol. 16, September, 1979, pp. 102–129.
    (b) Robert H. Brown, "Graveyard Clocks: Do They Really Tell Time?", *Signs of the Times,* June 1982, pp. 8–9.
58. (a) Robert V. Gentry, "Radiohalos in Coalified Wood; New Evidence Relating to the Time of Uranium Introduction and Coalification," *Science,* Vol. 194, October 15, 1976, pp. 315–317.
    (b) Robert V. Gentry, "On the Invariance of the Decay Constant Over Geological Time," *Creation Research Society Quarterly,* Vol. 5, September 1968, pp. 83–84.
59. (a) B. Sahni, "Age of the Saline Series in the Salt Range of the Punjab," *Nature,* Vol. 153, April 15, 1944, pp. 462–463.
    (b) J. Coates, et. al., "Age of the Saline Series in the Punjab Salt Range," *Nature,* Vol. 155, March 3, 1945, pp. 266–267.
60. (a) John Woodmorappe, "The Essential Nonexistence of the Evolutionary-Uniformitarian Geologic Column: A Quantitative Assessment," *Creation Research Society Quarterly,* Vol. 18, No. 1, June 1981, pp. 46–71.

(b) Clifford Burdick, *Canyon of Canyons* (Caldwell, Idaho, Bible-Science Association, 1974).

(c) Edgar Nafziger, "The Grand Canyon and Creation," *Repossess the Land* (Bible-Science Association, Minneapolis, Minnesota, 1979) p. 162.

61. (a) John Morris, *Tracking Those Incredible Dinosaurs,* (San Diego: Creation Life Publishers, 1980).

(b) Fredrick P. Beierle, *Man, Dinosaurs, and History* (Perfect Printing, 1977).

(c) John Morris, "The Paluxy River Tracks," *ICR Impact Series,* No. 35 (San Diego: Institute for Creation Research).

(d) Taped interview with Dr. Carl Baugh, *Bible-Science Newsletter,* Vol. 20, No, 8, August 1982, p. 1.

62. (a) Rene Noorbergen, *Secrets of the Lost Races* (New York: The Bobbs-Merrill Company, Inc., 1977), pp. 40–62.

(b) Harry V. Wiant, Jr., "A Curiosity From Coal," *Creation Research Society Quarterly,* Vol. 13, No. 1, June 1976, p. 74.

(c) J. R. Jochmans, "Strange Relics from the Depths of the Earth," *Bible-Science Newsletter,* January, 1979, p. 1.

(d) Frederick G. Wright, "The Idaho Find," *American Antiquarian,* Vol. II, 1889, pp. 379–381, as cited by William R. Corliss in *Ancient Man, A Handbook of Puzzling Artifacts* (Glen Arm, Maryland: The Sourcebook Project, 1978) pp. 661–662.

(e) Frank Calvert, "On the Probable Existence of Man During the Miocene Period," *Anthropological Institute Journal,* Vol. 3, 1873, as cited by William R. Corliss in *Ancient Man, A Handbook of Puzzling Artifacts* (Glen Arm, Maryland: The Sourcebook Project, 1978), pp. 661–662.

63. (a) A Cook, "William J. Meister Discovery of Human Footprints with Trilobites in a Cambrian Formation of Western Utah," *Why Not Creation?* (New Jersey: Presbyterian and Reformed Publishing Co., 1970), pp. 185–193.

(b) "Human-Like Tracks in Stone are Riddle to Scientists," *Science News Letter,* October 29, 1938, pp. 278–279.

64. (a) Henry M. Morris, *King of Creation* (San Diego: Creation Life Publishers, 1980), pp. 152–153.

65. (a) Robert H. Brown, "Can We Believe Radiocarbon Dates?", *Creation Research Quarterly,* Vol. 12, No. 1, June 1975, pp. 66–68.

(b) Robert H. Brown, "Regression Analysis of C-14 Age Profiles," Unpublished Manuscript, July 28, 1980.

66. (a) Thomas G. Barnes, *Origin and Destiny of the Earth's Magnetic Field* (San Diego: Institute for Creation Research, 1973).

68. (a) Stuart E. Nevins, "Evolution: The Ocean Says No!" *ICR Impact Series,* No. 8 (San Diego: Institute for Creation Research).

69. (a) "What Happened to the Earth's Helium?", *New Scientist,* Vol. 420, December 3, 1964, pp. 631–632.

(b) Melvin A. Cook, *Prehistory and Earth Models* (London: Max Parrish, 1966) pp. 10–14.

71. (a) Nevins, pp. ii–iii.

(b) George C. Kennedy, "The Origin of Continents, Mountain Ranges, and Ocean Basins," *American Scientist,* pp. 491–504.

72. (a) Cook, p. 341.

73. (a) Fritz Heide, *Meteorites* (Chicago: University of Chicago, 1964) p. 119.
    (b) Peter A. Steveson, "Meteoritic Evidence of a Young Earth," *Creation Research Quarterly,* Vol. 12, June, 1975, pp. 23–25.
74. (a) Henry M. Morris, editor, *Scientific Creationism* (San Diego: Creation Life Publishers, 1974), pp. 151–153.
    (b) Steveson, pp. 23–25.
    (c) Hans Peterson, "Cosmic Spherules and Meteoritic Dust," *Scientific American,* Vol. 202, February 1960, p. 132.
77. (a) "Analyses of Historical Data Suggest Sun Is Shrinking," *Physics Today,* September 1979, pp. 17–19.
    (b) David W. Dunham, et. al., "Observations of a Probable Change in the Solar Radius Between 1715 and 1979," *Science,* Vol. 210, December 12, 1980, pp. 1243–1245.
78. (a) Thomas D. Nicholson, "Comets, Studied for Many Years, Remain an Enigma to Scientists," *Natural History,* March 1966, pp. 44–47.
    (b) Harold Armstrong, "Comets and a Young Solar System," *Speak to the Earth,* ed. George F. Howe (New Jersey: Presbyterian and Reformed Publishing Co., 1975) pp. 327–330.
    (c) Steidl, pp. 58–59.
    (d) Lyttleton, p. 110.
79. (a) II. H. Aumann and C. M. Gillespie, Jr., "The Internal Powers and Effective Temperature of Jupiter and Saturn," *The Astrophysical Journal,* Vol. 157, July 1969, pp. 169–172.
    (b) "Close Encounter with Saturn," *Time,* November 10, 1980, p. 78.
    (c) M. Mitchell Waldrop, "The Puzzle That Is Saturn," *Science,* September 18, 1981, p. 1351.
    (d) Andrew P. Ingersoll, "Jupiter and Saturn," *Scientific American,* December 1981, p. 93.
    (e) Steidl, pp. 51–52, 55.
80. (a) Steidl, pp. 60–61.
81. (a) Harold S. Slusher, *Age of the Cosmos,* ICR Technical Monograph No. 9 (San Diego: Institute for Creation Research), p. 16.
    (b) F. Hoyle and J. V. Narlikar, "On the Nature of Mass," *Nature,* Vol. 233, 1971, pp. 41–44.
    (c) William Kaufmann III, "The Most Feared Astronomer on Earth," *Science Digest,* July 1981, p. 81.
    (d) Geoffrey Burbidge, "Redshift Rift," *Science 81,* December 1981, p. 18.
82. (a) Gerardus D. Bouw, "Galaxy Clusters and the Mass Anomaly," *Creation Research Society Quarterly,* September 1977, pp. 108–112.
    (b) Steidl, pp. 179–185.
    (c) Joseph Silk, *The Big Bang* (San Francisco: W. H. Freeman and Co., 1980), p. 188–191.
83–91. (a) Violet M. Cummings, *Noah's Ark: Fact or Fable?* (San Diego: Creation-Science Research Center, 1972).
    (b) Tim LaHaye and John D. Morris, *The Ark on Ararat* (San Diego: Creation Life Publishers, 1976).
    (c) John Warwick Montgomery, *The Quest for Noah's Ark* (Minneapolis, Minnesota: Bethany Fellowship, Inc., 1972).

(d) John D. Morris, *Adventure on Ararat* (San Diego: Institute for Creation Research, 1973).

(e) Rene Noorbergen, *The Ark File* (California: Pacific Press Publishing, 1974).

92–104 (a) Walter T. Brown, Jr., taped lectures, available on request, 1319 Brush Hill Circle, Naperville, IL 60540.

106. (a) Byron C. Nelson, *The Deluge Story in Stone* (Minneapolis, Minnesota: Bethany Fellowship, Inc., 1968), pp. 169–190.

108. (a) George F. Howe, "Seed Germination, Sea Water, and Plant Survival in the Great Flood," *Scientific Studies in Special Creation* (New Jersey: Presbyterian and Reformed Publishing Co., 1971), pp. 285–298.

Part 4
Responding to the Creationist Challenge

Scientific Creationism and Its Critique of Evolution

by V. Elving Anderson

Everyone has a 'batch of beliefs' (more or less recognized and organized) which are used to interpret the mysteries of personal existence. For the most part, these are brought to one's interpretation
of scientific data, rather than derived from science. Serious difficulty
arises when the origin of such beliefs is over-simplified. Thus it
becomes necessary to guard against a scientific exclusivism and
dogmatism that sometimes are found in the writings of science enthusiasts.

PART I: THE CRITIQUE

For many people in the general public, one of the most puzzling aspects
of science is the way in which scientists can disagree on many topics
ranging from the use of nuclear energy to the fluoridation of water
supplies. Different groups of scientists come to quite opposite conclusions, even though presumably they are using the same information. To
a bystander the controversy between evolution and scientific creationism must seem to be just another illustration of the same familiar
problem.

When we are confronted with puzzles like these, we do well to ask
questions such as the following:

1. Are the two groups really asking the same question? It is sometimes possible for opponents to talk past each other without even realizing it.

2. Are the groups using the same sets of data? Personal biases can
lead to selectivity in the choice of information and the emphasis placed
upon various aspects of the problem.

3. Are the groups using the same rules for evaluating the evidence
and coming to a conclusion?

It should be apparent that a public debate is not a satisfactory way to resolve questions of this type. The outcome is too much dependent upon the way in which the question is phrased or upon the personality and delivery of the speakers. More effort goes toward audience appeal than to a thoughtful and fair evaluation of the issues.

On the other hand it is essential that the major arguments used by scientific creationists should be stated clearly, as a basis for further discussion. Therefore a preliminary version of this paper was submitted to a few scientific creationists and some changes were made in response to their comments.

It must be recognized, of course, that there is a considerable variation of opinion among creationists. There are many, for example, who believe in God as creator and sustainer but who do not accept any of the arguments for a young earth. Many creationists accept micro-evolution, but others reject any change in species. Also, there is considerable disagreement about the advisability of laws concerning science teaching in this area. Some, for example, hold that the positive actions of education and persuasion are more appropriate and more effective in the long run than coercion. To be sure, those who accept the concept of evolution have their own areas of disagreement.

The criticisms of evolution are divided into three groups in order to assist our consideration of them. The first consists of those that are broad in nature and that depend significantly upon areas outside of science itself. The second group is more strictly scientific but makes no claim concerning the age of the universe. Many of these points have been made for some decades by proponents of special creation. The third set consists of those points that emphasize recent origin (a young earth). These have been emphasized by advocates of scientific creationism (also known as creation science). My own evaluation follows in a later section.

Broad Issues

1. Exclusion of Religion

Evolution is a mechanistic, atheistic theory and therefore is a basic dogma of agnosticism, humanism, and atheism in general. Furthermore (it is argued by some) evolutionary philosophy has led to communism, racism, and many other questionable movements. (Several respondents felt that this is not a primary point and should not be used by creationists. If such outcomes result, they are more a misuse of evolution than an inherent part of the concept.)

A variation of this theme is the claim that science in general is atheistic, since it excludes consideration of supernatural factors.

2. Ultimate Origins

Questions of ultimate origins cannot be handled by science at all. Laboratory experiments related to the origin of life have not remotely approached the synthesis of life from non-life. The limited results that have been obtained depended upon laboratory conditions that were artificially imposed and extremely improbable.

In contrast, a creationist model would state that the basic systems of nature were developed by supernatural creative processes which were different from, and existed prior to, the present natural laws and processes. This postulate by definition removes origins from scientific inquiry.

A somewhat more modest claim is that scientists might be able to suggest possible scenarios for origins but will never be able to decide what actually occurred.

3. Historical Interpretations

An extension of the same reasoning is that proposed steps in evolution involve unique unrepeatable historical events. Neither creation nor evolution has ever been observed by human witnesses, neither is subject to the experimental method, and neither is capable of falsification. Thus neither is a valid scientific theory.

Astronomers do not have observational data about the actual formation of the universe. Biochemists do not have observational data about the actual first living substance on the earth. Paleontologists do not have observational data about the first human beings. Therefore assumptions commonly made by scientists on these points should be termed "supranatural."

4. World View

Evolution is a world view and thus constitutes a religion which is opposed to Christianity and other forms of theism. Many famous scientists such as George Gaylord Simpson, Julian Huxley, Carl Sagan, and E. O. Wilson have presented their own personal world views as flowing directly and inevitably from scientific research.

Scientific Issues Not Depending on Time

5. The Second Law

The second law of thermodynamics states that matter and energy always tend to change from complex and ordered states to disordered states. Therefore, the universe could not have arisen by chance and could not have existed forever.

Things tend to go from order to disorder (entropy tends to increase) unless added energy is provided by a conversion mechanism (such as photosynthesis). The consequences of increasing entropy are seen in the loss of usable energy and of order. In this context, evolution (which involves increased complexity) is seen as impossible.

Finally the argument is stated in terms of information. The DNA code is too complex to have arisen through completely mechanistic processes. The system at the time of origin must have been open to information from outside of the system (in the form of the thought or will of the creator). Matter is the substrate, but not the source of the DNA code.

6. Fossils

There are no transitional fossils between major categories. Variation has always been within limits. Thus the missing links are still missing. In reference to the tree diagram often used for evolutionary models, it is admitted that variation may have occurred in the horizontal sense, but never in a vertical manner.

The fossil "record" contains no incipient or transitional kinds between major categories of biological organisms, and this is exactly what would be predicted by the creation model. The fossil "record" is essentially barren and then there is a sudden great outburst of life at a high level of complexity.

The "horizontal" variation remains within the limits of a "kind." A "kind" has been defined as a group of organisms that possesses variant genes for a common set of traits, but that does not interbreed with other groups of organisms under normal circumstances. Others think that the outer limits of variation might correspond approximately to the taxonomic category of "family."

A final corollary is that hominoid fossils are always clearly human or ape, never intermediate. Thus it is claimed that Neanderthal man was really human, but that its primitive features resulted from nutritional deficiencies and pathological conditions. On the other hand, *Ramapithecus* and *Australopithecus* are clearly ape-like.

7. Mutation and Selection

The joint action of mutation and selection could not have brought about the variety of present living kinds from a simple primordial organism. The effect of rare mutations in producing "horizontal" variation is acknowledged. It is the extrapolation to evolution on a grand scale that is denied.

Mutations are nearly always harmful in an organism's natural environment. Therefore, it is unlikely that this mechanism could have provided the many beneficial mutations required for progressive evolution.

Natural selection involves circular reasoning. The survival of the fittest simply means that the fittest organisms leave the most offspring. What is needed is a criterion of natural selection that is independent of survival. Meanwhile, natural selection should not be considered a sufficient explanation of how mutations produce more fit organisms.

There is a disagreement among evolutionists on a number of details of evolutionary theory. This implies that there is less certainty about Darwinian evolution than is generally assumed. Of particular interest is the attention given to "punctuated equilibria," a concept which involves more rapid change and more discontinuity than is usually admitted in discussions of evolution.

A final argument is based on probability estimates. The probability that random mutation and natural selection could have produced complex kinds from a simpler kind is infinitesimally small, even over a period of billions of years. This is the kind of point that was recently made by Sir Fred Hoyle, who is not known to be a creationist. He argued that the information content of the higher forms of life is represented by the number $10^{40,000}$—representing the specificity with which some 2,000 genes, each of which might be chosen from 10^{20} nucleotide sequences of the appropriate length, might be defined. Evolutionary processes would, Hoyle said, require several Hubble times to yield such a result. The chance that higher life forms might have emerged in this way is comparable with the chance that "a tornado sweeping through a junk-yard might assemble a Boeing 747 from the materials therein."[1]

8. Uniformitarianism and Catastrophism

The earth's geological features were fashioned largely by rapid catastrophic processes that affected the earth on both a global and regional scale. This is in contrast to a central postulate of evolution, namely that all major geological formations can be explained by present processes acting essentially at present rates without resorting to any world wide catastrophe. This argument would include, but is not limited to, flood geology. Furthermore, it can be associated with claims for a young earth but it does not need to be so restricted.

Another way of stating this point is to refer to such geological phenomena as the following: the formation of mountain ridges, the thick deposition of thick sequences of sedimentary rocks including fossils, the initiation of glacial ages, and the extinction of dinosaurs and other animals. It is claimed that these features can be explained satisfactorily by catastrophic events such as the following: huge floods, massive asteroid collisions, large volcanic eruptions, devastating landslides, and intense earth quakes.

Particular attention is given to situations in which inversions of the rock layers appear to exist, in the sense that older layers lie on top of younger ones. In order to explain this circumstance, evolutionary geologists are said to postulate vast "over-thrust-faults." Creationist geologists accept the evidence for local folding and over-thrusting on a small scale but argue that physical evidence along the contact line, such as brecciation, gouge, and slickensides, does not exist for the supposed large thrust-faults.

Scientific Issues Related to Time

9. Astronomic Time

In a very large universe it would take a long time for the light to arrive from the most distant point. Thus the estimates of distance become an indirect indicator of time.

One of the estimates of astronomic distance is based on "redshift," a change in the frequency of the light waves towards the red end of the spectrum. Creationist writers have focused on instances of discordant redshift in which the estimated distance to some body is not as large as the theory would require. Some bodies are too bright to be as far away as the redshift theory would imply. Furthermore, some bodies which are grouped in the same location have different redshifts.

A somewhat different argument is the claim that there are no direct (geometric) means of estimating distances to astronomical objects that are more than about 330 light years away from the earth. Any greater distances have to be derived by secondary means from theories that may have questionable postulates.

Creationists also object to the "big-bang" hypothesis for the origin of the universe. This would be paralleled by comments from Sir Fred Hoyle, who acknowledged that "steady state theories of cosmologies, of which he was one of the chief exponents in the 1950s, are not now tenable because of the evidence for evolutionary galactic and stellar processes. But the big-bang view is similarly not tenable because of the way in which it implies the degradation of information."[2]

10. The Age of the Sun

Here again there are a series of creationist arguments for a recent origin of the sun. For example, all theories of modern evolutionary cosmogony assume that hydrogen fusion is the energy source of the sun and stars. The conversion of hydrogen into helium produces neutrinos, but a search for solar neutrinos has failed to detect the predicted number. This is consistent with evidence for a shrinking sun and casts doubt on hydrogen fusion as a major energy source. The sun may actually be burning by means of gravitational energy.

11. The Age of the Earth

A very long list of arguments has been advanced to support the concept of a young earth. Once again I will select one which has been explored by creationists more recently.

The value of the earth's magnetic dipole moment decays at about 5% per 100 years. 7,000 years ago the earth's magnetic field was calculated to be 32 times its present value. If we take the maximum value of the earth's magnetic field to be less than that of a magnetic star, calculations would show that the earth's magnetic field could not have been in existence for more than 10,000 years.

12. The Age of Rocks

The main methods for assigning ages to rocks use radiometric determinations. This assumes a decay of uranium through a number of steps to lead and a similar decay of potassium to argon.

In general these methods assume that: (a) no decay product was present at the outset (or at least the initial quantities can be estimated accurately), (b) radioactive materials or products do not move in or out of the rock, and (c) the decay rate was constant over time. Creationists claim that each of these assumptions is questionable.

A further criticism is based on the claim that estimates of age (even from the same rock) vary so widely that one cannot have any confidence in the research methods.

A third criticism has to do with the methods used to plot the data. The radiometric data from the rock sample are plotted on a graph. A straight line (or isochron) is then drawn to give the best fit to the points. The age of the rock is then estimated from the slope of this line. Creationists claim that the slope of an isochron may simply reflect the occurrence of a mixture of components in the rock. It is hard to establish independently whether or not such mixtures have occurred, and thus the slope of the line has no meaning with reference to age.

A final argument has to do with radioactive halos. In some minerals, rings of discolored areas can be detected around radioactive inclusions. Certain examples (Polonium halos) show no inner ring. This is taken as evidence that the process did not start with the first element in the decay chain. There was no parent element, and therefore the material must have been created in an intermediate form. This is taken to be a signature of divine activity.

PART 2: A CRITIQUE OF THE CRITIQUE

A thorough response to these points would require more time and space than are available in this brief review. The issues do need to be addressed, however, and I would hope that the list might serve as an agenda for extended discussion. Furthermore, some guidelines for constructive debate can be outlined. It should be obvious, for example, that the words "evolution" and "creation" have different meaning to different people and in various situations.

The Meanings of Evolution

1. Evolution as Special Theory (microevolution).

This includes the view that many animals and plants have undergone changes over the course of time so that new species have been formed. The relevant facts are partly derived from direct experimental tests and are partly historical in nature. Much of the discussion of evolution in biology textbooks is at this level.

2. Evolution as General Theory (macroevolution).

Here it is assumed that the development of new classes of organisms (such as reptiles and mammals) involved the same processes as the formation of new species. As the theory of evolution becomes more comprehensive and involves the more distant past, however, the theory becomes farther removed from the basic data and thus less subject to direct test. Simplifying assumptions must be introduced to handle the extremely large volume of data, but the issues are still testable in principle, and (in my opinion) thus scientific.

3. Evolutionism as Religion.

Some scientists have attempted to construct an entire world view on the basis of evolutionary theory. This approach can be called evolutionism and often represents a form of scientism. It is important to realize that, when a famous scientist such as Julian Huxley denies the existence of God, he is stating his religious beliefs rather than a necessary and logical conclusion from his scientific investigations.

The Meanings of Creation

1. Special Creation.

This view of creation holds that a number of "kinds" were brought into being by God and that further limited changes over time have taken place only within these boundaries. Some of those holding this view would accept what I have described above as a special theory of evolution but would call it variation. Special creation can fit either a short or a long time span.

2. Creation as General Doctrine.

For many people creation is an important theological doctrine. Humankind and all things that exist are dependent upon God who is Creator and Sustainer. The created order is good, and purpose can be seen in God's providence. Humans stand in a special relationship to God, in their capabilities to respond toward God. It should be noted that these views are not dependent upon the presence or absence of scientific explanations. Nor is either a short or a long time scale an essential element. Such a belief in God as Creator is independent of whether the theory of evolution is true or false, in whole or in part.

3. Scientific Creationism (or Creation Science).

This view has been described in the first part of this paper and is the one generally involved in creation/evolution controversies.

The Nature of Science

The methods of science are limited to space/time and can neither prove nor disprove the existence of God. This stance, however, is described better as non-theism than as atheism.

On the other hand, serious problems arise when this objectivity is violated by extravagant statements such as E. O. Wilson's claim that "The final decisive edge enjoyed by scientific naturalism will come from its capacity to explain traditional religion, its chief competitor, as a wholly material phenomenon."[3] This should be recognized as a matter of personal opinion that is beyond the scope of scientific investigation.

Another illustration of over-claim appears in a recent book advertisement. "The *ultimate* test of the hypothesis that the origin of life is *inevitable* will be to examine life that has arisen *independently* elsewhere." The words that I have italicized cannot be justified in this context as part of a serious scientific claim.

The historical nature of the evidence used to support evolution is readily admitted, but this fact cannot be used to deny the validity of any scientific study of the past. A crude analogy is seen in mystery stories in which detectives often are called in to solve a crime after the fact, even if none of them was present and sometimes even if there were no surviving witnesses.

What we need is some sense of perspective about the past, such as that suggested by Bube:

> In the teaching of evolution, what we can say about the processes going on at present is the most solidly based, what we can say about processes in the immediate past is probably largely valid, what we can say about processes in the distant past becomes increasingly speculative, and we can say nothing scientific at all about absolute origins. Yet the typical discussion of evolution starts the other way round, presenting theories about origins as if they were the foundation of our evolutionary knowledge and established beyond the shadow of a doubt. What evolution teaching needs is not the introduction of an alternative non-scientific "creation theory," but the reformation of the present courses so that they are faithful to the potentialities and limitations of the scientific method.[4]

Another statement of the same principle is seen in a recent letter by Hildemann to *Science:* "The integrity of science is not compromised by stating that the *ultimate* origins of matter and life are unknown and open to conjecture. Indeed, evolutionary scientists, among whom I count myself, could well take greater care in separating facts from conjecture."[5]

Science and Faith

Science and religious faith represent two distinct ways of knowing which are described more accurately as complementary than as mutually exclusive. It is impossible to consider fully the meaning of science for man without asking religious questions. It is fruitless to ask whether the topic of "origins" is scientific or religious, since both facets become involved eventually. Mechanism and meaning are two aspects of one and the same reality.

Everyone has a "batch of beliefs" (more or less recognized and organized) which are used to interpret the mysteries of personal existence. For the most part, these are brought to one's interpretation of scientific data, rather than derived from science. Serious difficulty arises when the origin of such beliefs is over-simplified. Thus it becomes necessary to guard against a scientific exclusivism and dogmatism that sometimes are found in the writings of science enthusiasts. Hooykaas has made this clear:

> As long as we recognize that the world of science is not the *whole* reality we live in, but only one particular aspect of it, and that our religious or our aesthetic experience is as *real* and as *fundamental* as our scientific experience, no harm is done. . . . If we do not clearly realize these things, our silence about God in nature will end in the

deification of nature. . . . Materialism, evolution, scientism—that is, pseudo-religion under the disguise of science—will take the place of religion.[6]

This is the single most important issue in the current controversy. As long as evolution is treated as science (open to investigation and leading to testable hypotheses) I have no basic argument. (To be sure, there are unresolved questions, but in science we never run out of questions.) But when evolution is used as the basis for a comprehensive world-view which denies the existence of God as Creator, then the discussion has been changed from science to religion. Such a belief might well be considered as a sophisticated form of nature worship. An important aspect of nature has become elevated to an organizing principle, not only for science as such, but for all of life's questions.

This is essentially the point made by C. S. Lewis in his essay "The Funeral of a Great Myth":

> The central idea of Myth is what its believers would call 'Evolution' or 'Development' or 'Emergence,' just as the central idea in the myth of Adonis is Death and Re-birth. I do not mean that the doctrine of Evolution as held by practising biologists is a Myth. It may be shown, by later biologists, to be a less satisfactory hypothesis than was hoped fifty years ago. But that does not amount to being a Myth. It is a genuine scientific hypothesis. But we must sharply distinguish between Evolution as a biological theorem and popular Evolutionism or Developmentalism which is certainly a Myth.[7]

The Second Law of Thermodynamics

I cannot understand why creationist writers have continuously misinterpreted the second law of thermodynamics as *prohibiting* evolution. That law has a rigorous mathematical definition that restricts its application. Note the following points:

1. The second law applies only to closed systems for which the parameters are carefully defined. A general increase of entropy in the universe as a whole does not prevent localized decreases.

2. If an apparent exception to a scientific "law" is noted, there are several options to be explored. The situation may be one to which the law does not apply, or the law is not sufficiently general and should be modified. Apparent exceptions to laws are not illegal or prohibited; they are problems to be considered and resolved. Thus the case for or against evolution at any level is made on other grounds.

3. The second law of thermodynamics (which concerns energy) has served as a model for the development of analogous concepts in information theory. How life could arise without prior information is a very

interesting question that is not resolved easily, but it is a different one. It cannot be settled by reference to thermodynamic concepts.

Is Darwinism Dead?

Recent articles have given the impression that evolutionary theory is in total disarray and that many prominent scientists have given up their earlier acceptance of evolution. The facts in the case are quite different.

Was Darwin wrong? Yes, in many ways. He accepted pangenesis, the view that elements from body cells migrate to the gonads and can be transmitted to offspring. He accepted Lamarckianism, the concept that external environmental influences also can be transmitted. He accepted blending inheritance, the idea that transmissible factors from parents blend and are permanently altered en route to the next generation. He did not understand the nature of genetic variability.

Yet that is just what should happen to scientific theories. They are altered continuously by new discoveries. Many thought that Darwinism would disappear when Mendelian genetics arrived on the scene, but the basic concept of natural selection was enriched, not abolished. Similarly, new evidence about the nature of genetic mutations and the effects of population size and population structure have been incorporated.

The current argument involves the question as to whether all evolutionary change must be small and gradual or whether episodes of more rapid change may have occurred. The answer already appears clear, namely that change may be slow at times and rapid at others. This is not a new idea, however. It is only that more adequate data have been accumulated.

Is the "survival of the fittest" a tautology? Yes, but natural selection is not, and that is the major theme of evolutionary theory. The point is that differential reproduction (some organisms surviving and reproducing better than others) can produce changes in gene frequencies. Natural selection has been demonstrated so often in experimental tests that it should be considered firmly established.

The "Young Earth" Model

A belief in the recent origin of the universe is not a majority view among Christians, even among those who accept the Bible as divinely inspired. A careful review of the way in which the doctrine of creation is interpreted within the Bible as a whole shows that the emphasis is primarily upon the meaning of existence and very little about time. Thus the term "creation model" arbitrarily fixes upon one of several interpretations. As a result, persons who reject recent origin may reject any other view of creation without understanding the distinctions.

I also have serious difficulties with the scientific arguments advanced for a young earth, since they seem to fall into one of three categories:

1. They fail to take other data into account. For example, there is evidence indicating that the earth's magnetic field fluctuates and that straight line extrapolations into the past are not appropriate.

2. They misrepresent the methodology. It is true that isochrons can be produced by mixing, but independent tests can be used to detect whether or not mixing has occurred.

3. They assume that any disagreement at the cutting edge of scientific investigation totally invalidates all other approaches to a given question.

Summary

Most of those who write about the creation/evolution controversy have failed to recognize that there are many research scientists and science teachers who read the Bible and believe in God as creator and sustainer, but think that neither the scientific nor the biblical evidence supports a recent origin of the universe. From their point of view there are problems with both extreme poles of the controversy.

Some evolutionists claim that science has made religious faith obsolete, but this misrepresents the nature and limitations of science. Such an arrogant attitude provides an interesting contrast with studies in bioethics about the relevance of values for our understanding and use of science and technology.

At the other extreme, it is a serious theological mistake (in my opinion) to think that creation can be discussed adequately without reference to religious sources. The attempt to make creation strictly scientific appears to reflect the belief that questions about the meaning and purpose of human existence are best resolved within a scientific context, another dubious assumption.

It is for these reasons that laws for teaching the "creation model" will cause serious dilemmas for many teachers who really do believe in creation. Furthermore, whenever the two models are compared thoroughly, it seems inevitable that some students will feel that criticisms of the creation model are anti-religious in nature.

The aspect of the "creation model" that is most easily tested by scientific evidence (and that is a central point of most proposed laws) is the argument for recent origin. The problems mentioned above would be reduced if in public discussions and in the classroom we would use the more descriptive term, the "young earth model."

REFERENCES

1. Reported in *Nature,* 294 (November 12, 1981): 105.

2. *Nature,* 1981.

3. E. O. Wilson, *On Human Nature* (Cambridge, MA: Harvard University Press, 1978).

4. R. H. Bube, "Creation and Evolution in Science Education," *Journal of the American Scientific Affiliation* 25 (1973): 69–70.

5. W. H. Hildemann, "Letter," *Science* 216 (1982): 1048.

6. R. Hooykaas, *The Christian Approach in Teaching Science* (London: Tyndale Press, 1966).

7. C. S. Lewis, "The Funeral of a Great Myth," in *Christian Reflections* (Grand Rapids, MI: W. B. Erdmans Publishing Co., 1967).

Answers to the Standard Creationist Arguments

by Kenneth Miller

The reason evolutionary science does not make reference to a creator is for the same reason that mathematics, cell biology, organic chemistry, and hydraulic engineering do not make reference to a creator: none of these are theological subjects. They are *nontheistic*, as all scientific and mathematical systems must be. Imagine how ridiculous Dr. Gish would have sounded had he declared, 'Thus, while not all those who do long division are atheists, the practice of long division *is* an atheistic practice'.

In the spring of 1982, a pre-recorded debate between Dr. Russell Doolittle of the University of California at San Diego and Dr. Duane Gish of the Institute for Creation Research was aired over national television. In this debate, produced by Jerry Falwell and taped in October 1981, Dr. Doolittle made an excellent case for the exclusion of creationism from science classrooms. He argued its religious nature and its failure to meet the standards of scientific investigation. Dr. Gish, in a stunning presentation, made an effective summary of the standard creationist debate arguments. Because his performance was so widely viewed, the points he made have become the creationist arguments most familiar to millions of television viewers. We will see them crop up again and again in school board controversies, legislative battles, and court cases. It would be practical, therefore, that answers to these standard arguments be made available. The purpose of this article is to provide them.

Such debates, of course, are neither part of the scientific process nor a contribution of anything to scientific understanding. Their purpose is political; so scientists participate only in the hopes of making them educational. The Doolittle-Gish debate was no exception. While many in the overwhelmingly one-sided audience may have been delighted with Dr. Gish's performance, it fell sadly short of anything that could be recognized as scientific argument.

THE CREATION MODEL

The most remarkable failure—and the most obvious—was Gish's lack of a single sentence during the entire debate which described "the creation model." He made his whole presentation a game of "hide the ball," never once revealing what his "theory" or "model" was. The closest he came was when he said:

> According to the concept of creation, or, as it may be called, *the creation model,* the origin of the universe and all living forms came into being through the designed purpose and deliberate acts of a supernatural creator. The creator, using special processes not operating today, created the stars, our solar system, and all living types of plants and animals.

Dr. Gish did not say *when* that creation event occurred, and he did not say whether all the animals and plants were created in their present forms at the same time or whether they were created in different forms and at different times. These are not trivial points because, without them, "creation science" does not make a single scientific statement. Without any details on the creation "theory" being presented, Dr. Doolittle had no way to discuss it.

While this tactic may be an excellent debating strategy, one that keeps one's opponent on the defensive due to a focusing of the attack on *his* ideas alone, it is very bad science. A theory that is kept hidden from discussion cannot be analyzed on its own merits. Therefore, if we were to declare that Dr. Gish had "won" the debate, we would only be saying that evolution had been questioned, not that a case had been made for creation.

Gish, of course, would disagree. His opening statement was, "There are two fundamentally different explanations for the origin of the universe and the living things it contains." This statement implies that, if he can disprove or cause people to doubt evolution, he has proved creation. But such a view constitutes cultural arrogance. There are a number of different hypotheses concerning origins that have been postulated by scientists in the past. One could name spontaneous generation, for example. The hypothesis of panspermia is the suggestion that life originated elsewhere and came to this planet through space. Various cyclical hypotheses propose fluctuation or change back and forth. And the number of religious ideas are legion.

The only time we find ourselves limited to just two "fundamentally different explanations" is when we compare naturalism and supernaturalism. But Gish is foolish if he thinks that he represents the infinite number of supernatural explanations and that Dr. Doolittle was to represent all the naturalistic possibilities. Furthermore, for Gish to take such a position, he would have to deny that creationism is a part of natural science. This would effectively bar it from any natural science class and thereby end the debate. Supernatural science must depend on supernatural evidence—not evidence from the natural world. To the extent that creationists argue from natural evidence and propose naturalistic mechanisms for their creation model (the model Gish did not state in the debate), they place themselves in the naturalistic camp with the evolutionists.

After misstating the controversy in his first statement, Dr. Gish went on to misstate the theory of evolution in his second and following statements. He said:

> According to the theory of evolution—or, as we should more properly call it, "the evolution model"—everything in our universe has come into being through mechanistic processes, which are ascribed to properties, inherent in matter. No supernatural intervention of any kind was involved. In fact, by definition, God is excluded. Thus, while not all evolutionists are atheists, the theory of evolution *is* an atheistic theory.

Such a clear and effective misrepresentation of his opponent's position was a beautiful rhetorical maneuver—one which was almost guaranteed to win the approval of Gish's audience while at the same time placing Dr. Doolittle in a very awkward position. This is a classic case of winning an argument by distorting the idea you are challenging. The key issue in this debate should have been whether living organisms on earth have changed (evolved) through the hundreds of millions of years for which science has excellent fossil records and other evidences or whether living things have remained unchanged from an initial creation event which occurred no more than about ten thousand years ago. Instead, Dr. Gish made the key issue of the debate a theological question over whether or not God exists. His arguments for a creator involved appeals to the second law of thermodynamics, design, the supposed mathematical improbability of things arising naturalistically, and "gaps" in the fossil record. He seemed to maintain the viewpoint that, if he could prove the existence of God, he would thereby have disproved the theory of evolution. Since Dr. Doolittle did not come to discuss theology and as that is not his specialty, the result was that the two debaters found themselves talking about two different issues.

The reason evolutionary science does not make references to a creator is for the same reason that mathematics, cell biology, organic chemistry, and hydraulic engineering do not make references to a creator: none of these are theological subjects. They are *nontheistic,* as all scientific and mathematical systems must be. Imagine how ridiculous Dr. Gish would have sounded had he declared, "Thus, while not all those who do long division are atheists, the practice of long division *is* an atheistic practice." After all, "no supernatural intervention of any kind" is involved. It must be that elementary school teachers who instruct our children in nonmiraculous math are teaching "a basic dogma of agnosticism, humanism, and atheism."

I would like to add that Dr. Gish's suggestion that evolution and creation are mutually exclusive ideas is insulting to me personally (I am a Roman Catholic) as well as to the great majority of scientists of Christian, Jewish, Moslem, Hindu, Buddhist and other faiths who understand quite well that biological evolution is a scientifically supported fact. The theory of evolution is not inconsistent with the belief in a created universe *per se.* However, it *is* inconsistent with the creationist belief in a universe that was created no more than ten thousand years ago in which all living things were created at the same time in essentially the same form they take today. But this is the very "creation model" that Dr. Gish would not discuss.

THE NATURE OF SCIENCE

After misrepresenting the controversy and evolution, Dr. Gish then went on to misrepresent science, which he accomplished admirably.

> Let us dispense, once and for all, with the notion that this is a debate between science and religion. Each concept of origins is equally scientific and each is equally religious. In fact, neither qualifies as a scientific theory. The first requirement of science is observation. Obviously there were no human observers to the origin of the universe, the origin of life, or, as a matter of fact, to the origin of a single living thing. These events were unique, unrepeatable historical events of the past. . . . Ultimately then, no theory of origin can be considered a scientific theory in the strict sense.

It is crucial for creationists that they convince their audiences that evolution is not scientific, because both sides agree that creationism is not. So, Dr. Gish proposed this ingeniously stringent set of requirements for a scientific theory. He seems to say that not only is science based on observation (which is true) but that it requires eyewitnesses to all events (which is false). This is a strange suggestion. No one has ever *seen* an atom, just its effects. Do atoms therefore not exist? The wave and particle aspects of electrons have only been determined by the images they leave on film when certain experiments are performed. These images record *past* events, not present realities. Is subatomic physics then a faith? The same questions could be asked of astronomy, chemistry, and geology—not to mention much of the rest of science. Dr. Gish's overly limited interpretation would wipe away most of the world's evidence for anything.

In fact, even creationism could prove nothing. This is why Dr. Gish had to contradict himself in the debate by saying:

> Although there were no human witnesses to any of these events [of creation], creation can be inferred by the normal methods of science: observation and logic.... Creation and evolution and *inferences* based on circumstantial evidence and predictions based on each model can be tested and compared with that circumstantial evidence.

So which is it? Or perhaps it is neither. Perhaps the creation-evolution controversy should really be a debate over which act of faith is best supported by the circumstantial evidence. This is a strange mixture of religion and science—a mixture that denies we can ever attain knowledge of historical events. Imagine what would happen if Gish's requirements were followed in our courts of law. We could only convict criminals who were directly observed committing their crimes. But since crimes are rarely committed in full view of others, our courts have to take this into account. In both law and science there is a common-sense precedent to use circumstantial evidence carefully to resolve questions about natural events, even when they are historic, unique, and unwitnessed. Dr. Gish's narrow definition of science is simply self-serving. It is a way of promoting confusion about evolution and bringing the acquired data of hundreds of years of scientific research down to the level of Dr. Gish's brand of faith. Only by such questionable thinking can creationism be seen as an equal and alternate model.

If we ignore the creationist's arbitrary rules of science and compare the two models in the normal way, we find that evolution is scientifically testable, right along with many aspects of creationism. For example, there is observable evidence for evolution. This evidence is found in the fossil record, the phylogenetic trees for living and extinct animals, the geographic distribution of organisms, the phases of embryologic development, observed mutations, observed natural selection, observed geological changes, and laboratory experiments in biology, among other things. Both evolution (which predicts that the evidence will show life has changed through time) and creationism (which predicts an absence of change, except for extinction) are scientifically testable. Dr. Gish would like to pretend this is not true because creationism fails the test of evidence while evolution passes it.

Evolution also predicts a consistent pattern of relationships between animals. This prediction is also testable. For example, if humans appeared to be most closely related to chimpanzees by one criterion, but to butterbeans by another, to chickens by a third criterion, and to bullfrogs by a fourth, there would be no consistent pattern, and evolution would thereby be disproved. But all techniques for determining relationships have consistently given results that fit with the evolutionary prediction. Creationists have recently tried to claim that some data go against the prediction (which shows that creationists also see this prediction as significant), but their arguments are all based on incorrect data. After a century and a quarter of strenuous questioning and testing in many fields, the theory of evolution stands stronger than ever. It could be falsified if it were wrong, but efforts to falsify it have continually failed. Evolution unites genetics, physiology, paleontology, embryology, biogeography, systematics, and geology into a coherent whole. And this is another reason why evolution is a good scientific theory.

THE NATURE OF THE UNIVERSE

Dr. Gish next implied that the theory of evolution says that the universe created itself. Nothing could be further from the truth. Notions of how the universe originated are altogether outside the province of science. Such questions of first cause properly belong to the realms of philosophy and theology. Evolution speaks only of change through time. The universe could have begun in any number of ways, and yet we would still have to separately learn whether or not biological evolution takes place and existing life forms evolved from ancestral ones.

Nonetheless, Dr. Gish spent a great deal of time insisting that the universe could not have created itself and that a creator must therefore exist. He simply ignored the possibility that a creator might have formed a universe in which evolutionary processes then brought about the formation and development of living things. Yet, if such events actually did take place, Dr. Gish's particular brand of creationism would be falsified—a possibility he refused to consider.

"According to evolution theory," he said, "disorder spontaneously generated order" by means of the "Big Bang." That is to say, a cosmic explosion created the orderly cosmos we see today in a manner that is actually contrary to the second law of thermodynamics. Dr. Gish was wrong again, but this argument went over well because people naturally visualize an explosion as *disorderly* and the present state of the universe as *orderly.* Yet, in a thermodynamic sense, *order* means "energy available for work" and *disorder* means "energy unavailable for work." Therefore it is actually true that the universe was more orderly at the time of the "Big Bang" but has grown progressively more disorderly as it has expanded. Dr. Gish was simply playing on the popular meaning of these words while speaking of the science of thermodynamics which uses them differently.

Astronomers are well aware that the universe, taken as a whole, is "running down" in accord with the second law of thermodynamics. Evolution harmonizes with that. There is nothing in evolutionary theory that states the universe has ordered itself. *Cosmic evolution* is what happens as the universe runs down. It is the second law in action. *Biological evolution* is what happens in pockets of the universe where the process temporarily reverses itself due to greater losses of energy elsewhere. For example, in our pocket of the universe there is an increase in complexity associated with living organisms and their evolution. This is made possible by the *decrease* in available energy in the sun. The energy loss of the sun provides thousands of times the energy demanded by the second law to account for the increase in complexity on our planet. Dr. Gish therefore set up a straw man with his claim that the second law of thermodynamics prohibits evolution.

In his rebuttal, however, Gish argued that receiving energy from the sun was not sufficient to create life. He claimed that there must also be an "energy conversion machine," much like a car's motor, and a "control system," much like a car's driver, if there is to be an evolutionary increase in complexity. He argued that life has these properties infused into it by the creator but inanimate matter does not.

However, in actual fact, the raw, uncontrolled application of energy *does,* under certain conditions, cause the formation of complex molecules (although not automobiles!). Stanley Miller and Harold Urey demonstrated this in their famous experiment nearly thirty years ago. Furthermore, inanimate matter can often increase in complexity in nonbiological ways. Snowflakes form from water and dust all by themselves, and complex and energetic whirling wind storms also arise spontaneously from random converging wind systems. Where are the divine "energy conversion machines" and "control systems" in these phenomena?

THE LAWS OF PROBABILITY

The next argument was an old standard. Dr. Gish noted the great complexity of living cells and the various other forms of life on earth. He argued that the mathematics of probability would render it impossible for life to develop from nonlife all by itself, no matter how much time was allowed:

> Most proteins consist of several hundred amino acids, each arranged in precise sequence, and DNA and RNA usually consist of *thousands of nucleotides* also arranged in precise order. The number of different possible ways these subunits can be arranged is so incredibly astronomical that it is literally *impossible* for a single molecule of protein or DNA to have been generated by chance in five billion years.

He backed up this claim by citing calculations by Hubert Yockey. But these calculations are based on two false assumptions which stack the deck against evolution: first, that a *particular* nucleotide or amino acid sequence must assemble completely by chance—and only that *specific* sequence will be accepted—and, second, that no small nucleotide chains are capable of self-replication.

Yet, in the globin protein sequence (the polypeptide part of hemoglobin) only seven amino acids, out of more than one hundred, are always the same when we examine the many globins which are used by different organisms. If the creationist calculations are done with this fact in mind, we would discover that such sequences form very quickly. Second, the sequences would not have to assemble from scratch. Recent work by Orgel and Eigen and others has shown that RNA nucleotides can spontaneously form small chains. Furthermore, these small chains can procede to self-replicate. Often when such organic molecules get to be twenty to twenty-five amino acids long, they can spontaneously double their lengths through this replication process. (Indeed, many of the molecules found in living things bear evidence of having evolved in exactly this way.) The net result is thousands and thousands of variant copies being produced quickly. Therefore, the sequences that Dr. Gish says could *never* form would in fact self-assemble in a few months or years, given the whole earth as a laboratory. Since Yockey's calculations do not allow for this replication, his mathematical results are light years away from the truth.

Gish argued next that hundreds of different functional proteins would have had to form simultaneously. He assumed that this also would be another impossibility. Yet, there are numerous papers with copious data showing that the many modern proteins appear to have derived from a few ancestral proteins. He also assumed that, if modern cells have two hundred proteins, the earliest protocells also had two hundred proteins. A wealth of experimental results refutes that assumption as well. However, in spite of the open availability of all this data, the creationists go right on making these same tired old statements.

Reading these creationist impossibility calculations always brings to mind other impossibility calculations, some made by eminent scientists of their day, which were also based on erroneous assumptions. Lord Kelvin calculated that powered aircraft could never fly. Others calculated that steamships could never carry enough fuel to cross the Atlantic. One should always keep in mind the computer-age dictum: "Garbage in—garbage out."

THE FOSSIL RECORD

The key claim of evolution is descent with modification, the idea that animals alive today evolved from earlier forms. All the previous talk about the supposed impossibility of life evolving from nonlife says nothing about descent. Evolution is not really a concept of origins. A creator could have created life and then everything could have evolved from there. Such a fact would still falsify Dr. Gish's unstated creation model.

In order to defeat the notion of descent, Dr. Gish claimed that "the missing links are still missing," that there are gaps in the fossil record so severe that the record simply does not show evolution. This is a shocking set of untruths.

The fossil record not only documents evolution but the very existence of the fossil record was the force that drove unwilling scientists to admit nearly two centruies ago that living forms had changed (evolved). This record shows intermediate form after intermediate form. There is a long series of intermediates linking reptiles with mammals. There are evolutionary sequences showing the evolution of the horse, the elephant, sea urchins, snails, major groups of plants, and many other animals now extinct. Furthermore, these fossils show an orderly succession which fully documents the evolutionary tree of life.

The reason Gish says that intermediate forms do not exist is because his model requires that he explain them all away. For example, *Archaeopteryx,* a clear intermediate between reptiles and birds which in some ways is more closely linked with the little dinosaurs of the period than with later birds, is declared by Gish to be "100 percent bird." Why? Because it has feathers. This is where he draws the line. Yet, if one really wanted to discuss the *Archaeopteryx* fossils in detail, one should be aware that several fossilized *Archaeopteryx* skeletions were discovered before one was found with feathers preserved. How were these specimens first classified? They were thought to be reptiles and were placed in museums alongside other small dinosaurs. In short, *Archaeopteryx* was an animal whose skeletal structure was reptilian but upon whose skin the first feathers had appeared. Just how much more intermediate does something have to be?

Since the fossil record is actually very complete and is getting better all the time with continuing new discoveries, it is only by refusing to see what is plain that creationists can deny that the fossil record supports evolution. And even if these gaps were as profuse as Gish claims, the fossil record would still reveal an impressive lineage for animals living today. It would still reveal that the further back one goes in time, the more numerous the extinct forms and the less similar they are to modern forms.

Nonetheless, Gish made an impressive-sounding case by citing "authorities" supportive of his claims. In a classic out-of-context quote he voiced the words of Dr. Corner, a Cambridge botanist, who wrote, "Much evidence can be adduced in favor of evolution, but I still think that to the unprejudiced the fossil record of plants is in favor of creation." However, what Dr. Corner actually said was that ". . . the fossil record of *higher* plants is in favor of *special* creation" (emphasis added). What did Corner mean by that? He meant that the major form of higher plant (the angiosperms or flowering plants) appeared on earth about 135 million years ago, and we have no good fossil evidence as to what forms they evolved from. Corner meant to emphasize in his statement just that lack of ancestral evidence and pointed out that the higher plants appear so suddenly that one could almost believe that they had been *specially* created—just as if a creator had said, "Let there be angiosperms," and so they appeared.

One might get the impression that Dr. Gish's creation model suggests exactly that: that the appearance of the angiosperms represents a specific and individual creative act in which they were formed from scratch by a creator 135 million years ago. Although Dr. Gish seemed quite willing to leave that false impression with his listeners, he in fact holds to a radically different view.

His real position is that all animals and plants were created at the same time (or in six solar days) only about ten thousand years ago. Such a view means that angiosperms were *always* present and their fossils should be found in the oldest rocks available. However, there is no evidence of their existence prior to 135 million years ago, while other land plants appear in the record hundreds of millions of years earlier. The fact that various life forms appear in various places along the geologic column is actually deadly evidence against Gish's notion of a single creation event. But he gets away with implying this evidence is consistent with his creation model because he never really presents this model.

HUMAN EVOLUTION

The big emotional issue among creationists is human evolution. It might be safe to say that all their previous arguments exist only to support the notion that humans are in no way linked to the other animals. To this end, Gish quoted Sir Solly Zuckerman in order to claim that *Australopithecus* did not walk upright. The quote is dated 1970. Since then, several pelvic fossils and one nearly complete *Australopithecus* skeleton have been found. There is now not the slightest doubt that this animal walked upright, much as we do. But Dr. Gish quoted from a decade-old source and therefore ignored the latest findings.

His information on Lucy is no better. Gish declared, "Since Johanson describes this creature as totally ape from the neck up, the only basis for the idea that this creature was a link between man and ape is a notion that it did walk upright."

But Johanson never claimed that Lucy was an ape. He simply stated that from the neck up she was essentially a hominid with a number of apelike *features.* And, from the neck down, she should be linked with the human family due to her fully upright stature. She and her colleagues walked just as we do today. This is clear from the detailed anatomy of the hip, knee, and ankle, not to mention the 3.7 million-year-old footprints in the volcanic ash at Laetoli, Tanzania. Extensive comparative anatomy and biomedical analysis render this judgment of Lucy's locomotion to be far more than a guess. The value of this discovery is that it shows how hominid bipedalism preceded both tool use and the modern human cranial capacity.

To conclude his attack on human evolution, Dr. Gish reminded his audience of the Piltdown Man hoax. This is surprising since the hoax was revealed and exposed not by anti-evolutionists but by scientists. The same techniques that exposed the Piltdown hoax now verify the authenticity of the work done by Johanson and others. However, Dr. Gish refuses to accept in one case the same sort of dating evidence he is delighted to use against evolutionists in another.

Gish also mentioned Nebraska Man, for which the evidence turned out to be a number of fossilized pig's teeth. However, what he failed to mention was that since the discovery of Nebraska Man in 1922, it was contested by scientists worldwide. In fact, in every case that creationists have pointed out that scientists made errors, the errors were originally discovered by scientists themselves—not by creationists who have made no significant contribution to the literature of evolution.

THE AGE OF THE EARTH

In his rebuttal to Dr. Doolittle's remarks about "scientific creationism" requiring a young earth and universe, Dr. Gish declared, "This debate is not about the time of origins, but about the 'how' of origins. These are separate questions." Not only was there no agreement to ignore the question of time made prior to the debate but the idea that the earth and universe are only a few thousand years old is a major plank in Gish's model.

It is true that some creationists accept the theory of an old earth and universe. But are they the creationists who are pushing for equal time in the public schools? It doesn't seem so when one reads the definitions of creationism that appear in the Arkansas law. Creationism is defined there as including "explanation of the earth's geology by catastrophism, including the occurrence of a worldwide flood and a relatively recent inception of the earth and living kinds." This view is considered the only valid form the creation model can take. It is seconded in the major creationist public school text books, particularly *Origins: Two Models* by Richard Bliss and *Scientific Creationism* edited by Henry Morris. These books argue for an earth and universe that are only ten thousand years in age. To gain admittance into the Creation Research Society, one must swear to a statement that includes the words: "All basic types of living things, including man, were made by direct creative acts of God during Creation Week as described in Genesis."[1]

Creationists should therefore be willing to answer critiques of this aspect of their model, even if it is the weakest plank in their platform. It is to Gish's credit that he did make a concession and speak about the age of the moon. He declared that, when scientists dated the moon rocks, "they got ages of all kinds—from over a few thousand to many multiplied billions of years. They simply selected the date that had to be right, which had to be 4.6 billion."

But Gish was wrong on two counts. First, *every single rock* from the moon for which rubidium-strontium isochrons could be determined (the most sensitive and reliable way of radiometric dating) showed an age of formation of *billions* of years. Second, the "picking and choosing" of dates, which he criticizes, is not to find dates that fit with evolution. The "picking and choosing" is really over which rocks have not been altered by outside factors in such a way that they would yield inaccurate dates. Just as you don't give up on the notion of ever knowing what time it is because some watches are broken, those who do radiometric dating don't give up determining the age of the earth and moon just because some rocks are known to be unreliable measures.

THE PUBLIC SCHOOLS

In a stunning close that appealed to the audience's sense of fair play, Dr. Gish compared creationists to Galileo facing opposition from the "stifling dogma" of the establishment. He claimed that there could only be two reasons why scientists were against equal time for creationism in public school science classes: either they were practicing an insulting form of paternalism designed to protect students from error and indoctrinate them in evolutionary ideas, or they were fearful that evolution could not survive in the free marketplace of ideas.

What the audience may not have realized is that this appeal is common to every pseudoscientific group. The very same arguments could be used to intimidate the schools into giving equal time to astrology, hollow-earth theory, "ancient astronauts," and the search for Atlantis. Furthermore, if creationism is being proposed in the name of academic freedom, why is legislation involved? In *not one* of the fifty states has evolution been *legislated* into the classroom. Evolution is taught for the same reason that the cell theory and germ theory of disease is taught: each theory successfully fought it out in the scientific arena and convinced the scientific community (including the teachers of science in public schools). What Dr. Gish is trying to encourage is the use of public pressure to determine what is and what is not science, and he is trying to force creationism into the schools through the back door without first winning the scientific debate in the way that all past theories have had to do.

Creationists are not being persecuted by scientists; they have deliberately *avoided* the scientific community. And here we could reverse Dr. Gish's claim: creationists must be fearful that creationism cannot survive a careful scientific scrutiny in the free marketplace of ideas. This must be why creationism is the only hypothesis in need of special legislative protection. Most scientists, on the other hand, support the freedom of local school boards to determine the scientific content of their instruction. It is ironic that the only state in which citizens are *not* free to make such choices is Louisiana, where a law supported by creationists has taken that freedom away.

The reasons, then, that scientists are against equal time for creationism are that it would remove academic freedom and local control from the public schools and that it would unconstitutionally promote sectarian religion.

That religion is the real issue behind the scenes is made plain by a statement by Dr. Henry Morris, director of the Institute for Creation Research, of which Dr. Gish is associate director. In a February 1979 cover letter mailed with the Institute's publication *Acts & Facts,* Dr. Morris wrote:

> Although our message to the educational world necessarily and properly stresses the scientific aspects of creationism, we can never forget we are actually in a spiritual battle and need always to be clothed in God's whole armor (Ephesians 6:11) if the creation witness is to continue to grow in its ministry to a world that needs desperately to know its Creator and Savior.

Dr. Gish stated in *Evolution: The Fossils Say No!:*

> By creation we mean the bringing into being of the basic kinds of plants and animals by the process of sudden, or fiat, creation described in the first two chapters of Genesis.[2]

This is the hidden creation model. So now we see why Dr. Gish didn't wish to mention it in debate. It would have revealed the real purpose behind the creation movement: to bring bibilical fundamentalism into the science classroom.

Dr. Gish's audience was made up of sincere and well-meaning Christians who desired to defend God and promote fairness. They were not aware of how his appeals would effectively misdirect their energies in ways harmful both to science and religious freedom. Yet, this is how far creationists must go in order to buoy up a discarded and disproved theory of science and a minority position in religion. Citizens should not be misled into subsidizing sectarian religious pseudoscience in the public school science classroom.

ACKNOWLEDGEMENTS

I give special thanks for the valuable input, ideas, and arguments provided by Frank Awbrey, Frederick Edwords, Donald Johanson, William V. Mayer, Wayne Moyer, Philip Osmon, Robert Schadewald, and William Thwaites in the preparation of this article.

REFERENCES

1. This statement is included in J. A. Moore, "On Giving Equal Time to the Teaching of Evolution and Creation." Included in this Volume.

2. Duane T. Gish, *Evolution: The Fossils Say No!* (San Diego, CA: Creation-Life Publishers, 1978), p. 24.

The Transition between Reptiles and Mammals

by Robert E. Sloan

For any given time during the evolution of mammal-like-reptiles, we have at our disposal rocks containing their fossils from less than 2 percent of the world's land surface. Because of the global distribution of land faunas, this is not a great handicap. We either have the precise species involved in the origin of mammals or very close relatives in our collections.

INTRODUCTION

A major problem in science education today is the political resurgence of a fundamentalist Christian set of beliefs currently known as "Scientific Creationism." This politico-religious movement attacks the data and concepts of evolution and geologic time on the grounds that they are only theories and that the first eleven chapters of Genesis provide a more correct theory. All levels of teachers of biology, geology, and astronomy are under attack by this pseudo-science in the classroom, school board, editorial letter columns and state and federal legislatures.

Proponents of Scientific Creationism attack conventional science not in the usual scientific journals where the battle of scientific ideas is normally fought, but entirely within the political arena where they can ignore the conventional rules of logic. A definitive statement of their position on paleontology is Gish's *Evolution: The Fossils Say No!*[1] On 67 of the 97 text pages I found at least one error of fact, logical error, or quotation out of context, all chosen carefully to mislead the reader. On checking a standard college logic text with a list of logical fallacies, I found that Gish did not manage to miss a single one! Their works have the appearance of scholarship, but not the substance.

One of the most often repeated remarks of the creationists states that "None of the intermediate fossils that would be expected on the basis of the evolution model have been found. . . . between amphibians and reptiles, between reptiles and birds or mammals. . . ."[2] This is an outrageously false statement. Space will not permit the documentation here of the many known transitions between major classes of organisms. The detail in which such transitions are known is increasing rapidly as exponentially more paleontologists look for them. Most of these transitions took place relatively rapidly (in less than 5 million years) and in small geographic areas, requiring some serendipity and more work than usual to document. However, a fossil record of the transition between reptiles and mammals has been well-known since about 1878! While the creationists deny it to this day,[3] the fact remains that this is the most thoroughly documented transition between major classes in the whole fossil record. Without looking very hard I found over 1000 technical papers and books written on the subject. The mammal-like-reptiles were the biggest and most common land animals of the world for about 120 million years. About 400 genera have been described, more are described every year. Found in every continent, they are not rare. A casual phone survey to some of my colleagues suggests that there are over 8000 specimens of the early forms—pelycosaurs—in North American museums and over 20,000 specimens of the later forms—therapsids—in North American and African museums. The Geological Survey of South Africa alone collected over 2000 specimens in about 5 years as an aid to geologic mapping for Uranium deposits.

These fossils show a complete and continuous transition between the most primitive reptiles known and the first mammals. The sequence is normally omitted from most elementary texts in biology and geology simply because most scientists don't need to know it, and perhaps because dinosaurs are more glamorous. But in the present religio-political controversy, documentation against the outrageous statements of the creationists must be available.

WHAT IS A MAMMAL?

There are no living intermediates between reptiles and mammals. Even the egg-laying mammals of Australia, the platypus and echidna, are definitely mammals. Linnaeus defined mammals as warm-blooded, fur-bearing vertebrates with mammary glands; most give birth to live young and provide parental care. Reptiles on the other hand are cold-blooded, scaled or naked skinned and rarely provide parental care. Unfortunately, these features are rarely preserved in the fossil record

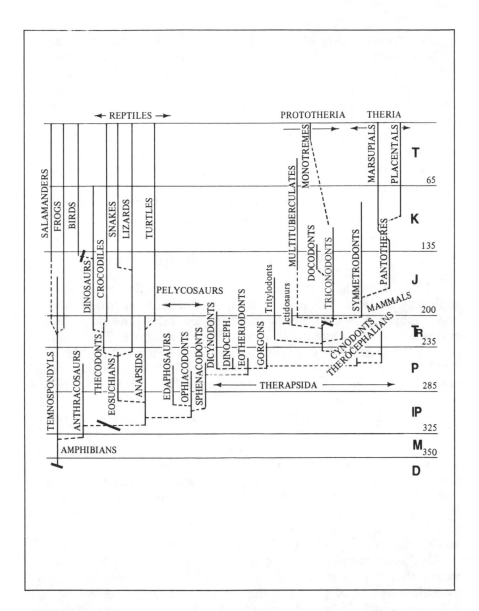

FIGURE 1. Phylogenetic chart of the recent tetrapods and their ancestors designed to show their latest common ancestors. Solid vertical lines represent the known time extent of a group or taxon, usually a suborder, order or subclass. Dashed line represents the transition from one taxon to another. Heavy diagonal solid bars represent the conventional class boundaries where one class originated from another.

which consists mostly of bones and teeth. On occasion they can be deduced from the fossil record. Some skeletal features that mammals have and most reptiles do not include: 2 occipital condyles, 3 ear bones and a single lower jaw bone, dentary-squamosal jaw joint, a secondary palate, terminal growth—growth stops at maturity—2-rooted cheek teeth, limited tooth replacement, and a division of vertebrae, into thoracic and lumbar associated with the diaphragm. These and other characteristic features were not acquired at the same time in the fossil record. Instead each successive group of mammal-like-reptiles is more like a mammal than the last. The big difference between the two classes is completely blurred in the fossil record; to a great degree the definition of a mammal becomes scientific nit-picking!

HISTORY OF STUDY OF MAMMAL-LIKE-REPTILES

In the 120-plus years since the first mammal-like-reptile was described by Owen in 1859, over 1000 papers and books have been written on the subject. Few are obsolete. In this review I shall cite mostly major review articles that summarize the previous literature. The origin of reptiles from amphibia and pelycosaurs from the earliest reptiles have best been summarized by Carroll (1964, 1969, 1970), Clark and Carroll (1973), Carroll and Baird (1972) and Reisz (1972).[4] The pelycosaur radiation is best covered by Romer and Price (1940).[5] Olson (1962) and Boonstra (1963, 1969) summarize the early therapsid radiation and the Middle Permian of the U.S., South Africa and Russia.[6] Broom (1932) is a classic example of the state of therapsid knowledge 50 years ago and should have been enough to convince the creationists;[7] it clearly wasn't. Keyser and Smith (1977–78) summarize the Late Permian and Early Triassic therapsid faunas of the Karoo basin of South Africa.[8] Bonaparte (1970) summarizes the South American Triassic faunas.[9] Romer (1970), Hopson (1969), Hopson and Crompton (1969), Hopson and Kitching (1972), Allin (1975), and Bonaparte and Costa Barberena (1975) discuss various aspects of the transition between the most advanced therapsids and the earliest mammals.[10] "Mesozoic Mammals" by Lillegraven *et al* (1979) is the single most useful review of that subject and includes a useful chapter on the origin of mammals.[11]

WORLD GEOGRAPHY AND CLIMATE DURING THE ORIGIN OF MAMMALS

Few of the lay person's general ideas about world geography and climate during the late Paleozoic and early Mesozoic eras are correct.

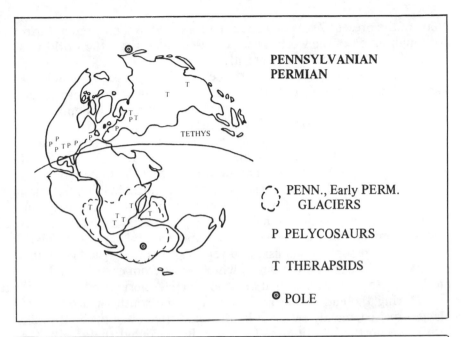

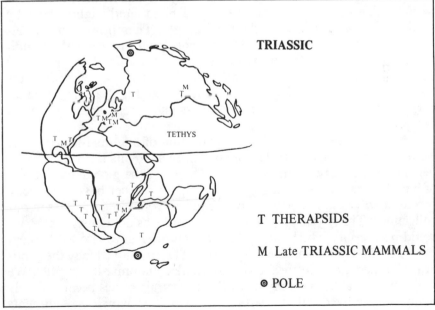

FIGURE 2. Maps of the world continent Pangaea during the Pennsylvanian, Permian and Triassic periods showing the position of the poles and the equator and the areas where fossil mammal-like-reptiles and mammals are known.

None of the present 7 seas or 7 continents existed in their present form. Although much of the world's coal was deposited then, the world was not uniformly tropical. Far from it!

There was a single major world ocean, Panthalassa, of which the Pacific Ocean is only about half. There was a single southern hemisphere continent, Gondwana, which drifted about near the South Pole. During this time, the combined continents of North America and Europe collided with the African part of Gondwana to form the Atlas-Appalachian mountains and slightly later Siberia and China collided with Europe to produce the Ural mountains. While North America and Europe were located on the equator and were tropical in climate, much of Gondwana was covered by an ice cap during the first half of the history of the origin of mammals. A large equatorial sea, Tethys, extended from Panthalassa between Asia and Gondwana. Amphibians arose from crossopterygian fish, and reptiles (including the first mammal-like-reptiles, the Pelycosauria) which arose from early amphibians on the equatorial continent that became North America and Europe.[12]

During the end of the Early Permian the southern hemisphere glaciers melted, mainly because Gondwana drifted off of the South Pole into more northerly latitudes. This greatly expanded the area of the world which could be colonized by land plants and slightly later by reptiles. The later mammal-like-reptiles (the Therapsida), mammals and dinosaurs all evolved on the global continent Pangaea. Many individual kinds of Middle Permian to Early Jurassic land animals were distributed over several of the modern continents. At any one time there was a more or less uniform land fauna. After the Triassic, Pangaea began to break up into smaller continents and spread apart, a process not yet over.

For any given time during the evolution of mammal-like-reptiles, we have at our disposal rocks containing their fossils from less than 2 percent of the world's land surface. Because of the global distribution of land faunas, this is not a great handicap. We either have the precise species involved in the origin of mammals or very close relatives in our collections. Thirty years ago our fossil record was not dense, that is, we could not specify which family of one suborder gave rise to the next suborder. Today we can always specify the family and usually the genus of the immediate ancestor for any particular mammal-like-reptile. We may never be able to specify all of the ancestral species because surely many species lived only in parts of the world in which no sediments were deposited or from which all the sediments were eroded. This does not make the origin of therapsids from pelycosaurs, or the origin of mammals from therapsids, less sure.

EXTINCTIONS, REPLACEMENTS, AND PARALLEL EVOLUTION DURING THE ORIGIN OF MAMMALS

At any one time during the evolution of mammal-like-reptiles, the precise ancestry of mammals rests only in one species. Which species, of course, was continually changing, but almost always it was a small sized carnivorous or insectivorous member of its group (never an herbivore!), with skull length from 30 to 200 mm, usually about 100 mm in skull length. Periodically most of the mammal-like-reptiles would become extinct, sometimes for climatic reasons, sometimes by competition with more advanced mammal-like-reptiles, and at the end by competition with rapidly evolving dinosaurs. When this occurred, one of the surviving species would undergo an adaptive radiation and give rise to many species of a new more advanced family or suborder, varied in size, diet, and habits which would replace their ecological predecessors. Since each such radiation started from a more advanced and mammal-like heredity than the last, and the advances of mammals over reptiles are beneficial in most habitats, the later members of each order commonly evolved exactly similar mammal-like features independently. This is parallel evolution by definition. As recently as 15 years ago it was still possible to suggest that several groups of advanced mammal-like-reptiles independently gave rise to several different groups of primitive mammals. The data now available suggest this is not the case, that the first family of mammals, the Morganucodontidae, arose once from a single species of the cynodont family Chiniquodontidae.[13]

In succession then the first species of the Ophiacodontia, Sphenacodontia, Eotheriodonta, Gorgonopsia, Therocephalia, Cynodontia, Chiniquodontidae and Morganucodontidae each represent the ancestry of mammals. Thankfully we do not have to have the first species of each of these groups to understand what happened; a slightly later species will do as well.

TIME AND ROCK STRATA

A short perusal of any standard historical geology text will show the reader the basis for the geologic time scale. The named systems of rock and periods and epochs of time are easily recognized on the basis of common fossils. The age in millions of years of most fossils and rock units are based on interpolation from rocks whose approximate stratigraphic age is known and which contain one or more radioactive decay series that can be dated. The named periods and epochs are known

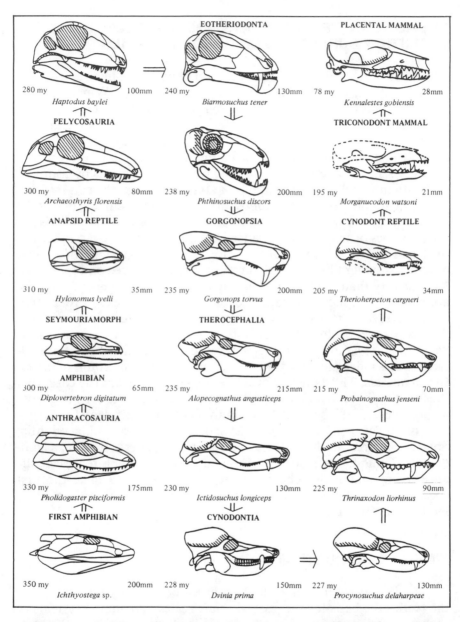

FIGURE 3. Drawings of skulls of animals in or very close to the direct line of ancestry from the earliest amphibian to the earliest reptile through the mammal-like reptiles to mammals. Follow the arrows up the left column, down the center column and up the right column. Approximate age in millions of years and skull length in millimeters are listed above the name of the animal; class, order or suborder are in bold capital letters over the skull drawing.

APPEARANCE OF MAMMALIAN CHARACTERS	ANAPSIDA	PELYCOSAURIA	EOTHERIODONTA	GORGONOPSIA	THEROCEPHALIA	CYNODONTIA	ADVANCED CYNO.	PROTOTHERIA	PLACENTALS
AMNIOTE EGG	X	X	X	X	X	X	X	X	X
TEMPORAL OPENING		X	X	X	X	X	X	X	X
CANINE TOOTH		X	X	X	X	X	X	X	X
CORONOID PROCESS				X	X	X	X	X	X
ERECT GAIT					X	X	X	X	X
2-3-3-3-3 TOE BONES						X	X	X	X
SECONDARY PALATE						X	X	X	X
DOUBLE CONDYLE						X	X	X	X
ENLARGED DENTARY						X	X	X	X
COMPLEX JAW MUSCLES						X	X	X	X
DIFFERENTIATED TEETH						X	X	X	X
TYMPANIC BONE						X	X	X	X
QUADRATE/ARTICULAR HEARING						X	X	X	X
LOSS OF THIRD EYE						X	X	X	X
DIAPHRAGM						X	X	X	X
LONG MATERNAL CARE (MILK?)						X	X	X	X
HAIR						X	X	X	X
FORWARD ILIUM						X	X	X	X
SINGLE NARES							X	X	X
DENTARY/SQUAMOSAL JAW JOINT							X	X	X
LOSS OF POST ORBITAL BAR							X	X	X
FUSED NECK RIBS								X	X
2 ROOTED TEETH								X	X
EPIPHYSES/TERMINAL GROWTH								X	X
BIG BRAIN									X
LIVE BIRTH									X
SINGLE LOWER JAW BONE									X

FIGURE 4. Table to show in which of the suborders of mammal-like-reptiles or primitive mammals a particular mammalian characteristic first appeared.

exactly for most fossiliferous rocks. The number of million years ago is subject to errors of 2 to 3 percent, due to interpolation errors or to small errors of analysis. As new data accumulate, decay constants are subject to periodic revision, slightly changing the reported ages for a given rock. All ages given here are consistent and are based on the Phanerozoic Time-Scale.[14] Expect them to be revised in the future, but do not expect a change in the order of the fossils.

ECOLOGICAL HISTORY OF THE MAMMAL-LIKE-REPTILES AND THE ORIGIN OF MAMMALS

Amphibians arose from crossopterygian fish by modification of fins to legs and loss of gills in adults during the Late Devonian period near the equator in North America and Europe.[15] By the Early Mississippian, amphibians had radiated into 3 main groups, one of which, the Temnospondyli, has frogs and salamanders as living descendants. Another, the Lepospondyli, has the legless caecilians as living descendants, while the third, Anthracosauria, has the reptiles, birds and mammals as its only living descendants. The more terrestrial anthracosaurs gave rise to the first known anapsid reptiles by the Early Pennsylvanian. Very shortly after reptiles appeared, the first species of the Synapsida, an ophiacodont pelycosaur named *Protoclepsydrops,* arose from the Anapsida. All of this took place in less than 10 percent of the world's land area, within 20 degrees of the equator in North America and Europe. The total time involved from the origin of amphibians to the first pelycosaur was about 40 million years. While this was happening, during the Late Mississippian to Early Permian, another third of the land surface was covered by a continental ice cap. Pelycosaurs were the largest and dominant land animals of their time. Most were carnivores; only a few species were herbivores. Carnivore/herbivore ratios in the faunas were 1 to 1, suggesting they were all cold blooded.

At the end of the Early Permian, the southern hemisphere glaciers melted. As the climate warmed, a small carnivorous sphenacodont pelycosaur, *Haptodus,* which had lived in the cooler uplands of the tropics, gave rise to the Eotheriodonta, the earliest suborder of the later order of synapsids, the Therapsida. Over the next 1 or 2 million years most of the pelycosaurs became extinct, probably by competition with the warmer blooded eotherodonts. The smallest carnivorous pelycosaurs, the varanopsids, survived through the end of the Middle Permian, while the herbivorous caseid pelycosaurs lasted until large, efficient, more and more warm blooded herbivorous Dinocephalia evolved from the eotheriodonts. The pelycosaur radiation lasted some

70 million years, with a few ecologically unusual stragglers surviving another 5 million years. The carnivore/herbivore ratio in eotheriodonts and dinocephalians was 1 to 8, suggesting they were more warm blooded than modern reptiles, but much less so than modern mammals.

In addition to giving rise to the herbivorous dinocephalians, the eotheriodonts also gave rise in succession to two groups of initially small more advanced suborders, the herbivorous Dicynodontia and the carnivorous Gorgonopsia. In turn the gorgons gave rise very rapidly to another suborder, the carnivorous Therocephalia, all during the early Middle Permian. As the climate in the southern subcontinent Gondwana warmed up, all of these animals were able to invade the developing temperate zone communities in present-day South Africa. Here they and their descendants for the next 20 million years are preserved in a 3-mile thick pile of river flood plain and stream channel sediments called the Karoo series (often spelled Karroo). These sediments came from the rapidly rising Heritage mountains in then adjacent Antarctica.

As a result of some unknown ecological change, the last of the large carnivorous eotheriodonts, the large herbivorous dinocephalians and the last 2 pelycosaurs became extinct at the end of the Middle Permian. The small gorgons, therocephalians and dicynodonts then rapidly diversified in size and adaptation to fill the ecological vacuum left by the extinction. The Late Permian faunas of South Africa are very rich in species and ecologically complex. Depending on the size of the herbivores, the carnivore/herbivore ratio varies from 1 to 20 to 1 to 32, implying a continuing increase in average body temperature to near mammalian conditions.

A few million years before the end of the Permian—probably in the tropics since the earliest member occurs in Russia—the last therapsid suborder, the Cynodontia, arose from small advanced therocephalians. This, the most mammal-like of all therapsid suborders, rapidly spread over the whole world. Cynodonts took care of their young until they were about 60 percent of the adult size. Several associated family groups have been found preserved together.

At about the end of the Permian, most of the gorgons, most of the therocephalians and most of the dicynodonts became extinct. This extinction is associated with a climatic and vegetational change. The *Glossopteris* flora of small trees, shrubs and bushes is replaced by one dominated by *Dicroidium* trees. It is also approximately contemporary with a global major marine extinction perhaps due to a drastic lowering of sea level. This would have greatly increased aridity.

The major survivors of this extinction were small cynodonts and large dicynodonts. The small carnivorous cynodonts gave rise in the next few million years to large carnivores as well as to the large herbivorous gomphodont cynodonts. Meanwhile a group of diapsid reptiles that had been very minor members of the world fauna during the Permian became large predators. These Eosuchian thecodonts diversified into the ancestors of crocodiles and dinosaurs and replaced the large cynodont carnivores at the end of the Early Triassic. Giant dicynodonts continued to be the large herbivores throughout the Triassic until dinosaur herbivores appeared at the very end of that period. By the beginning of the Late Triassic, the only cynodonts left were small insectivores or small to middle-sized herbivorous derivatives of the gomphodonts, both very close to being mammals. Dinosaurs were extremely varied in size and diet, ranging from 3 to 10 meters in length. A small cynodont such as *Therioherpeton* gave rise to both the "ictidosaur" *Pachygenulus* and to the first family of mammals, the Morganucodontidae, before the end of the period. The cynodonts had held sway over the earth for some 30 million years before being replaced by dinosaurs. Dinosaur communities were then typical of the world fauna until the end of the Cretaceous, some 130 million years later. Throughout this time mammals were all small, most with skulls of 25 to 35 mm length. A very few giants are known with skulls 75 mm in length. Only when dinosaurs became extinct about 65 million years ago did mammals recover the diversity in size and variety that their ancestors, pelycosaurs and therapsids, had in the Permian and Triassic.

STANDARD CREATIONIST ARGUMENTS AGAINST THE TRANSFORMATION OF REPTILES TO MAMMALS

Gish argues that reptiles did not give rise to mammals on the following grounds: "There are no transitional forms showing, for instance, three or two jaw bones, or two ear bones. No none has explained yet, for that matter, how the transitional form [as if there were only one] would have managed to chew while his jaw was being unhinged and rearticulated, or how he would hear while dragging two of his jaw bones up into his ear."[16]

About half of the species of Jurassic mammals have 2 bones in the lower jaw, the dentary and the coronoid, and a few even have the splenial! The first certain placental mammal, *Kennalestes,* from the Late Cretaceous, is the latest mammal to have the coronoid bone as well as the dentary in the lower jaw. None has 2 ear bones because 2 won't work. What actually happened was that the rear-most jaw bones functioned both as ear bones and jaw joint in late therapsids and early mammals. These small bones never moved much. As the animal chewed, the pressure of the muscles on the jaws locked them tightly to the skull and lower jaw and hearing was temporarily suspended. (We still don't hear very well when chewing noisy food like celery). When the animal was not chewing, those small bones were free to move on a different axis of rotation and amplified sound. Hearing and chewing were complementary activities; the animal did one or the other at any one time, never both. Gish states: "No explanation is given how the intermediates managed to hear while this was going on."[17] Yet he cites references that discuss precisely that! The best reference with most detail on the subject is Allin (1975), but the 1973 and 1979 references by Crompton and Jenkins that he does cite discuss this point adequately.

Gish continues: "As insuperable as this problem appears to be, it pales into relative insignificance when we consider the fact that the essential organ of hearing in the mammal is the organ of Corti, an organ not possessed by a single reptile, nor is there any evidence that would provide even a hint of where this organ came from. The organ of Corti is an extremely complicated organ. It is suggested that the reader consult one of the standard texts on anatomy for a description. One cannot but marvel at this complex and wondrously designed organ. It has no homologue in the reptile from which it could be derived. It would have to be created de novo, since it was entirely new and novel."[18] Flowery words, so I checked. In Romer (1956), the first place I looked, I found that the basilar papilla of lizards and crocodiles is exactly homologous with the organ of Corti, a little smaller perhaps, but then mammals hear better.[19]

Gish continues: "The structure of the thoracic girdle of the mammal [as if they were all the same] differs fundamentally from that of the reptile. In the reptile it articulates with the breastbone by means of the coracoid and forms part of the thorax [?]. This is not the case with mammals."[20] Opossums and the egg-laying mammals all have coracoids. Cynodont shoulder girdles are almost identical to those of the monotremes. He continues: "There is no structure in a reptile that is in any way similar to the mammalian diaphragm."[21] But all of the cynodont reptiles for which we know the skeleton had a diaphragm! Earlier in that paper Gish refused to admit that *Morganucodon* and *Kuehneodon* were mammals on the grounds that they had the extra jaw bones of reptiles. By inference he now claims that they were mammals since their close relative *Megazostrodon* had a diaphragm! He can't have it both ways!

SUMMARY

The mammal-like-reptiles are known from tens of thousands of specimens of hundreds of genera of rat-to-cow-sized animals found over the last 120 years in all continents of the world. They were the ecological dominants of land communities for some 120 million years until supplanted by dinosaurs. Each successive suborder of carnivorous mammal-like-reptile has more mammalian characters than the preceding suborder. Later members of any suborder are more mammalian than early members of the same suborder. The earliest pelycosaurs differ from the most primitive and contemporary anapsid reptiles only in having a temporal opening, a feature which permitted them to snap their jaws shut faster. The latest cynodonts differ from the earliest mammals only in lacking two-rooted cheek teeth, and in having more than one set of replacement teeth, but do have teeth similar in shape to those of the first mammals. The largest remaining morphological gaps in the fossil series amount to less than one percent of the morphological distance from the most primitive reptiles to the most primitive mammals. The gaps will be filled with future collecting, as a comparison between what was known to Broom in 1932 and what is known in the more recent references will show.

REFERENCES

1. Duane T. Gish, *Evolution: The Fossils Say No!* (San Diego, CA: Creation-Life Publishers, 1978).

2. D. T. Gish and R. B. Bliss, *Summary of the Scientific Evidence for Creation,* 1981. Included in this volume.

3. Gish and Bliss, 1981.

4. R. L. Carroll, "The Earliest Reptiles," *Journal of the Linnean Society. Zoology,* 45 (1964): 61–83; R. L. Carroll, "Problems of the Origin of Reptiles," *Biology Review* 44 (1969): 393–432; R. L. Carroll, "The Ancestry of Reptiles," *Philosophical Transactions, Royal Society of London* (B) 257 (1970): 267–308; J. Clark and R. L. Carroll, "Romeriid Reptiles from the Lower Permian," *Bulletin of the Museum of Comparative Zoology* 144 (1973): 354–407; R. L. Carroll and D. Baird, "Carboniferous Stem-Reptiles of the Family Romeriidae," *Bulletin of the Museum of Comparative Zoology* 143 (1972): 349–63; R. Reisz, "Pelycosaurian Reptiles from the Middle Pennsylvanian of North America," *Bulletin of the Museum of Comparative Zoology* 144 (1972): 27–61.

5. A. S. Romer and L. W. Price, *Review of the Pelycosauria, Geological Society of America Special Paper* 28 (1940): 538 pp.

6. E. C. Olson, "Late Permian Terrestrial Vertebrates: U.S.A. and U.S.S.R.," *Transactions, American Philosophical Society* 52, pt. 2 (1962): 1–224; L. D. Boonstra, "Early Dichotomies in the Therapsids," *South African Journal of Science* 59 (1963): 176–95; L. D. Boonstra, "The Fauna of the *Tapinocephalus* Zone (Beaufort Beds of the Karoo)," *Annals, South African Museum* 56 (1969): 1–73.

7. R. Broom, *The Mammal-Like Reptiles of South Africa and the Origin of Mammals* (London: H. F. and G. Witherby, 1932).

8. A. W. Keyser and R. M. H. Smith, "Vertebrate Biozonation of the Beaufort Group with Special Reference to the Western Karoo Basin," *Annals, South African Geological Survey* 12 (1977–78): 1–35.

9. J. F. Bonaparte, "Annotated List of the South American Triassic Tetrapods," *Proceedings, Gondwana Symposium,* Capetown and Johannesburg, 1970: 665–82.

10. A. S. Romer, "The Chanares (Argentina) Triassic Reptile Fauna. VI.A Chiniquodontid Cynodont with an Incipient Squamosal-dentary Jaw Articulation," *Breviora,* no. 344 (Cambridge, MA: Harvard University) 1970, pp. 1–18; J. A. Hopson, "The Origin and Adaptive Radiation of Mammal-Like Reptiles and Non-Therian Mammals," *Annals, New York Academy of Science* 167 (1969): 199–216; J. A. Hopson and A. W. Crompton, "Origin of Mammals," in *Evolutionary Biology,* ed by T. Dobzhansky, M. K. Hecht, and W. C. Steere (New York: Appleton-Century-Croft, 1969), vol. 3, pp. 15–72; J. A. Hopson and J. W. Kitching, "A Revised Classification of Cynodontia (Reptila, Therapsida)," *Palaeont. Afr.* 14 (1972): 71–85; E. F. Allin, "Evolution of the Mammalian Middle Ear," *Journal of Morphology* 147 (1975): 403–38; J. F. Bonaparte and M. Costa Barberena, "A Possible Mammalian Ancestor from the Middle Triassic of Brazil (Therapsida-Cynodontia)," *Journal of Paleontology* 49 (1975): 931–36.

11. J. A. Lillegraven, Z. Keilan-Jaworowska, and W. A. Clemens, eds., *Mesozoic Mammals* (Berkeley: University of California Press, 1979).

12. See A. S. Romer, *Vertebrate Paleontology,* 3d ed. (Chicago: University of Chicago Press, 1966), and C. Scotese et. al., "Paleozoic Base Maps," *Journal of Geology* 87 (1979): 217–78.

13. See Romer, 1970, and Hopson and Kitching, as cited above.

14. See W. B. Harland, A. G. Smith, and B. Wilcock, *The Phanerozoic Time-Scale* (London: Geological Society, 1964).

15. See Romer, 1966.

16. Gish, 1978.

17. D. T. Gish, *The Mammal-Like Reptiles*, ICR Impact Series no. 102, December 1981.

18. Gish, 1981.

19. A. S. Romer, *Osteology of the Reptiles*. Chicago: University of Chicago Press, 1956; see especially pp. 33–38, fig. 37.

20. Gish, 1981.

21. Gish, 1981.

ADDENDUM

A major new reference published after this paper was written is: Kemp, Thomas S. *Mammal-like Reptiles and the Origin of Mammals*. New York: Academic Press, 1982. 362 pp.

Defining "Kinds"—
Do Creationists Apply a
Double Standard?

by Frank T. Awbrey

Accepting that a body of evidence infers common ancestry for flies, horses, cats, and dogs, but claiming that exactly analogous evidence infers nothing about human-ape ancestry is not sound scientific reasoning. It is blind prejudice. Creationists clearly reject evolution in this case not because there is scientific evidence against it but because it conflicts with a cherished belief.

Creationists long ago gave up on their original idea of fixity of species. One reason is because simple calculation can show that Noah's Ark could not possibly have held pairs from each of some two to five million species (there would be less than one-half cubic foot per pair), nor could Noah and his family have possibly taken care of them all. A second reason is that the evidence for adaptive change and species formation is overwhelming. Therefore, they had to develop another concept.

ORIGINAL CREATED KINDS

The current creation science stand on this matter is very nicely summed up by Dennis Wagner, editor of *Students for Origins Research,* in his answer to a letter by Dr. C. A. Zimmerman of Aurora College. Zimmerman asked whether or not creationists are "opposed to any and all evolution for any and all cases." Wagner defined three levels of evolution and stated that creationists object only to the third level— macroevolution—which leads to the formation of higher taxonomic categories such as genus, family, order, and so forth. He then said:

> The main thrust of the creationist belief is that there is a set of originally created kinds that were designed with a vast but limited

potentiality for variation. This variation was gradually released through the degenerative process of inbreeding in which the offspring species never again reach the hereditary variability of the parent.[1]

He proposed that the term *genus* should apply to the original created kind and finished with:

Is the creationist opposed to macroevolution then? Well according to the definition of macroevolution, sometimes yes, sometimes no. If by macroevolution it is meant descent with modification of the coyote, wolf, dingo, pampa fox, asiatic jackal, and domestic dog from the originally created Canid kind, then the answer is no, the creationist is not opposed to macroevolution. If, however, the definition of macroevolution means the descent with modification of the bird kind from the reptilian kind, then the answer is yes, the creationist is opposed to macroevolution.[2]

Creationists do not have an exact definition of the original created kind for the same reasons that taxonomists cannot precisely define species: every imaginable gradation between species exists. Gish, a leading creation scientist, says that a basic animal or plant kind would include all species that have truly shared a common gene pool. Furthermore, although no new kinds have arisen since the original creation, "the concept of special creation does not exclude the origin of varieties and species from an original created kind."[3] In an article that first appeared in *Creation Research Society Quarterly* in 1971, F. L. Marsh says:

The descriptions of kinds in Genesis I give us ground for hypothesizing that the individuals of any particular Genesis kind would have chemistries sufficiently alike to make them fertile *inter se,* but sufficiently different to make them incompatible with individuals of every other kind. If this hypothesis is valid, then ability to cross would demonstrate membership in the same basic type.[4]

He then lists examples of plant and animal kinds, based upon "true fertilization," that is, whether or not "both reduced parental sets of chromosomes join and participate in the first division of the fertilized egg."[5] In cases such as horses and asses, or dogs, coyotes, and wolves, the genus is the kind (or the *baramin,* as creationists call it). If members of different families within an order can be crossed, that order is the created kind, and so on. He admits that in some cases mutation and chromosomal rearrangements may have occurred that prevent interbreeding, but membership in a kind can be determined from external appearance as in fruit flies.

As biologists have recognized for over half a century, this is an objective and testable approach to the question of origins. An enormous amount of experimental hybridization has been done and is being done to help taxonomists assess relatedness in classifying many plant and animal species. Evolutionists interpret the results as strongly supporting the theory of descent with modification. Although creationists also appear to agree, within the limits stated by Wagner (1980) and Marsh (1971),[6] they reject or ignore a large body of straightforward evidence that relates directly to the question of common descent and hybridization. Evidence is accepted by creationists only when it does not conflict with their beliefs.

DETERMINING GENETIC RELATIONSHIP

An overwhelming body of experimental evidence clearly shows that the ability to form hybrids is strongly tied to the amount of genetic relatedness between species. Genetic relatedness can be directly measured by a number of techniques.[7] Members of a local population exchange genes freely and are genetically very similar. On a scale of zero to one, their "genetic similarity" (one measure of the proportion of shared genes coding for essentially identical proteins) is 0.90 or higher. Some barriers to hybridization may exist between subspecies whose genetic similarity is about 0.8. Full species have genetic similarities ranging from 0.8 to about 0. 3 and usually have substantial, if not complete, barriers to hybridization, and so on.[8]

Numerous studies show that chromosomal similarity is also a good measure of genetic relatedness. The fine structure of the gene-bearing chromosomes is an extremely complex pattern of bands and lines. The probability that two different chromosomes would independently arrive at identical banding patterns is essentially zero. Wallace details the evidence that identical chromosomes in two different species prove common ancestry just as surely as identical scratch patterns on two bullets prove both came from the same gun.[9] Even so, related species do not always have highly similar chromosome numbers or banding patterns because extensive chromosome rearrangements sometimes occur during speciation.[10] Chromosome differences, when great enough, can cause the hybrids to be sterile, as in mules,[11] or to die as embryos, as in the cross between the domestic dog and the red fox.[12] If the differences are too great, the sperm will not even penetrate the egg. Conversely, the more similar the chromosomes of a sperm and an egg, the more likely their combination will result in Marsh's "true fertilization."

Evolutionists and creationists alike realize that all living species of a "kind" have inherited their genes from the same original ancestral species. Therefore, their body forms, chemistry, physiology, chromosomes, and genes are very similar. Conversely, the more similar any pair of species is for any and all of these features, the more likely it is they are descended from a close common ancestor and the more likely it is they can hybridize. If two species appear similar and are known to share most of their genes, creationists are usually perfectly willing to accept them as one "kind." If they are all interfertile, they are certainly one "kind." For example, asses, horses, zebras, and onagers all look rather similar. Their habits, behavior, diets, digestive systems, the proteins in their bodies, and the genes that manufacture those proteins are also similar. Moreover, the species in these four groups differ almost equally from other animals, such as rhinos. They form an obvious "natural group": the family *Equidae.* The final proof of their close relationship is that all are more or less interfertile, in spite of some differences in chromosome numbers.[13]

The general rule is that the higher the genetic similarity and/or the more similar the chromosomes of two species are in number and structure, the higher the probability that they can hybridize.[14] Thus, even in cases where hybridization experiments have not yet been conducted, the likelihood of successful hybridization can be objectively predicted.

These objective criteria, for example, can be used to determine the relatedness of two ape species—the gibbon and the siamang—and, thus, how likely they are to be one "kind." They are easily distinguished, and taxonomists have placed them in different genera of the same family. These two species live together in Southeast Asia, but are not known to interact. The gibbon has twenty-two pairs of chromosomes, the siamang has twenty-five pairs. Their chromosome banding patterns have been so extensively rearranged that only one chromosome still bears a recognizably similar banding pattern in both species.[15] Their genetic identity, another measure of the proportion of their genes coding for essentially identical protein, is 0.76.[16] No natural hybrids have ever been reported. Their separation into different genera seems morphologically and behaviorally justified. However, they are genetically as closely related as most rodent species belonging to one genus.[17] They appear, chromosomely, to be at least as closely related as horses (thirty-two chromosome pairs) and onagers (twenty-seven to twenty-eight pairs)[18] or domestic dogs (thirty-nine chromosome pairs) and the red fox (eighteen to twenty pairs).[19] Scientist view these observations as strong evidence for close evolutionary relationship. Since 1975, two hybrids have been born in the Atlanta, Georgia, Zoo.[20] Gibbons and siamangs unquestionably are highly modified descendants of a recent common ancestor, according to both evolutionist and creationist criteria.

Let us now apply these principles to another pair of mammal species. Their genetic identity is 0.70, about equal to the gibbon and siamang. Unlike the gibbon and siamang, their chromosomes are virtually identical even though one species has one pair of chromosomes more than the other. It is remarkable that 99 percent of the chromosome banding sequences of one species are clearly discernible in the chromosomes of other species. The banding sequences are mostly in the same locations in the two chromosome sets, but in one species nine short segments are inverted, eighteen chromosomes have other minor changes, and one long chromosome has split to form two short ones, accounting for the different number of chromosomes. No hybrids have been found in nature, and no one has reported producing them in the laboratory. Nevertheless, if the proven criteria for genetic relatedness are objectively applied, these two species are merely one more example of close common descent or "variation without a created kind."

Few—whether evolutionist or creationist—would object to this interpretation if the species in question were fruit flies, horses, dogs, or even monkeys; but the two species involved are actually humans and chimpanzees.[21] Evolutionists are not surprised, because these observations simply agree with previous fossil, anatomical, and embryological evidence. The picture suddenly changes for scientific creationists, however, because they are irrevocably committed to the tenet that humans are unique. All members of the Institute for Creation Research (ICR) must sign a statement of faith affirming their belief that humans were separately and specially created. The statement that humans did not evolve from an animal ancestor is ICR tenet number 4 and is incorporated in the ICR bylaws.[22] This belief forces them to deny that all this evidence, which would be compelling proof of close relationship for almost any other species pair, has any relevance at all. A typical creationist reply might be that, at most, they show a common designer.

A DOUBLE STANDARD

The important question here is whether or not changing the meaning of evidence in cases where it conflicts with a belief is scientifically legitimate or intellectually honest. Accepting that a body of evidence infers common ancestry for flies, horses, cats, or dogs, but claiming that exactly analogous evidence infers nothing about human-ape ancestry is not sound scientific reasoning. It is blind prejudice. Creationists clearly reject evolution in this case not because there is scientific evidence against it but because it conflicts with a cherished belief.

When creationists finally acquiesced to the voluminous evidence that species had not remained absolutely fixed and unchanged since creation, they lost their war against the concept of evolution. Because the direct evidence for species divergence (that is, speciation) is so abundant and straightforward, "scientific" creationists had to accept it or appear as irrational as those who used the Bible to argue that the earth is flat. Creationists now argue that new species may arise within kinds, but that no species may change into a new kind. The question is: what limits divergence? Sheep and goats can hybridize. If one "kind" can diverge this far, why not as far as sheep and cattle or sheep and camels? If that far, why no further? Gene products and the genes themselves show no boundaries between kinds. All available evidence suggests that, as long as they reproduce, as long as their genes mutate, and as long as they are subjected to selection, species will continue to diverge, essentially without limit.

I suggest that creationists made a tactical error when they began to pay attention, albeit selectively, to scientific evidence. This is because that evidence actually contradicts their most sacred belief: the belief that humans are uniquely created. Even if "true fertilization" between human and ape were reported, would scientific creationists reject special creation and accept the evolutionary relationship of humans and apes? If they were rational practitioners of science they would.

REFERENCES

1. Dennis Wagner, *Students for Origins Research* (Fall 1980): 3.

2. Wagner, 1980.

3. Duane T. Gish, *Evolution: The Fossils Say No!* Public School ed. (San Diego, CA: Creation-Life Publishers, 1978), pp. 32, 40.

4. F. L. Marsh, "The Genesis Kinds in the Modern World," in *Scientific Studies in Special Creation,* ed. by W. E. Lammerts (Grand Rapids, MI: Baker Book House, 1971), pp. 136–55.

5. Marsh, 1971.

6. Wagner, p. 3; Marsh, 1971.

7. T. Dobzhansky et al., *Evolution* (San Francisco: W. H. Freeman, 1977), p. 572.

8. F. J. Ayala, "Genetic Differentiation During Speciation," in *Evolutionary Biology,* ed. by T. Dobzhansky, M. K. Hecht, and W. C. Steere (New York: Plenum Press), vol. 8, pp. 1–78.

9. B. Wallace, *Chromosomes, Giant Molecules, and Evolution* (New York: W. W. Norton, 1966), p. 166.

10. M. J. D. White, *Modes of Speciation* (San Francisco: W. H. Freeman, 1978), p. 455.

11.. M. Kaminsky, "The Biochemical Evolution of the Horse," *Comparative Biochemistry and Physiology,* 63b (1979): 175–78.

12. A. B. Chiarelli, "The Chromosomes of the Canidae," in *The Wild Canids, Their Systematics, Behavioral Ecology, and Evolution,* ed. by M. W. Fox (New York: Van Nostrand Reinhold, 1975), pp. 40–53.

13. O. A. Ryder, N. C. Epel, and K. Benirschke, "Chromosome Banding Studies of the Equidae," *Cytogenet. Cell Genet.* 20 (1978): 323–50.

14. Dobzhansky, 1977; A. P. Gray, *Mammalian Hybrids* (Farnham Royal, England: Commonwealth Agricultural Bureau, 1971).

15. R. H. Myers and D. A. Shafer, "Hybrid Ape Offspring of a Mating of Gibbon and Siamang," *Science* 205 (1979): 308–10.

16. E. J. Bruce and F. J. Ayala, "Phylogentric Relationships between Man and the Apes: Electrophoretic Evidence," *Evolution* 33 (1979): 1040–56.

17. Ayala, 1975.

18. Ryder, et al., 1978.

19. Chiarelli, 1975.

20. Myers and Shafer, 1979.

21. See Bruce and Ayala, 1979; D. A. Miller, "Evolution of Primate Chromosomes," *Science* 198 (1977): 1116–24; M. C. King and A. C. Wilson, "Evolution at Two Levels in Humans and Chimpanzees," *Science* 188 (1975): 107–16; J. Yunis, J. R. Sawyer, and K. Dunham, "The Striking Resemblance of High Resolution G-Banded Chromosomes of Man and Chimpanzee," *Science* 208 (1980): 1145–48.

22. H. M. Morris, "The Tenets of Creationism," *Impact,* no. 85 (1980: Institute for Creation Research).

Creationist Misunderstanding, Misrepresentation, and Misuse of the Second Law of Thermodynamics

by Stanley Freske

Creationists are not showing that evolution contradicts the second law of thermodynamics; instead, they are saying that the second law, as accepted by conventional science, is incorrect and insufficient to explain natural phenomena. They insist that something else of their own making must be added—namely, a divinely created directing program or a distinction between different kinds of entropy.

One of the cornerstones in the crumbling foundation of creationist "science" is the notion that evolution contradicts the second law of thermodynamics. The classical version of this law may be stated as follows: *The entropy of an isolated system can never decrease.* (An isolated system is one that does not exchange energy or matter with its surroundings.) Creationists originally argued that a decrease in entropy is exactly what evolution requires, hence the conflict with the second law. This argument was used in an article by Dr. Morris of the Institute for Creation Research (ICR) as late as 1973.[1] As is the usual practice among creationists, he tried to support it with out-of-context quotations from the writings of respected scientists.

Actually, it is not difficult to find inaccurate statements regarding entropy in popular science literature. Ever since the time it was first defined, entropy has been recognized as a most elusive quantity as far as understanding its physical significance is concerned. Defining it mathematically in terms of other quantities is no problem; however, this cannot be done to advantage in popular debates, a situation that creationists have been quick to capitalize on. Entropy has been defined nonmathematically as a measure of disorder, equilibrium, uncertainty, and unavailability of energy. Actually, to consider only the entropy content of a system is not enough; a system can gain entropy and, at the same time, become more organized, unbalanced, and richer in information and available energy. (A few examples will be considered later on.) What is important is the entropy *deficiency* of the system. We define this as the difference between the system's entropy *capacity* (the maximum amount of entropy the system is capable of holding with its present energy content) and the amount of entropy it is actually holding. This deficiency may also be referred to as *negentropy* (short for negative entropy)—a concept which, had it been generally adopted, might have been less confusing than entropy. Negentropy, then, has been defined as a measure of order, information, lack of equilibrium, and the availability of energy for doing work. But most fundamentally, negentropy—or entropy deficiency—is a measure of the *improbability* of a system being in a given state. For this reason, when we discuss such things as the improbability of a certain nucleotide sequence, for example, we are also discussing entropy and the second law of thermodynamics.

A final warning: the word *order* in popular usage is highly ambiguous and should be scrupulously avoided in explanations of entropy for the benefit of anyone not already familiar with scientific jargon, lest it cause a great deal of confusion. (The mathematically inclined reader can refer to such works as Sears in 1959 and Brillouin in 1962 for more detail.)[2]

OPEN SYSTEMS

The creationist argument given in the first paragraph contains a gaping flaw, and evolutionist debaters wasted no time in pointing it out: While the classical version of the second law does indeed state that the entropy of an isolated system cannot decrease, evolving systems are not isolated! One might expect that at this point the issue would be considered settled and everyone would pack up and go home. However, such an expectation would never be entertained by anyone familiar with the peculiar tenacity of creationists.

Let us see how Morris responds after he has been confronted with the clear evidence that evolving systems are open. In 1976, he said: "The second law really applies only to *open* systems, since there is no such thing as a truly isolated system."[3] This statement suggests that he lacks the ability to distinguish between theoretical and practical concepts—an ability which is absolutely essential for the understanding of much of physics. It is certainly true that the second law applies to all thermodynamical systems; it wouldn't be much of a law otherwise. But the particular statement of the second law that Morris has in mind— namely, that the entropy cannot decrease—applies only to isolated systems. It is a purely theoretical statement, and in theory, any desired system can be postulated whether or not it can exist in practice. Let me mention another example: The concept of an ideal gas is utilized throughout thermodynamics and is extremely useful, even though no such substance actually exists. Just as real gases approximate an ideal gas, some better than others, there are real thermodynamical systems that are very nearly isolated. In these systems we do not expect the entropy to decrease. On the other hand, in a wide open system the entropy can either increase, decrease, or remain constant. The second law does not in any way prevent entropy decreases and the generation of entropy deficiencies in local systems so long as there is an equal or larger increase in entropy outside the system. This concept is easily grasped by most college and even high school students of science but not, apparently, by creationists, including those boasting Ph.D.s in the sciences.

It might now seem that all we have to do is give some examples of open systems in which the entropy decreases and *then* we can pack up and go home. But alas, no such luck. In an attempt to counter this, creationists have introduced a new device, which one creationist, Mr. Elmendorf, calls "The Creative Trinity,"[4] a properly descriptive phrase with an appropriate ring that I will therefore adopt.

THE CREATIVE TRINITY

According to this creationist concept, a system can become entropy deficient only if three conditions are satisfied.[5] (1) Free energy must be supplied to the system. This is actually incorrect, since a *loss* of energy can also generate an entropy deficiency; however, the need for the system to be open is universally recognized, so further discussion is unnecessary. (2) The system must contain an energy conversion mechanism. When creationists are pressed, we find that just about anything qualifies as having a "mechanism," including matter itself, so the statement becomes quite meaningless. (3) The system must contain a directing program. This is variously referred to as intelligence, information, control system, and so forth by creationists. The idea is that this directing program did not arise through natural processes but was created by God. The Creative Trinity can also be interpreted as a statement to the effect that there are different kinds of entropy which are not interchangeable.

We must take careful note of an elementary fact which is often missed in debates on evolution and the second law: In spite of what they claim, creationists are no longer talking about the second law. They wish to give the impression that science, in this case thermodynamics, is on their side in their opposition to evolution. But the fact is there is nothing in thermodynamics that contradicts the phenomenon of an entropy deficiency being produced in a system when energy flows through it. On the contrary, this is what thermodynamics leads us to expect, and nothing else is needed, such as a directing program, etc. It is interesting to note that, in his resolution of the long-standing paradox of Maxwell's demon, Brillouin showed that, to enable the demon to distinguish between fast and slow molecules, energy has to be supplied to the system, thus producing an entropy increase elsewhere in just the amount required by thermodynamics.[6] And it doesn't matter whether the demon is an intelligent being or a simple mechanism.

Creationists are not showing that evolution contradicts the second law of thermodynamics; instead, they are saying that the second law, as accepted by conventional science, is incorrect and insufficient to explain natural phenomena. They insist that something else of their own making must be added—namely, a divinely created directing program or a distinction between different kinds of entropy. Let us now look at several examples to see how creationists attempt to support their claims and to show that their notions are wrong and unnecessary.

CRYSTAL GROWTH

The example of crystal growth is particularly interesting, because it has been misunderstood and misused by evolutionist debaters as well as by creationists. While the growing crystal is certainly an example of an open system in which entropy is decreasing, there is an important thermodynamical difference between it and a living system. In the crystal, the entropy is always at a maximum. In other words, while it is true that the entropy decreases as the liquid changes into a solid, this happens because the entropy *capacity* of the system decreases. The living system, on the other hand, contains an entropy *deficiency,* and this deficiency increases as the system grows or evolves. It should now be obvious that a debater who tries to draw too close a parallel between crystals and living systems will be in trouble.

Nevertheless, creationists have expended a great deal of effort attempting to explain the entropy decrease inherent in crystal growth. Elmendorf claims that there is no decrease in entropy, because liquids are more orderly than crystals.[7] When I pointed out to him in an exchange of letters that gases turn into liquids by a similar removal of heat, he decided that gases are the most orderly of all. I might have asked him why we observe changes of state in nature which proceed in the opposite direction by means of the simple addition of heat, such as snowflakes melting, however, I did not pursue the matter any further.

It is more interesting to examine the claim by both Elmendorf and Morris that crystals grow because of the divinely created directing program built into matter. Elmendorf simply tells us that "the molecules are pre-programmed,"[8] while Morris, with somewhat greater sophistication, explains that crystals are able to form only because of "the electrochemical properties of the molecules in the crystal."[9] This quotation from Morris may sound perfectly reasonable (or should I say conventional?), but only because it is out of context. He subsequently informs us that these properties "could never arise by chance" or "within the constraints imposed by the second law," and finally concludes that they must be the work of "an omniscient programmer."[10]

Two points should be noted here. First, Morris confuses the origin of matter and its properties with the process of evolution. This undoubtedly is done intentionally, since it is a common obscuring tactic among creationist debaters. Second, the divine programs built into matter are claimed to be capable of bringing about such entropy-reducing processes as crystal growth, development of a seed or egg into a mature organism, growth of populations, evolution of complex technologies, and so forth, but *not* capable of bringing about biological or even comparatively simple astronomical evolution. Creationists have nothing but contempt and ridicule for theistic evolutionists, an attitude made possible only by this severe inconsistency in their own belief system.

CONVECTIVE SYSTEMS

In their attempts to prove their version of the second law, creationists often use the example of a pile of bricks lying in the sun. This is supposed to represent an open system that, although it is receiving an abundance of high-grade energy, is not exhibiting any reduction in entropy. Creationists gloatingly draw our attention to the fact that such bricks have never been observed to organize themselves spontaneously into a building. What they apparently fail to understand is that under the given conditions, an entropy deficiency is in fact generated in the pile. After several hours of exposure to the sun, the temperature will be higher at the top than at the bottom. If we were to measure the temperatures throughout the pile, it would be a fairly simple matter to calculate the entropy deficiency. Useful energy could actually be extracted from the pile by means of a thermocouple, for example. Creationists should tell us where in this mundane pile of bricks we find the divine directing program and conversion mechanism, supposedly necessary for an entropy deficiency to be generated in the system.

Incidentally, this pile of bricks, absorbing heat at the top only, is an example of a system that becomes entropy deficient even though the entropy in the pile actually *increases*. This seeming paradox results from the fact that, as heat is added, the entropy capacity of the pile increases faster than the amount of entropy contained in it. If we began again with a uniform temperature throughout the pile and then allowed heat to be removed from the top, as when cooling at night, the entropy would in fact decrease in addition to an entropy deficiency again being generated. We may also note that in this case the cause is a *loss* of energy. When discussing crystal growth, we saw that a loss of energy produced a decrease in entropy, but not a deficiency. Almost any combination is possible and we have to be extremely careful in making general statements concerning entropy.

Other, more impressive convective systems, in which large entropy deficiencies develop spontaneously as a result of the simple influx of solar energy, are meteorological systems such as hurricanes, tornados, and lightning storms. And consider the water cycle: Heat from the sun evaporates water from the ocean; the vapor is carried over the land by winds, which are also generated by solar heat, and is forced up by mountains, where it precipitates; the water eventually forms rivers with waterfalls and finally flows back into the ocean to close the cycle. The waterfalls, of course, constitute a well-known source of available energy. Where, creationists, are the directing programs in these highly organized, entropy deficient systems?

MUTATIONS AND THE GENETIC CODE

The growth of a seed or egg into a mature organism constitutes an observable process involving a large and spontaneous increase in the entropy deficiency of a localized system. Creationists naturally claim that the genetic code making this possible is just the directing program included in their Creative Trinity. It is certainly true that the genetic program determines just what the egg will grow into. But it is *not* true that this program is what enables the system to develop an entropy deficiency. In the course of a year, the earth receives 1.6×10^{21} watt-hours of energy from the sun and reradiates almost the same amount into space. But, because the incoming radiation originates on a high-temperature source (the sun) and the outgoing radiation on a low-temperature one (the earth), the whole process results in an outflow of entropy or inflow of negentropy. This negentropy flux can be calculated to be 3.2×10^{22} joule/°K per year.[11] A significant portion of this negentropy is used in biological processes directed by genetic programs, but a considerably larger portion is used to generate entropy deficient meteorological systems without the benefit of directing programs. Thus, the genetic program only insures that a small portion of the negentropy is used to develop a *particular type* of entropy deficient system. The only legitimate question left is whether the first bit of replicating genetic material could have come about naturally without violating the second law.

We may first note that all the information stored in a fertilized mammalian egg-cell is equivalent to only about 4×10^{12} joule/°K of negentropy. Ordinary everyday processes that we observe all around us spontaneously develop entropy deficiencies that easily amount to billions of times this amount. Thus, it is not the generation of the entropy deficiency that constitutes the problem, although this is what creationists imply when they say that a natural origin of the genetic code would violate the second law.

Experiments of the type first performed by Stanley Miller have shown that the basic building blocks of life—amino acids and nucleotides—are generated spontaneously in a reducing atmosphere, consisting of compounds of carbon, hydrogen, oxygen, and nitrogen, when energy in the form of electrical discharges or high-energy radiation is supplied. We are unable to choose at this time the particular mechanism whereby these units assembled themselves into proteins and DNA (or RNA) respectively; there are several possibilities. A more important question is the probability of the spontaneous formation of such a chain with sufficient autocatalytic properties so that, once formed, it would promote its own duplication. Once this hurdle has been overcome, evolution can be expected to proceed through the combination of mutations and natural selection, as discussed later. For years creationists have been indulging in calculations intended to prove that the formation of the original functional chain is statistically impossible. Let us examine one such attempt by Dr. Gish, also of ICR.[12]

Gish begins by assuming that a functional chain would need to consist of 100 amino acids of the 20 different kinds found in living organisms. He then states that there are 10^{130} different varieties of such a sequence, which is correct. He then assumes arbitrarily and, he thinks, generously that 10^{11} of these variations might be functional. Stated more directly, he has assumed, entirely without justification, that only 1 out of 10^{119} combinations is useful. But, to show what an extremely generous man he is, Gish then assumes that 10^{21} varieties are formed every second during a period of 5 billion years. He is still perfectly safe, of course; with his assumption of 1 in 10^{119}, the useful chain would never form. Gish doesn't mention whether anyone has systematically examined the properties of any significant number of such sequences. But even if thousands had been investigated, this would be nowhere near 10^{119}, and it would be just as reasonable to assume that 1 in a trillion (10^{12}), 1 in a billion (10^9), or even 1 in a million (10^6) has the desired characteristics. Actually, the evidence we have points in this direction. For example, examination of hemoglobins of different species shows that only 7 out of a total of 140 sites always have the same amino acid.[13] The probability of these 7 sites being correctly occupied, assuming again 20 different amino acids, is 1 in a little over a billion (1.3 X 10^9).

Now, if we go by what little evidence we have and make the far more reasonable assumption that 1 in 10^9 is functional, and assume further that *only one* sequence forms each second (anywhere on earth), a functional one could be expected to form in about 32 years! On the time scales we are dealing with, even 32 million years is nothing, so we too can be generous and assume that only 1 out of 10^{15} randomly generated 100-member sequences is sufficiently autocatalytic. Let us see Gish or anyone else prove this impossible!

Perhaps the greatest unanswered question in biological evolution concerns the manner in which proteins and DNA (or RNA) became associated with each other. Creationists maintain that because we don't now know how this happened naturally, it could only have happened through divine design, and it is useless to investigate it further. We are fortunate that such attitudes have not prevailed universally at all times or science would never have evolved out of the Dark Ages.

We may speculate on whether evolution could at one time have proceeded through mutations and natural selection involving chains of amino acids only, but in the present discussion we will leave aside these early developments, of which enough is not yet known. Let us look, instead, at the evolution of the genetic program from that of primitive organisms even simpler than (and different from) modern viruses, to that of complex ones such as mammals. Although we recognize the enormous amount of variation possible in the normal genetic mixing associated with sexual reproduction, the only way in which something entirely new can be introduced is through mutations, including such phenomena as gene duplication. Creationists contend that, because of the second law, only detrimental mutations are possible. An examination of the mechanism involved will show that this contention is absurd.

Four nucleotides constitute the characters in the genetic code, and, for convenience, they are designated *A, C, G,* and *T* in the case of DNA. They are read in groups of three called codons, each of which codes for an amino acid. A simple type of mutation is one in which one nucleotide is replaced by a different one, and, as a result, a different amino acid is coded for. (Because of a redundancy in the code, this does not always happen.) Since the genetic program has already been brought to near perfection through natural selection, a mutation is usually detrimental to the organism. It therefore tends to be weeded out of a population or, if it gives rise to a recessive gene, is limited in its spread. But there is, of course, no natural law which prevents an occasional mutation from benefiting the organism, especially if the latter exists in a changing environment. Such a mutation would tend to become more common and spread throughout the population. (An example is the acquisition of drug resistance on the part of asexually reproducing organisms, where variability due to genetic mixing does not play a part.) The important point here is that, as far as the second law is concerned, it makes no difference which nucleotide substitution occurs. *The entropy content of the genetic message does not depend on whether the substitution turns out to be beneficial or detrimental to the organism.*

We might profit from an examination of the fallacy that an accumulation of beneficial mutations would contradict the second law. It undoubtedly derives from the fact that, if such an accumulation were the result of a totally random process, it would indeed be contrary to the predictions of the second law. However, if each beneficial mutation is favored over an indifferent one, which in turn is favored over a detrimental one, then the process is by no means random, and we cannot invoke the second law to predict its outcome. The selective process just described is, of course, what we commonly refer to as natural selection.

In order for the complexity of the code to increase, a simple nucleotide substitution is not enough; instead, nucleotides need to be added to the existing sequence, perhaps through the process of gene duplication. Such an addition does constitute a minute negentropy increase, but, as we have seen, this does not at all violate the second law, since there will be a corresponding entropy increase elsewhere. In other respects, the addition is like the simple substitution discussed earlier; in particular, the entropy change in the genetic material is in no way dependent on whether the organism is helped or harmed, and the few beneficial mutations will be favored and accumulate, for added complexity.

SUMMARY

In their first and crudest attempt at creating the illusion of a contradiction between evolution and the second law of thermodynamics, creationists simply ignored the fact that evolving systems are not isolated. Their next endeavor consisted of altering the second law by maintaining that it precludes entropy decreases in all systems, not just isolated ones. Although they still occasionally make either or both of these claims in debates, they apparently realized at some point, presumably after having been confronted with examples proving them wrong, that a new device was needed. So, they invented the "Creative Trinity." This actually *replaces* the second law, but they still refer to it as the second law of thermodynamics in order to maintain the air of scientific respectability.

There is a virtually unlimited number of examples of natural systems in which entropy deficiencies develop spontaneously, provided only that energy is allowed to flow across their boundaries, thus disproving the creationist requirement for a divine directing program or different kinds of entropy. We are awaiting coherent responses from the creationists dealing with these examples.

This leaves only the task of examining the validity of the claim by creationists that genetic programs could not have developed naturally and must therefore have been intelligently created. A simple calculation of the probability of formation of a sufficiently autocatalytic chain of amino acids and an elementary examination of the process of evolution through mutations and natural selection from simple organisms to complex ones show that, whatever difficulties occur in the natural origin of life, they do not involve any violations of the second law of thermodynamics.

REFERENCES

1. Henry M. Morris, *Acts & Facts, ICR Impact Series,* no. 3 (April 1973).

2. F. W. Sears, *Thermodynamics, the Kinetic Theory of Gases, and Statistical Mechanics* (Reading, MA: Addison-Wesley, 1959); Leon Brillouin, *Science and Information Theory* (New York: Academic Press, 1962).

3. Henry M. Morris, *Acts & Facts, ICR Impact Series,* 40 (October, 1976).

4. R. G. Elmendorf, *How to Scientifically Trap, Test, and Falsify Evolution* (Bairdford, PA: Association of Western Pennsylvania, 1978), p. 38.

5. Morris, 1976.

6. W. Ehrenberg, "Maxwell's Demon," *Scientific American,* 217 (5) (November 1967): 103–10.

7. Elmendorf, 1978.

8. Elmendorf, 1978.

9. Morris, 1979.

10. Morris, 1979.

11. See Myron Tribus and Edward C. McIrvine, "Energy and Information," *Scientific American,* 225 (3) (September 1971): 179–84, 86, 88.

12. Duane T. Gish, *Acts & Facts, ICR Impact Series,* 58 (April 1978).

13. See M. F. Preutz and H. Lehmann, "Molecular Pathology of Human Haemoglobin," *Nature* 219 (August 31, 1968): 902–09.

Finding the Age of the Earth: By Physics or By Faith?

by Stephen G. Brush

At the present time, the evidence is overwhelming that the earth is several billion years old; many different determinations by radiometric dating methods give an age close to 4.5 billion years. The criticisms of radiometric dating published by creationists have no scientific basis, and can be justified only by arbitrarily rejecting well-established results of modern physical science.

SUMMARY

For several decades, scientists have generally accepted, on the basis of radiometric dating, that the earth is three to five billion years old. The recent revival of "creationism" in the United States has raised a challenge to this conclusion; creationists prefer a much shorter time scale, in part to support their biblical cosmogony, and in part because they realize Darwinian evolution, which they oppose, requires a very long time scale. The creationist critique of radiometric dating, and their argument that the earth is less than 10,000 years old, may appear plausible to those who are not familiar with modern physical science. In fact, the creationist demand for "equal time" would involve not only presenting an alternative to biological evolution but also replacing some of the best-established theories of physics by obsolete or speculative hypotheses which would ultimately have to be taught in order to justify a young earth.

According to the current "scientific creationism" model which has been proposed for teaching in public schools, not only the earth but the entire Universe was created only a few thousand years ago. This model was refuted more than a century ago, when the distances of a few nearby stars were first determined directly by trigonometric parallax measurements. It was pointed out by Serres (1843) that at least some of the stars in nebulae must (on the basis of their relative brightness) be several hundred thousand light years away, so they must have been in existence long before the supposed creation of the Universe 6000 years ago. The creationists can answer this argument only by postulating that God created the light rays en route to earth in such a way as to make it *appear* that they came from distant stars. Such a postulate —that the world was recently created but looks very old—can be neither proved nor disproved, and has no place in a science course where all postulates must be testable.

The argument that there has not been enough time for the slow process of Darwinian evolution to produce the present forms of life goes back to the nineteenth century. Lord Kelvin estimated that a homogeneous solid sphere as big as the earth would have cooled down from an initial molten state to its present temperature in less than 100 million years, if no *sources* of heat were present. Since Darwin had mentioned 300 million years as a geological time period when discussing evolution, Kelvin's estimate was generally considered an argument against evolution.

The discovery of radioactivity showed that Kelvin's basic assumption was wrong—radioactive substances in the earth's crust might easily generate enough heat to balance a substantial part of that lost by conduction out into space. Moreover, Rutherford and Boltwood suggested that the relative proportions of uranium, lead and helium in rocks could be used to estimate the time since they had crystallized, and by 1907 ages of more than a billion years had been found in this way.

The radioactive decay of a nucleus is a random process, described by a decay constant λ which gives the probability that a nucleus will decay in a definite time interval. Half of the original nuclei will have decayed, on the average, in a time $T_{1/2}$ known as the "half-life." (According to the mathematical theory, the number of non-decayed nuclei decreases exponentially with time, as $e^{-\lambda t}$, and the half life is $T_{1/2} = 0.693/\lambda$.) The randomness of this process is a direct consequence of quantum theory, and is well established experimentally. Since the nucleus is a very small part of the atom and is insulated from the effects of other atoms by electrical repulsion, external conditions such as pressure and temperature have almost no effect on the decay constant, except in extreme cases such as inside stars.

One method frequently used to estimate the age of the earth relies on the ratios of lead isotopes, determined in a mass spectrometer. Two of these isotopes may be produced by alpha-decay of uranium isotopes. A uranium 238 nucleus, containing 92 protons and 146 neutrons, may emit 8 alpha particles (nuclei of helium containing 2 protons and 2 neutrons) and 6 electrons, thus becoming a lead 206 nucleus, containing

82 protons and 124 neutrons. Similarly, uranium 235 decays to lead 207. Ordinary lead also includes lead 204, which is not produced by any radioactive decay of another element, and lead 208, some of which is produced by the decay of thorium. As far as is known, *chemical* or *geological* processes cannot change the relative abundances of these isotopes.

If we knew the "primeval abundances" of the lead isotopes, i.e. the relative amounts of isotopes 204, 206, 207 and 208 at the time the earth was formed, we could then subtract these from the present abundances to estimate the amount formed by decay of uranium and thorium, the "radiogenic component." If there had been no chemical separation of lead from uranium since the time of the earth's formation, we could estimate the age of the earth by comparing the amount of radiogenic lead with the amount of uranium present in a rock.

Of course it is not valid to assume that no chemical separation of lead from uranium has occurred, since all rocks were probably molten at some time after the formation of the earth. The assumption *may* be valid for a period of time since the rocks were last crystallized (the "time of mineralization," t_m).

In 1946, Arthur Holmes and F. G. Houtermans pointed out that the age of the earth could be estimated from the relative abundances of lead 206 and 207 in rocks even though those rocks had crystallized more than a billion years after the formation of the earth. The key to their method is that there are *two* radioactive decays involved, with different decay constants, and therefore the end products (206 and 207) accumulate at different rates. Suppose one has several rocks formed at the same time (t_m) but with different proportions of uranium and lead. Then a plot of the amount of radiogenic lead 207 against the amount of radiogenic lead 206 should be a straight line. This is the so-called "isochron" (graph for rocks formed at the *same time*). From the slope and intercept of this line one can obtain two relations between t_m, the age of the earth t_0, and the primeval isotopic abundance ratio for lead 206 and 207. If one additional piece of information about these parameters can be found, then all three can be estimated.

One possible way to apply the isochron method, suggested by Holmes and Houtermans, is to construct isochrons for collections of rocks with different values of t_m. The calculation also assumes that the primeval isotopic abundance ratio was similar to that of certain rocks with a very high relative abundance of the non-radiogenic isotope 204. In this way Holmes and Houtermans estimated that the age of the earth is about 2.9 billion years.

In 1953, this estimate was revised when Patterson, Brown, Tilton and Inghram measured the abundance of lead isotopes in some meteorites which contained very little uranium. (Thus almost all of their lead 206 and 207 should be non-radiogenic.) Taking this as the primeval abundance, Patterson and Houtermans arrived at the figure 4.5 ± 0.3 billion years.

Although there was originally some doubt as to the validity of the assumption that the primeval abundance of lead isotopes in the earth is the same as that now found in some meteorites, subsequent research has confirmed it. While an age based on any single set of rock or meteorite data may be subject to criticism, the fact that many independent determinations give the same value makes it extremely unlikely that they could all be seriously wrong. The isochron method is self-checking in the sense that if the group of rocks selected does not yield a straightline plot, then one knows they were not formed at the same time and can reject them.

The best current estimates of the age of the earth all lie well within the limits of error of the original Patterson-Houtermans value. It is remarkable that this value has been stable for twenty-eight years, especially in view of the radical changes that have taken place in the earth sciences during that time.

Other methods for determining the ages of very old rocks have been used to check and reinforce those found by the uranium-lead method. One method uses the beta-decay of rubidium 87 to strontium 87; another uses the decay of potassium 40 to argon 40 by capture of an orbital electron.

A systematic examination of criticisms of radiometric dating by creationists (H. M. Morris and H. S. Slusher) shows that they are based on ignorance or misunderstanding. For example:

1. *Rocks cannot be dated by radiometric methods alone—these methods must be combined with assumptions about the fossils they contain, which depend on the validity of evolution.* This is simply not true for the methods used to determine the oldest rocks and the age of the earth (see above).

2. *Lead 206 may be converted into lead 207 by free-neutron capture so their abundance ratio cannot reliably be used to estimate the age of the earth.* The only evidence the creationists present for this process is based on misreading of published data. In one case they assume that no lead 204 was found in a certain rock whereas in fact it was not measured. In the other case they drew conclusions from the amount of lead 208 found in a rock without realizing that none of it is radiogenic.

3. *The decay constant may depend on cosmic rays and thus could have been affected by supernovae in the past.* The only technique this applies to is carbon dating (where it is not the decay constant itself that would be affected) and here the result would be to make the dates too

young rather than too old. An experiment by L. R. Maxwell showed that changes in cosmic ray intensity have no effect on the rate of alpha decay.

4. *Radioactive decay rates may be affected by pressure, temperature, or chemical state.* The observed changes are generally less than one percent and would have no significant effect on the accuracy of radiometric dating, according to the authority cited by the creationists.

5. *A variation in decay dates would be expected from Dudley's "neutrino sea" theory, which suggests that radioactivity is not a random process but can be triggered by events in the medium surrounding the nucleus.* Dudley's theory is not supported by any evidence; its acceptance would require rejecting not only established principles of nuclear physics but also quantum theory and relativity. (Dudley's "neutrino sea" is essentially an ether based on "hidden variables" of the type known to be inconsistent with quantum mechanics.) Dudley himself states that radioactive decay constants vary by less than ten percent and that his theory therefore does not entail any substantial inaccuracy in radiometric dating methods.

6. *Experiments have shown that the decay rates of cesium 133 and iron 57 vary, hence there may be similar variations in other decay rates.* There are no such experiments; both are stable isotopes, not subject to radioactive decay at all!

To support their short time scale, creationists rely on (1) the Bible, (2) estimates based on various processes such as the influx of certain elements to the ocean, (3) decay of the earth's magnetic field. The first argument cannot be used in a public school science class so we ignore it. The second is unreliable according to the creationists themselves, since we cannot assume that the rates of those processes were constant in the past. The third is the only one that seems to be given much confidence by the creationists in their "scientific" publications.

Thomas G. Barnes developed the argument that the decay of the earth's magnetic field implies its recent creation. He assumed that the field is due to electric currents which are not maintained by a dynamo or any other energy source but are decaying exponentially with time as their energy is dissipated into heat. This assumption is flatly contradicted by almost all current research on geomagnetism, and is not even supported by the observational data which Barnes himself presents.

Barnes determined the parameters of his exponential decay curve from measurements of the earth's magnetic dipole in 1835 and 1965 (ignoring the fact that the decay is very nearly linear, *not* exponential, between those dates) and then extrapolates the curve backwards to 20,000 B.C. Since the field would thus have the incredibly large value of 18,000 gauss, he concludes that the earth must have been created after that time. But recent research in archaeomagnetism shows that the extrapolation is invalid.

To justify his calculation, Barnes has to reject the Elsasser theory (now generally accepted, at least qualitatively, by geophysicists) that the earth's magnetic field is maintained by dynamo action of the fluid motions in the core. He must also reject the conclusion (an integral part of the modern theory of plate tectonics or continental drift) that the earth's magnetic field has a long and well-documented history of polarity reversals. The one skeptical authority on whom Barnes relies, J. A. Jacobs, has recently decided that the reversals are now firmly established, thus undermining the basis of Barnes' calculation.

Conclusion

At the present time, the evidence is overwhelming that the earth is several billion years old; many different determinations by radiometric dating methods give an age close to 4.5 billion years. The criticisms of radiometric dating published by creationists have no scientific basis, and can be justified only by arbitrarily rejecting well-established results of modern physical science. Their argument for a young (10,000 year) earth from decay of the earth's magnetic field is completely refuted by empirical data and is incompatible with all currently accepted principles of geomagnetism.

TIME SCALES FOR THE TWO MODELS

The recent revival of "creationism" has raised an issue that most scientists thought was settled decades ago, the validity of the multi-billion year time scale for geological history. Indeed, by insisting that not only Man but the earth and the entire Universe were created in six days no more than about 6000 years ago, the creationists have adopted a position that has not been scientifically respectable for the last 150 years. Whereas a few professional scientists 100 years ago continued to deny that the human race had evolved from "lower" forms of life, even they had already abandoned the Mosaic chronology, i.e. the doctrine that the world was created in 4004 B.C. One has to go back to about 1830 to find a period when a significant number of geologists reckoned the age of the earth in thousands rather than millions of years.[1]

According to the textbook prepared by the Creation Research Society for use in high school biology courses, "Most creationists believe that the age of the earth can be measured in thousands rather than millions or billions of years."[2] However, in some of their publications the creationists are ambiguous about the extent to which their "model" depends on a short time scale. Thus in *Scientific Creationism* we read that "the creation model does not, in its basic form, *require* a short time scale. It merely assumes a period of special creation sometime in the past, without necessarily stating when that was." Apparently this flexibility is needed so that certain aspects of the theory of human evolution can be criticized by quoting experts who rely on fossils more than a million years old.[3] Nevertheless "it is true that it [the creation model] does fit more naturally in a short chronology. Assuming the Creator had a purpose in His creation, and that purpose centered primarily in man, it does seem more appropriate that He would not waste aeons of time in essentially meaningless caretaking of an incomplete stage or stages of His intended creative work."[4] It is interesting to note that this statement appears in the "public school edition" which claims to treat the subject "solely on a scientific basis, with no references to the Bible or to religious doctrine."[5]

The "theological necessity of a young Universe" is explained by creationist T. Robert Ingram as follows:

> to suppose a Creation untold ages ago is really to dismiss the notion of Creation, as a serious matter; and to do that, in turn, is to play down, and eventually ignore or deny, the difference between the Creator and his Creation.[6]

Note that this "necessity" is not perceived by most religions or even by most Christians.[7] Although they seem to have some disagreement among themselves, the important point for our purposes is that the creationist materials most likely to be introduced into public school science classes do insist on the validity of the short time scale. The law recently passed in Arkansas, requiring "balanced treatment" for creationism and evolution, includes "a relatively recent inception of the earth" as part of its definition of the former (Act 590, signed into law 19 March 1981).[8]

After devoting about ten pages to a critique of radiometric dating methods (to which we return in a later section), *Scientific Creationism* presents a review of "evidence for a young earth." The definitive statement here is that

> 10,000 years seems to be an outside limit for the age of the earth, based on the present decay of its magnetic field.[9]

A more recent summary of the creation model states:

> the age of the earth appears to be about 10,000 years . . . Extrapolating the observed rate of apparently exponential decay of the earth's magnetic field, the age of the earth or life seemingly could not exceed 20,000 years.[10]

Another publication, *The Scientific Case for Creation,* asserts that "the decay of the earth's magnetic field, of all processes, probably most nearly satisfies the necessary uniformitarianism assumptions and so probably yields the best physical estimate of the earth's age," namely "the earth almost certainly was created less than 10,000 years ago."[11] Each of these publications also mentions estimates by other methods yielding hundreds of thousands or even millions of years, not as if they should be accepted but simply as objections to the billion-year time scale.

Despite this ambivalence, which seems to be mainly a smokescreen generated to throw doubt on the evolutionist theory without committing the creationists to a definite alternative, the Director of the Institute for Creation Research has made at least one unequivocal statement on the subject. In his book *The Remarkable Birth of Planet Earth,* Henry Morris wrote:

> The only way we can determine the true age of the earth is for God to tell us what it is. And since He *has* told us, very plainly, in the Holy Scriptures that it is several thousand years in age, and no more, that ought to settle all basic questions of terrestrial chronology.[12]

I have not yet found a creationist who can point out such a statement in the Bible, other than Bishop Ussher's seventeenth-century addition to the King James version.

If the creationists only wanted to attack evolution, they could have saved themselves much trouble and embarrassment by ignoring the time-scale problem. After all, it was quite common for pious scientists in the nineteenth century to assert that an indefinite period of time intervened between the First and Second Days of Creation, or that each Day corresponded to 1000 years or an indefinite geologic epoch. In this way they could accept the results of geology (and astronomy) indicating a long time scale without letting this affect their belief in the Special Creation of Man. But modern-day creationists have made it quite clear that they reject any such compromise with (what they consider) a literal interpretation of *Genesis,* and therefore they feel compelled to argue for a "young earth" in many of their publications. Since they have chosen to do so, I will assume, in comparing the "creation model" with the "evolution model," that the former entails creation of the earth and the rest of the Universe no more than 10,000 years ago, while the latter entails formation of the earth several billion years ago and a Universe which is at least ten billion years old. (To be more specific, evidence of life on earth 3.5 billion years ago has recently been reported; the earth is probably about 4.5 or 4.6 billion years old; and the "Big Bang" may have occurred about 18 billion years ago.) Since the difference between the time scales of the two models is *six orders of magnitude*—i.e., the evolution model considers that the world is about a *million times as old* as the creation model does—we don't have to pin down either of them very precisely. It seems fair to say that any good evidence for an earth older than a million years would be extremely damaging to the creation model, while any good evidence for an earth younger than 100 million years would be extremely damaging to the evolution model. Historically, just such evidence was considered damaging to Darwin's theory when it was brought forth by Lord Kelvin in the 1860s, but it was based on assumptions later found to be wrong.[13]

Aside from the age of the earth itself, radiometric dating methods support evolution by showing that, in general, simple forms of life were present on earth before the more complex forms. If one can arrange a sequence of fossils, starting with the most primitive and ending with those similar to humans, and reliably assign dates going back millions or hundreds of millions of years, this would be strong evidence for evolution against creationism. The problem of "gaps" in the fossil record would then be completely irrelevant. Many of the same arguments about the reliability of methods for dating the age of the earth apply also to the dating of fossils. (Note that radiocarbon dating, which applies mainly to objects formed more recently than about 50,000 years ago, depends on additional assumptions about the earth's atmosphere, and is considerably less reliable than the other methods I will be discussing here (see Suess [1980] for a recent review of radiocarbon dating.)[14]

The creationists reject not only radiometric dating but also the entire picture of earth history developed by geologists and geophysicists over the last two centuries. They claim that fossils and rocks in a geological column were not gradually deposited over millions of years but were laid down in a few months or years after a worldwide flood or "great hydraulic cataclysm."[15] As we will see in section 8, in order to support their argument that the earth is only a few thousand years old, creationists reject the evidence for reversals of the earth's magnetic field, on which much of modern plate tectonics is based. And, since they claim that not only the earth but the entire Universe was created a few thousand years ago, the creationists reject some of the major results of modern astronomy.

The Arkansas law applies not only to biology courses but all "lectures, textbooks, library materials or educational programs [that] deal in any way with the subject of the origin of man, life, the earth, or the Universe."[16] The "creation-science" that must be inserted to balance "evolution-science" includes "sudden creation of the Universe, energy, and life from nothing" and "explanation of the earth's geology by catastrophism, including the occurrence of a worldwide flood," as well as "relatively recent inception of the earth and living kinds."[17] In Arkansas or any other state that has passed a similar law it would be illegal to teach an earth science course based on any of the currently-used texts unless a substantial amount of creationist material were added. For example, the text *Investigating the Earth,* sponsored by the American Geological Institute, based on the original Earth Science Curriculum Project,[18] has chapters on "Earth's biography" and sections on magnetic reversals, age of the moon and solar system, and cosmology, which make up about twenty-five percent of the total; these would all be considered "evolution-science" which must be balanced by giving equal time to creationist doctrines.

TIMES FROM STELLAR DISTANCES

The creationists' claim that the entire Universe was created only a few thousand years ago was thoroughly refuted by the middle of the nineteenth century, in a rather simple way. In 1838, the German astronomer Friedrich Wilhelm Bessel reported the first accurate determination of the distance of a star (other than the sun) from the earth. By measuring the parallax (change in apparent position as seen from earth at different places in its orbit around the sun) of the star 61 Cygni, he found that its distance is approximately 10^{14} kilometers. (This is within ten percent of the accepted modern value.) It is known that the speed of light is about 3×10^5 km/sec, so that in a year light would travel nearly 10^{13} kilometers. (More precisely, a light-year is 9.46×10^{12} km.) Thus light starting from 61 Cygni would take a little more than ten years to reach us.

By this time William Herschel and other astronomers had already estimated that the distances of some nebulae must be hundreds or thousands of times greater than those of stars such as 61 Cygni. The basis of such estimates is the physical law that the *apparent* brightness of a light source decreases as the square of its distance from the observer, and is proportional to its *intrinsic* brightness (how bright it would appear at a standard distance). Several nebulae could be resolved into separate stars of very low but measurable brightness. If one assumed that such a star had the same intrinsic brightness as 61 Cygni, then one could determine its distance.

The first scientist to perceive the implications of these astronomical discoveries for creationism was Marcel de Serres, Professor of Mineralogy and Geology at the University of Montpellier in France. In his book on the creation of the earth and the celestial bodies, Serres (1843) pointed out that some nebulae must be at least 230,000 light

years away, on the basis of the results mentioned above.[19] It is important to note that this conclusion does not depend on the assumption
that all stars have the same intrinsic brightness (which is now known
to be false) but only that *at least one* star in a nebula be as bright as
61 Cygni. The only way Serres' argument could fail is if 61 Cygni just
happened to be millions of times brighter (intrinsically) than *every other
star* in the sky, a rather unlikely circumstance. We do not have to make
any assumptions about the accuracy of other methods of measuring
stellar distances, or about the rate of processes such as the expansion
of the Universe, in order to accept Serres' conclusion, which is simply
that *some stars in the sky must have existed much more than 6000 years
ago,* contrary to the creationist doctrine.

Serres went even further than this, to point out an absurd consequence of the creationist assumption that the entire Universe was
created at the same time only a few thousand years ago. In that case
Adam would have seen no stars in the sky (other than the sun) for ten
years; after that they would have started to appear, one by one, as their
light first reached the earth. (To refine this argument in the light of
more recent discoveries, we might say four years rather than ten, but
we would still be granting for the sake of argument the creationist
assumption that no stars now visible are more than 6000 light years
distant.) Throughout recorded history, according to this hypothesis,
the number of stars seen in the sky would have increased every year
until the present multitude is visible. Such a remarkable phenomenon,
if it had really taken place, could hardly have gone unnoticed, especially
by the early seafaring people who relied on the stars for navigation. A
new star, rather than being such a rare event that it undermined the
credibility of Aristotle's cosmology in 1572, would have been almost an
everyday occurrence according to the creationist model.

How do the creationists answer this argument? Their explanation
[by Henry M. Morris in *The Remarkable Birth of the Planet Earth*] is
worth quoting at length, and should be read by anyone who still thinks
it is possible to defend "a scientific creation model" without falling back
on theology:

> If the stars were made on the fourth day, and if the days of creation
> were literal days, then the stars must be only several thousand years
> old. How, then, can many of the stars be millions or billions of
> light-years distant since it would take correspondingly millions or
> billions of years for their light to reach the earth?
>
> This problem seems formidable at first, but is easily resolved
> when the implications of God's creative acts are understood. The

very purpose of creation centered in man. Even the angels themselves were created to be "ministering spirits, sent forth to minister for them who shall be heirs of salvation" (Hebrews 1:14). Man was not some kind of afterthought on God's part at all, but was absolutely central in all His plans.

The sun, moon, and stars were formed specifically to "be for signs, and for seasons, and for days, and years," and "to give light upon the earth" (Genesis 1:14, 15). In order to accomplish these purposes, they would obviously have to be visible on earth. But this requirement is a very little thing to a Creator! Why is it less difficult to create a star than to create the emanations from that star? In fact, had not God created "light" on Day One prior to His construction of "lights" on Day Four? It is even possible that the "light" bathing the earth on the first three days was created in space as en route from the innumerable "light bearers" which were yet to be constituted on the fourth day.

The reason such concepts appear at first strange and unbelievable is that our minds are so conditioned to think in uniformitarian terms that we cannot easily grasp the meaning of creation. Actually, real creation necessarily involves creation of "apparent age." Whatever is truly created—that is, called instantly into existence out of nothing—must certainly look as though it had been there prior to its creation. Thus it has an appearance of age.

The factor of created maturity obviously applies in the case of Adam and Eve, as well as of the individual plants and animals. There is nothing at all unreasonable in assuming that it likewise applies to the entire created Universe! In fact, in view of God's power and purposes, it is by far the most reasonable, most efficient, and most gracious way He *could* have done it.[20]

Readers familiar with the history of the subject will recognize that Morris has revived one of the most notorious methods for explaining away the evidence of the antiquity of the earth—the *Omphalos* (navel) theory of Philip Gosse, published in 1857.[21] Gosse proposed that Adam had a navel because God created him to look as if he had been born in the usual way with an umbilical cord; and that God thus created the entire Universe in such a way that it *appeared* to have a history of previous existence. In the same way we could of course assume that God created the Universe ten seconds ago and that our own memories of previous experiences were created at the same time. The creationists may themselves accept such a religious doctrine in order to explain away the overwhelming evidence for biological evolution and the antiquity of the world, but they have no right to inject it into the public schools under the guise of a "scientific" hypothesis.

DARWIN VS. KELVIN

Before 1905, one of the scientific objections to Darwin's theory of evolution was that there had not been enough time for such a slow process as natural selection to produce its effect, since Lord Kelvin had estimated the age of the earth to be substantially less than 100 million years. Creationists have recently revived this objection, despite the fact that the original basis for Kelvin's estimate is now known to be completely wrong, and a short time scale for earth history can be maintained only by denying the validity of radioactive and other dating methods. To understand this situation we must first review briefly the Darwin-Kelvin controversy. Further details may be found in the books by Burchfield (1975) and Brush (1978):[22] a good overview of the history of geological time estimates is given by Faul (1978).[23]

In the first edition (1859) of his *Origin of Species,* Charles Darwin conjectured that certain geological processes such as the gradual removal of solid material from chalk cliffs by water might have been going on for as long as 300 million years. This figure was introduced as a concrete example of the magnitude of the time periods involved in natural history, even though the theory of natural selection had no time parameters in it, and thus there was no direct *biological* way to estimate how much time evolution would take. An anonymous reviewer (presumably a geologist) criticized this estimate and Darwin removed it from later editions, but the unfortunate impression was created that he was "retreating" from his original position that there had been ample time for evolution.[24] He could just as well have started with an estimate of 30 million years, to which no one could have raised serious objections, and it would have made no difference to the biological aspect of evolution.

William Thomson (later known as Lord Kelvin) was at this time the best known of the younger generation of British physicists and was highly respected for his research on heat theory. He was also the only person in Great Britain who was thoroughly familiar with Fourier's mathematical theory of heat conduction. This theory allowed one to reconstruct the thermal history of a simple physical system, such as a sphere the size of the earth, provided certain assumptions were accepted. First, the entire sphere must be a solid (so that no heat flows by convection); second, it must have constant thermal conductivity and heat capacity throughout; third, it was initially all at the same high temperature, surrounded by an infinite space at a lower temperature; fourth, no heat is generated or destroyed anywhere in the system. Then, from data on the present temperatures at points just below the surface (average vertical temperature gradient), one can compute the amount of time that has elapsed since the sphere was at a specified initial temperature. In this way, Kelvin estimated that a period of 100 to 200 million years might have been required for the earth, assumed initially to have been at a uniform temperature of several thousand degrees (above the melting points of rocks), to reach its present state.[25]

Kelvin also attempted to estimate the age of the sun, assuming that its energy comes from converting the gravitational attraction of its parts into heat, using data on its present rate of heat loss. He concluded that it is "most probable that the sun has not illuminated the earth for 100,000,000 years, and almost certain that he has not done so for 500,000,000 years."[26]

Kelvin thought that his results should "sweep away the whole system of geological and biological speculation demanding an 'inconceivably' great vista of past time, or even a few thousand million years, for the history of life on the earth."[27] He did not reject the principle of evolution itself, and indeed the conflict between Kelvin and Darwin has been somewhat exaggerated by historians. It was primarily the randomness and lack of conscious direction in Darwin's theory of natural selection that was offensive. Evolution guided by Divine Wisdom—the "argument of design" as it was then called—would go much faster and thus would not be hampered by the limited time-scale imposed by heat-conduction calculations.[28]

The major public defender of evolution in nineteenth-century Britain was T. H. Huxley, known as "Darwin's bulldog." In a lecture in New York in 1876, Huxley pointed out that Kelvin's physical argument was entirely irrelevant to biology. It is up to the geologists and physicists, he declared, to decide on the age of the earth; once they have agreed among themselves, biologists will accept the decision. Biologists are interested only in whether it is a fact that evolution has taken place.

> We take our time from the geologists and physicists, and it is monstrous that, having taken our time from the physical philosopher's clock, the physical philosopher should turn round upon us, and say we are too fast or too slow.[29]

This is a perfectly reasonable argument, in view of the fact that evolutionary theory did not at that time *require* any particular time scale, in the absence of direct *biological* evidence of the *rate* of evolutionary change. But Huxley's point was ignored by those who looked for a decisive battle between physics and evolution. And most geologists, rather than defending their earlier statements that geological processes had been slowly acting over very long periods (hundreds of millions of years or more), revised them to conform to Kelvin's estimates. In the 1890s, new heat-conduction calculations by P. G. Tait and Clarence

King reduced Kelvin's original limit of 100 million years to somewhere between 10 and 20 million years, creating considerable strain on the geological time-scale.

At the beginning of the twentieth century the situation changed radically for a reason that neither geologists nor physicists could have anticipated—the discovery of radioactivity. Following the isolation of radium by Marie and Pierre Curie, it was announced by Pierre Curie and Albert Laborde (1903) that radium salts generate a substantial amount of heat.[30] Himstedt (1904) pointed out that, if there are widespread deposits of radium in the earth, then the heat they produce must be taken into account in studies of the thermal history of the earth.[31] Liebenow (1904) followed up this suggestion by estimating that the presence of 1/5000 of a milligram of radium per cubic meter, distributed uniformly throughout the earth's volume, would be sufficient to compensate for the observed loss of heat by conduction through the crust.[32] A similar suggestion was made about the same time by Rutherford and Strutt. Thus the possibility of continual generation of heat over long periods of time invalidated Kelvin's assumption that the earth is simply cooling down from an initial high-temperature state.

It was soon recognized that the relative proportions of lead, helium, radium, and uranium could be used to estimate the ages of those rocks (Rutherford, 1905, 1906).[33] Strutt (1905) obtained an estimate of 2.4 billion years and Boltwood (1907) found an age of 2.2 billion years for one rock sample; the inspiration for both studies seems to have come from Rutherford.[34] After some initial skepticism, the validity of this method for estimating ages was generally recognized, along with the thousandfold increase in the time scale over that which had previously been accepted. Historical accounts of the early period of radiometric dating may be found in the works by Burchfield (1975) and Badash (1968, 1969, 1979).[35]

Reviewing Kelvin's four assumptions on which he based his estimate of the age of the earth, we can see that his result was much too low because of the falsity of his fourth assumption, that no heat is generated or destroyed anywhere in the system. But in the twentieth century the other three assumptions were also found to be incorrect. Contrary to Kelvin's first assumption, the earth has a liquid core,[36] and even its mantle, which seems to be solid, transfers significant amounts of heat by convection.[37] Contrary to his second assumption, the physical and chemical state of the interior are different from that of the crust, so one cannot take the thermal conductivity and heat capacity as constant throughout. Contrary to the third assumption, there is no longer any good reason to assume that the earth was initially a hot fluid which has been cooling to its present state; it is just as likely that it was formed by the aggregation of cold solid particles and later warmed to its present temperature.[38] Thus any attempt to estimate the age of the earth from its thermal properties alone is likely to be completely wrong.

THEORY OF RADIOACTIVE DECAY

Several other methods for estimating geological time have been proposed. Each depends on the assumption that the rate of some process which we can measure at present has been constant (or changed in a known way) in the past. Thus if the process involves the accumulation of a physical quantity X (for example, the amount of salt in the oceans) and we assume that the time-rate of change of X has always been $R = dX/dt$, then according to a standard formula of calculus the net accumulation of X from time t_0 to t is

$$X = X_0 + \int_{t_0}^{t} \frac{dX}{dt}\, dt = X_0 + R(t - t_0). \tag{1}$$

The "age" is simply the time interval $(t - t_0)$ during which the process is supposed to have been operating, and is given by

$$A = t - t_0 = (X - X_0)/R, \tag{2}$$

or if we assume there is none of the quantity to start with ($X_0 = 0$), then

$$A = X/R. \tag{3}$$

Unfortunately, almost every process that has been suggested for this purpose is obviously affected by many physical, chemical, geological and biological factors that have *not* been constant in the past. The only apparent exception is radioactive decay. The early experiments by Rutherford and others indicated that decay involves the actual transmutation of one element into another, for example radium into lead with the emission of "alpha particles" (helium nuclei). The measured rate of emission was found to be proportional to the amount of radium present, but did not change with pressure, temperature or any other factors.

The process of radioactive decay could be described statistically by saying that an atom of radium has a certain probability of decaying during a small interval of time. If initially (t_0) there are N_0 atoms, and the probability that an atom decays during time Δt is λ, then the number of atoms that decay in that time will be, on the average,

λN_0. At any later time t, the rate of change of N will be proportional to the value of N at that time,

$$dN/dt = -\lambda\ N \tag{4}$$

or

$$dN/N = -\lambda\ dt. \tag{5}$$

Again using a standard formula of calculus to integrate this equation, we find that

$$\int_{N_0}^{N} \frac{dN}{N} = \ln(N/N_0) = -\lambda \int_{t_0}^{t} dt = -\lambda\ (t - t_0), \tag{6}$$

where "ln" means the "natural logarithm" to the base e (2.718. . .). Using the mathematical definition of this logarithm function, we then find that

$$N/N_0 = e^{-\lambda(t-t_0)}, \tag{7}$$

representing the well-known "exponential decay" curve. The ratio N/N_0 decreases from its initial value of one at $t=t_0$, eventually becoming indefinitely small. (While the mathematical curve extends to infinity, the physical curve would drop to zero at a random time when the last atom decays.)

The rate of radioactive decay is usually given in terms of the "half-life," defined as the time $T_{1/2}$ at which N/N_0 has declined to one-half. According to the above equation,

$$\ln(1/2) = -\lambda\ T_{1/2}.$$

Since

$$\ln\ (1/2) = -0.693,$$

we find that

$$T_{1/2} = -0.693/\lambda. \tag{8}$$

For example, the half-life of radium is about 1620 years; its decay constant (λ) is 1.36×10^{-11} per second. A gram of radium decays into about half a gram of radium and half a gram of lead after 1620 years; after 3240 years one would have .25 gram of radium and .75 gram of lead, and so forth.

To estimate the age of a rock, one has to recognize that the radium itself is produced by the decay of uranium, which has a much longer half-life. In fact there are a number of intermediate elements in the sequence that leads from uranium to lead, and moreover each element was found to have different varieties known as *isotopes.* While the early papers on radiometric dating of rocks generally arrived at the correct order of magnitude for their ages, all of this work had to be completely redone after the various radioactive decay series had been determined and it became possible to measure the amount of each isotope present. Unless one is familiar with the details of the history of the subject and can translate the nomenclature for isotopes into modern terms, little information can be obtained from papers published before 1930. I will therefore skip over this early period and consider only the most important research done after the present understanding of nuclear structure and transformations had been attained.

According to the Rutherford-Bohr model of the atom, almost all of the mass of an atom is concentrated in an extremely small nucleus, which has a positive charge Ze, where e is the magnitude of the electronic charge and Z is an integer equal to the atomic number of the element. There are Z electrons outside the nucleus, each having charge $-e$. Most observable properties of the atom are described with extraordinary accuracy by assuming that the nucleus behaves like a point charge while the electrons are governed by the equations of "quantum mechanics" as formulated in 1925–27 by Werner Heisenberg, Erwin Schrödinger, and Paul Dirac. Radioactive decay, on the other hand, involves (with one exception) only the nucleus, which is insulated from external influences by the electrons which define and maintain the size of the atom. The reason radioactive decay is unaffected by external conditions such as temperature and pressure is that every atom has a cloud of negatively charged electrons, and the electrostatic repulsion of these electrons prevents the central nuclei of the atoms from getting close enough to interact. If the electronic cloud does break down (for example at the extremely high pressures and temperatures found in stars) the positive electric charges of the nuclei themselves still supply a repulsive force which keeps them apart. This is one of the major reasons why the generation of nuclear energy from hydrogen fusion has been so difficult to accomplish. The electrical repulsive forces can be evaded by neutral particles such as neutrons, but such particles are themselves produced only under unusual conditions.

The one exceptional case in which radioactive decay involves particles outside the nucleus is the capture of an electron by the nucleus, yielding an element with atomic number one unit lower. An example of such a decay used in radiometric dating is the conversion of potassium to argon. One might expect that external conditions which change the electron distribution within the atom would affect the decay rate. Nevertheless, experiments show that the actual change produced by any plausible pressure or temperature is less than one percent.[39]

The random character of radioactive decay is a special case of the indeterminancy of quantum theory, as was pointed out in 1928 by George Gamow, Ronald Gurney and Edward Condon.[40] They showed that a particle held inside the nucleus by a "potential barrier" may be able to "tunnel through" the barrier and emerge on the other side, since if the barrier is finite the wave function of the particle is not completely localized and there is a finite probability that the particle will be outside the nucleus. Tests of the randomness of alpha decay have shown that any deviations are quite small and there is no indication that they could have a significant effect on the accuracy of the geological time scale.[41]

Prior to 1932 it was generally believed that the atomic nucleus consists of protons and electrons. Following the discovery of the neutron, a distinct particle which can be formed by putting together a proton and an electron, Heisenberg proposed that the nucleus contains only protons and neutrons. To a first approximation the proton and neutron have the same mass, and the mass-change resulting from the binding energy needed to hold the particles together is much smaller than the mass of a single proton. Thus the total mass of a nucleus will be close to Am_p, where m_p is the mass of the proton and A is an integer, known as the "mass number," equal to the total number of protons and neutrons. The number of protons is denoted by Z, the atomic number of the element. The number of neutrons is (A–Z); nuclei having the same Z but different A are called "isotopes" or "nuclides" of the element.

Since the chemical properties of an atom depend on the electrons outside the nucleus, and since the number of those electrons needed to attain electrical neutrality depends only on the number of protons in the nucleus (Z), it is generally impossible to separate isotopes of the same element by chemical methods. The mass spectrograph, developed by F. W. Aston in the 1920s, does allow a direct separation of isotopes by sending ions from a gas through an arrangement of electric and magnetic fields in such a way that ions of different mass will be deflected by different amounts. Modern methods of radiometric dating depend on such techniques for determining the isotopic composition of samples, since the decay process will generally result in an enrichment of one or more isotopes of an element over the other.

The most common types of radioactive decay which change one element to another are known as "alpha" and "beta." The alpha particle was found to be simply the nucleus of a helium atom, containing two protons and two neutrons. Thus a nucleus of uranium isotope 238, which contains 92 protons and 238–92 = 146 neutrons, decays by alpha emission to a nucleus of thorium isotope 234. This reaction can be written

$$_{92}U^{238} \xrightarrow{\alpha} {}_{90}Th^{234} + {}_2He^4. \tag{9}$$

In general the left subscript is Z and the right superscript is A, and these numbers must add up to the same values on both sides of the equation, if we assume that both charge (Z) and mass number (A) are conserved.

These two assumptions should be clearly identified. At present there is no evidence for violation of charge conservation. However, one of the theories of elementary particles currently under discussion does postulate that the proton is not stable but may decay in such a way that the total mass number of a system decreases.[42] Even if the hypothesis should be confirmed, it would not impugn the accuracy of radiometric dating methods since the predicted probability for the process is extremely small; the half-life for decay of a proton is thought to be greater than the estimated age of the universe by several orders of magnitude. The process could only be observed by studying a very large sample. Experiments are now in progress to test the hypothesis and may give an upper limit to the accuracy of the assumption that mass number is conserved.

[Uranium 238 may also undergo spontaneous fission to elements of medium atomic number; this process is used in some other dating methods.]

The next stage of the radioactive decay process which started with uranium 238 is the beta-decay of thorium 234 to protoactinium,

$$_{90}Th^{234} \xrightarrow{\beta} + {}_{-1}e^0 + {}_0(\nu^*)^0, \tag{10}$$

where the beta particle is identified as a negatively charged electron with mass close to zero ($Z = -1, A = 0$). The symbol ν^* stands for the antineutrino. Again, total charge and mass number are conserved. One may ask, where did the electron come from? The answer is that a neutron has decayed into a proton and an electron (and an antineutrino); this is why the resulting nucleus, protoactinium 234, has a *higher* atomic number than thorium.

The complete series starting from uranium 238 can be written as follows, omitting the symbols for helium, electrons and neutrinos but simply writing α or β above the arrows as appropriate:

$$_{92}U^{238} \xrightarrow{\alpha} {}_{90}Th^{234} \xrightarrow{\beta} {}_{91}Pa^{234} \xrightarrow{\beta}$$

$$_{92}U^{234} \xrightarrow{\alpha} {}_{90}Th^{230} \xrightarrow{\alpha} {}_{88}Ra^{226} \xrightarrow{\alpha}$$

$$_{86}Rn^{222} \xrightarrow{\alpha} {}_{84}Po^{218} \begin{Bmatrix} \xrightarrow{\alpha} {}_{82}Pb^{214} \xrightarrow{\beta} \\ \text{or} \\ \xrightarrow{\beta} {}_{85}At^{218} \xrightarrow{\alpha} \end{Bmatrix} {}_{83}Bi^{214}$$

$$\begin{Bmatrix} \xrightarrow{\beta} {}_{84}Po^{214} \xrightarrow{\alpha} \\ \text{or} \\ \xrightarrow{\alpha} {}_{81}Tl^{210} \xrightarrow{\beta} \end{Bmatrix} {}_{82}Pb^{210} \xrightarrow{\beta} {}_{83}Bi^{210}$$

$$\begin{Bmatrix} \xrightarrow{\beta} {}_{84}Po^{210} \xrightarrow{\alpha} \\ \text{or} \\ \xrightarrow{\alpha} {}_{81}Tl^{206} \xrightarrow{\beta} \end{Bmatrix} {}_{82}Pb^{206} \tag{11}$$

As can be seen there are some alternate paths, since a nucleus which is unstable against both alpha and beta decay may suffer either decay first and then the other.

Two other radioactive series end with lead:

$$_{92}U^{235} \rightarrow \cdots {}_{82}Pb^{207} \tag{12}$$

and

$$_{90}Th^{232} \rightarrow \cdots {}_{82}Pb^{208}. \tag{13}$$

In general the mathematical description of these processes would be quite complicated, but in the case of these particular radioactive series, a substantial simplification is possible because the rate of decay of the "parent" nucleus (uranium 238, uranium 235, or thorium 232) is extremely slow compared to those of all the "daughter" nuclei. Thus the half-life of uranium 238 is about 4.50×10^9 years which is more than 10,000 times as long as that of any other member of its series, and similar ratios hold for the other series. After a few million years, therefore, essentially all of the uranium which has decayed will have reached the end point of the series and we can assume that

$$N_{Pb} = N_U^0 \ (1 - e^{-\lambda t}) \tag{14}$$

where N_{Pb} is the number of lead 206 atoms and N_U^0 is the initial number of uranium 238 atoms. The number of uranium 238 atoms after time t will be

$$N_U = N_U^0 \ e^{-\lambda t} \tag{15}$$

so we have

$$N_{Pb} = N_U \ (e^{\lambda t}-1). \tag{16}$$

If t is still short enough so that $\lambda t \ll 1$ (i.e., in this case t is much less than a billion years) we can approximate the exponential function by

$$e^{\lambda t} \approx 1 + \lambda t, \tag{17}$$

which gives

$$N_{Pb} = N_U \ \lambda t$$

or $\tag{18}$

$$t = (N_{Pb}/N_U) \ 1/\lambda = (N_{Pb}/N_U)T \ _{1/2}/0.693.$$

Thus the age of a rock in which all the lead came from the decay of uranium 238 would be

$$Age = (N_{Pb}/N_U) \ 6.5 \times 10^9 \ \text{years} \tag{19}$$

or

$$Age = (7.5 \times 10^9 \ \text{years}) \ (\text{mass of Pb})/(\text{mass of U}). \tag{20}$$

As indicated above, this is an approximation valid only when the age is much less than $T_{1/2}$ but much more than the half-life of the next longest-lived member of the series (in this case uranium 234 which has

$T_{1/2} = 2.5 \times 10^5$ years). Since we will be considering the older rocks, we can remove the restriction $t \ll T_{1/2}$ by going back to the equation

$$N_{Pb} = N_U \ (e^{\lambda t} - 1) \tag{19}$$

and solving it exactly for t:

$$t = 1/\lambda \ln(1 + N_{Pb}/N_U). \tag{21}$$

Note that the age is proportional to the half-life, thus an error of ten percent in determining the half-life will lead to a corresponding error of ten percent in the estimated age.

There is a third kind of radioactivity, called "gamma" (γ) which involves the emission of a photon of electromagnetic radiation. This process changes the internal energy of the nucleus but does not change it to a different isotope. It should not be confused with the decays used for radiometric dating (see below, section 7, item e-5).

RADIOMETRIC DATING

One of the simplest methods for determining the age of a rock is to measure the amount of helium it contains, and compare it to the amount of uranium and thorium. This method depends on the assumptions that

(1) all of the helium present in the rock has been generated by radioactive decay through the series (11), (12) and (13), i.e. there are no other sources of helium,
(2) none of the helium generated in this way has been lost.

It is generally believed by scientists that assumption (1) is valid, so that the method would at least give a *minimum* age; the amount of helium present is not greater than the amount produced by radioactive decay. However, research in the 1930s indicated that assumption (2) is probably not valid; the ages found by the helium method were significantly less than those found by other methods. It is interesting to note that at least one rock—"Stillwater Norite" from Quad Creek, Montana— was found to have an age of more than 1.8 billion years by the helium method, so this would seem to be a *lower limit* for the age of the earth.[43]

A much more satisfactory method was eventually based on quantitative measurements of the abundances of the isotopes of lead, developed by Alfred Nier at the University of Minnesota starting in 1938. Nier found that the stable isotopes—lead 204, lead 206, lead 207 and lead 208—do not occur in the same proportions in all rocks. The last three of these isotopes can be formed by radioactive decay as indicated in the previous section, while lead 204 is not. If we knew the relative abundances of these four isotopes at some initial time, N_{204}^0, N_{206}^0, N_{207}^0, N_{208}^0, we could, for example, compare the amounts of the two isotopes generated by decay of uranium after a certain time t. (The abundances are usually computed relative to $N_{204}^0 = 1$.) Decay of uranium 238 yields lead 206 with a half-life of 4.5×10^9 years, while decay of uranium 235 produces lead 207 with a half-life of about 7.1×10^8 years. The lead produced by decay is called *radiogenic* lead, while the lead originally present (N_{206}^0 etc.) is called *primeval* lead. If the initial amounts of uranium 238 and uranium 235 were equal ($N_{238}^0 = N_{235}^0$), the radiogenic component of lead 207, $N_{207}-N_{207}^0$, would increase more rapidly than the radiogenic component of lead 206, $N_{206}-N_{206}^0$, because of the shorter half-life of uranium 235. Since the initial abundances of uranium isotopes were not in general the same, we must compare $(N_{206}-N_{206}^0)/N_{238}^0$ to $(N_{207}-N_{207}^0)N_{235}^0$ and we expect that the second ratio will increase with time more rapidly than the first.

In order to explain the observed fact that different rocks at the present time have different relative abundances of lead 206 and lead 207, Nier proposed that each consisted of a different mixture of primeval and radiogenic lead. At the time of its formation, each rock consisted of a certain amount of primeval lead and a certain amount of uranium and thorium which later decayed to produce radiogenic lead. If the primeval abundances were known, and the present abundances of all four isotopes are measured, then the time t since formation of the rock could be calculated:

let λ be the decay constant for $_{92}U^{238} \rightarrow _{82}Pb^{206}$

let λ' be the decay constant for $_{92}U^{235} \rightarrow _{82}Pb^{207}$

then

$$N_{238} = N_{238}^0 e^{-\lambda t}, \ N_{206} = N_{238}^0(1-e^{-\lambda t}) =$$
$$N_{238}e^{\lambda t}(1-e^{-\lambda t}) = N_{238}(e^{\lambda t}-1)$$
$$N_{235} = N_{235}^0 e^{-\lambda' t}, \ N_{207} = N_{235}^0(1-e^{-\lambda' t}) =$$
$$N_{235}e^{\lambda' t}(1-e^{-\lambda' t}) = N_{235}(e^{\lambda' t}-1)$$
$$(N_{207} - N_{207}^0)/(N_{206} - N_{206}^0) =$$
$$(N_{238}) \ (e^{\lambda' t}-1)/(N_{235})(e^{\lambda t}-1). \tag{22}$$

The problem then reduces to estimating the primeval abundances of the lead isotopes, N_{206}^0 and N_{207}^0. Nier and his colleagues assumed that the closest approximation to primeval lead would be a rock that has the highest proportion of lead 204, since all of that isotope is primeval. So they chose a galena from Ivigtut, Greenland, for which the abundance of the isotopes were in the ratios

$$1 : 14.54 : 14.60 : 34.45,$$

and assumed that was the ratio

$$N_{204}^0 : N_{206}^0 : N_{207}^0 : N_{208}^0.$$

The oldest rock sample, Huran Claim Monazite, gave an age of 2.57 billion years by this method, but this was substantially greater than the two billion years "usually taken as the earth's beginning" at the time, so they cautiously reported only that it "appears to have an age close to two billion years."[44]

In 1946 Arthur Holmes and F. G. Houtermans independently pointed out that Nier's method can be extended to give not only the ages of particular rocks but the age of the earth itself.[45] Suppose we take $t = 0$ to be the time of formation of the earth, and apply equation 22 to give the ratio of the amounts of lead isotopes generated between a time t_m when a particular rock was formed ("m" for "mineralization") and the present time t_0 after the earth's formation. Then we find

$$(N_{207} - N_{207}^0)/(N_{206} - N_{206}^0) =$$
$$(N_{238})(e^{\lambda't_0} - e^{\lambda't_m})/ (N_{235})(e^{\lambda t_0} - e^{\lambda t_m}) = Q. \qquad (23)$$

If we have a group of samples with the same age (t_m) but different admixtures of primeval and radiogenic lead, then they will have different abundances of the two lead isotopes (relative to lead 204) but if we plot N_{207} against N_{206} we should get a straight line,

$$N_{207} = QN_{206} - QN_{206}^0 + N_{207}^0, \qquad (24)$$

since all the other quantities in this equation should be constant. Houtermans called this line an "isochrone" since it displays the variation in isotopic composition for rocks formed from varying amounts of uranium and thorium at the same time.

Equation 24 is the fundamental equation for determining t_0, the present age of the earth, on the assumption that uranium has been decaying up to a time t_m after the earth's formation at $t = 0$. The particular rock being analyzed is assumed to have been formed at t_m from an arbitrary mixture of primeval lead whose isotopic composition is given by

$$N_{204}^0 = 1, \ N_{206}^0, \ N_{207}^0, \ N_{208}^0,$$

and uranium whose isotopic composition is given by

$$N_{235} = N_{235}^0 e^{-\lambda' t}{}_m, N_{238} = N_{238}^0 e^{-\lambda t}m.$$

How the equation is applied depends on which parameters are considered to be known and how much data are available to determine the others. In some cases the time of the mineralization of the rocks, t_m, can be estimated reliably from other methods; in that case the slope Q of the graph of N_{207} vs. N_{206}, and the present value of N_{238}/N_{235}, can be used in equation 23 to determine t_0. The intercept of the graph of equation 24 gives $N_{207}^0 - QN_{206}^0$; if Q can be determined for two different values of t_m, one can then estimate the primeval isotopic abundances N_{206}^0 and N_{207}^0. Conversely, if a reasonably good value of the age of the earth t_0 is known, this can be substituted into equation 23 to give t_m. The fact that the graph of N_{207} against N_{206} *does* give a straight line in most cases confirms the validity of the basic model, and allows one to exclude those rocks which have undergone further chemical differentiation so that they do not give st. aight-line plots.

When Holmes and Houtermans applied this method to the data available in 1946, they found $t_0 = 2.9 \pm 0.3$ billion years.[46] The primeval lead isotopic abundances were estimated to be $1 : (11.5 \pm 0.6) : (14.0 \pm 0.2) : (31.6 \pm 0.6)$.[47]

Although some scientists pointed out that the available data did not exclude a value for the age of the earth as high as 5 billion years,[48] the Holmes-Houtermans value of 3 billion years was generally accepted until 1953.[49] In that year a group of scientists at the University of Chicago and the California Institute of Technology reported that the abundances of the radiogenic lead isotopes in some meteoritic material were significantly lower than the figures previously considered "primeval" in estimating the age of the earth. The ratio of lead 204: lead 206: lead 207: lead 208 was found to be 1: 9.4: 10.3: 29.2. Moreover, the ratio of uranium to lead in these meteorites was extremely low, so little if any of the present abundance of lead 206 and lead 207 could be attributed to decay of uranium since the formation of the meteorite.[50] It seemed reasonable to suppose that this material was much less affected by chemical differentiation processes than minerals found in the earth's crust, so that these values were the most appropriate ones to use for the abundances at the time of formation of the earth, N^0_{206} and N^0_{207} in equation 23. Results based on these data were announced almost simultaneously in September 1953 by Clair Patterson and Friedrich Houtermans:[51]

Age of the Earth $= 4.5 \pm 0.3$ Billion Years

In discussing the discrepancy between the earlier result, based on inferring the primeval isotopic composition from several ores of different composition, and the new result based on meteoritic data, Patterson, Tilton and Inghram (1955) pointed out that each might be subject to criticism. The first, which they called the "ore method," leading to an age of 3.5 billion years, depends on the *scatter* of individual isotopic compositions about the curve representing the average radiogenic lead growth. The samples must be the same age (t_m) but must have different admixtures of primeval and radiogenic lead so the parameters of the curve of equation 24 can be determined. "Since the calculation is based on these second-order effects small errors in the underlying assumptions may be capable of introducing larger errors into the results."[52] Moreover, one has to assume that the uranium/ lead ratio has remained fixed after t_m, i.e. for billions of years, which seems unlikely. On the other hand, "the meteorite method uses a primordial lead of accurately known isotopic composition but makes use of extraterrestrial materials whose relationship to the earth is uncertain. There is no evidence that compels us to assume that when the earth was formed it contained lead with the same isotopic composition as that in iron meteorites."[53] Nevertheless both Houtermans (who had previously used the ore method) and Patterson's group leaned toward the meteoritic method, recognizing that further data must be collected in order to confirm its validity.

By 1956, Patterson thought that enough data were available to clinch the argument for the 4.5 billion year age. The meteorites used in the calculation had been found to have the same age by three independent radiometric methods, within the known limits of accuracy of each method: lead/uranium, potassium/argon, and strontium/ rubidium (see following sections for explanation of these methods). The most accurate method, based on the lead 207/lead 206 ratio, gave an age of 4.55 ± 0.07 billion years. Several terrestrial minerals were found to contain lead isotope ratios that fall on the same 4.55-billion year isochron as do the meteorites. Patterson concluded that the age of the earth is the same as that of the meteorites: "we should now admit that the age of the earth is known as accurately and with about as much confidence as the concentration of aluminum is known in the Westerly, Rhode Island granite."[54]

On the basis of a recent survey of ten determinations of the age of the earth, mostly based on the lead-uranium method, with results ranging from 4.43 to 4.59 billion years, Fouad Tera states that the best estimate is 4.53 to 4.56 billion years. He criticizes another determination (Manhes *et al.* 1979)[55] leading to a "low" result, 4.49 billion years; this indicates the degree of precision one can now expect in this area.[56] It is remarkable that in the 28 years since Patterson and Houtermans announced their estimate of 4.5 ± 0.3 billion years, almost every investigator has found a result lying within their original limits of error. Considering the major changes that have taken place in the earth sciences during this time, the stability of this result is quite impressive.

For further discussion of the lead/uranium dating method see Doe (1970), Hamilton (1965), Hamilton and Farquhar (1968, chapter by Kanasewich), and Jäger and Hunziker (1979, chapters by Köppel & Grünefelder and Gebauer & Grünefelder). A good textbook which covers radiometric dating as well as the background in geomagnetism needed for section 8 is Stacey (1977).[57]

OTHER RADIOACTIVE DECAYS USED FOR DATING

As often happens in science, our confidence in the presently accepted value of the age of the earth depends not on a single decisive experiment but rather on the mutual consistency of several independent measurements. The estimates from lead isotopic abundances are supported by two other age determinations which have been developed for rocks that crystallized hundreds of millions of years ago, the rubidium/strontium ratio, and the potassium/argon ratio. While it is conceivable that any one age determination could be erroneous because the rock experienced gains or losses of particular isotopes contrary to the assumptions of the method, it would be extremely unlikely that such processes affected the abundances of separate isotopes of three different elements in such a way as to make several rocks appear to have consistent ages of millions or billions of years when in fact they were formed only a few thousand years ago (unless, of course, one wants to assume that God arranged things just that way to fool us).

The use of the decay of rubidium to strontium for geological age determinations was suggested in 1937 by V. M. Goldschmidt and developed by Otto Hahn, Ernst Walling, and L. H. Ahrens.[58] The nuclear reaction is:

$$_{37}Rb^{87} \rightarrow {}_{38}Sr^{87} + {}_{-1}e^0$$

with a half-life of about 50 billion years. Since this is much greater than the age of most rocks, the amount of rubidium is essentially constant and equation 18 may be used in the form

$$Age = (N_{Sr} - N_{Sr}^0)/(N_{Rb}\lambda),$$

where N_{Sr}^0 is the amount of strontium 87 originally present in the rock.

In principle one would expect to have the same problem in estimating N_{Sr}^0 as one does in estimating the primeval abundances of the lead 206 and lead 207 isotopes in the lead-uranium method, but in practice there is much less difficulty and uncertainty because only one decay is involved, and the strontium 87 isotope forms only seven percent of common strontium. The most abundant isotope, strontium 88, can be measured along with strontium 87, and makes it easy to correct for the amount of non-radiogenic strontium 87 present. Alternatively one can use the "isochron" method described in the previous section, by taking three or more minerals known to have been formed at the same time (they may be in the same rock) but with different strontium/rubidium ratios. By plotting the amount of strontium 87 against the amount of rubidium 87, one can find the amount of strontium 87 that would be present if there were no rubidium 87 at all, from the intercept of the straight line, and this given N_{Sr}^0 directly. The fact that the points do fall on a straight line provides a check of the validity of the assumptions involved in this method, e.g., that no loss of radiogenic strontium 87 from the mineral has occurred since it was formed.

The rubidium-strontium method has been used to show that many rocks have ages of 2 or 3 billion years, and that many meteorites have ages of 4.6 billion years.[59] These results support the modern estimate of the age of the earth by confirming the ages of the rocks and meteorites used in these estimates, as mentioned above.

The potassium-argon method originated in a suggestion by C. F. von Weizsäcker in 1937, that potassium 40 might decay to argon 40 by capture of an orbital electron:

$$_{19}K^{40} + {}_{-1}e^0 \rightarrow {}_{18}Ar^{40}.$$

This suggestion was confirmed by experiment and it was found that the ratio argon 40/argon 36 in potassium-bearing minerals is correlated with their ages determined in other ways.[60]

Because of the widespread occurrence of potassium-bearing minerals, the potassium-argon method has become exceptionally useful in determining rock ages. Although argon is gaseous at normal temperatures, it is a much larger atom than helium and seems to be retained almost completely within the mineral; thus potassium-argon ages are more reliable than those estimated from the helium produced by alpha-decay of uranium. Although generally used for older rocks, the potassium-argon technique has been successfully applied to rocks as young as 5000 years, and the precision of the method makes it possible to establish the time scale for human evolution and for geomagnetic reversals.[61]

Although I will not discuss the "carbon dating" method here, because it cannot be used for dating objects older than about 70,000 years, it should be noted that it is essentially different from the three methods I have described. The method involves assuming that carbon 14 is produced in the upper atmosphere by cosmic ray neutrons striking $_7N^{14}$ atoms and ejecting protons according to the reaction:

$$_0n^1 + {}_7N^{14} \rightarrow {}_6C^{14} + {}_1p^1 .$$

The carbon 14 then decays back to $_7N^{14}$ by emitting an electron

$$_6C^{14} \rightarrow {}_7N^{14} + {}_{-1}e^0.$$

The carbon 14 is incorporated into organic substances where it is mixed with the stable isotope carbon 12 and carbon 13. It is assumed that after the organism dies it does not acquire any more carbon 14. Although the proportion of carbon 14 found in an organic substance will obviously be related to the number of years since it ceased to acquire this isotope, there is no reason to assume that the cosmic-ray flux and the composition of the earth's atmosphere have been fixed for the last 70,000 years. Thus the carbon dating method must be calibrated by comparing its results with ages determined in other ways, such as by counting tree rings. It is important to recognize that the errors involved in the radiocarbon method are due to factors that affect only this method, and do *not* apply to the uranium/lead, rubidium/strontium and potassium/argon methods.

Since creationists claim that not only the earth, but the entire Universe, was created a few thousand years ago, their model conflicts even more strongly with recent work in astrophysics. A survey of this work can be found in a review by Symbalisty and Schramm (1981), who conclude that "the long-lived radionuclides $^{234}Th/^{238}$ U and $^{187}Re/^{187}$ Os set a lower limit on the age of the universe of 8.7 X 10^9yr by their model-independent determination of the mean age of the elements."[62]

CRITICISMS OF RADIOMETRIC DATING

At the present time, the 4.5-billion-year time scale is accepted by the overwhelming majority of scientists. This fact in itself is remarkable since most other aspects of geophysics and planetary science are still in a state of lively controversy, and it would be difficult to find such nearly-unanimous agreement on a problem like, for example, the origin of the moon or the mechanism for continental drift. Aside from the creationists, only a handful of legitimate scientists have questioned the accuracy of the radioactivity estimate of the age of the earth. One of them is A. E. Mussett of the University of Liverpool, who argued that there is not sufficient reason to identify the primeval abundance of the radiogenic lead isotopes with the values determined from meteorites, as Patterson and others did. Because of the complex geological history of the earth which has affected the amounts of lead and uranium in the rocks now available for study, we may never be able to determine the precise age of the earth.[63] Nevertheless Mussett, who is the most severe critic among the scientists who have professional qualifications in this area, concluded that "the interval 5000 to 3500 million years ago almost certainly encompasses the formation of the earth as a separate entity within the solar system." Moreover, he points out that the objections to estimates of the age of the earth do not apply to the ages of meteorites, whose formation can be assigned quite confidently to a period from 4500 to 4700 million years ago.[64] Thus Mussett's criticism does the creationists no good at all, but only strengthens the evidence against their postulate that the entire universe was created less than 10,000 years ago.

It is noteworthy that no creationist has bothered to examine the precision of radiometric dating as critically as Donald McIntyre (1963) and the creationists do not even cite his work, suggesting they have no real interest in the scientific side of this subject.[65]

We now turn to the creationist criticisms, as presented in the books by Morris (1974) and Slusher (1981).[66] Many of them have been answered authoritatively by G. Brent Dalrymple, a geologist well known for his work on radiometric dating and geomagnetic reversals, so we need only refer to his article (Dalrymple, 1981).[67]

    (a) *"Rocks are not dated radiometrically.* Many people believe the age of rocks is determined by study of their radioactive minerals—uranium, thorium, potassium, rubidium, etc.—but this is not so. The obvious proof that this is not the way it is done is the fact that the geological column and approximate ages of all the fossil-bearing strata were all worked out long before anyone ever heard or thought about radioactive dating."[68]

This "proof" [offered by Morris] is easily refuted by looking at any nineteenth-century treatise on geology and comparing the statements made about the various geological ages with those in a modern textbook. For example, the American geologist James D. Dana stated that geology can determine only relative, not absolute lengths of these ages.[69] One sometimes finds vague statements about millions or hundreds of millions of years in the nineteenth-century books, but geological methods alone never provided any basis for a billion-year time scale before the introduction of radioactive dating. The literature on radiometric dating is so extensive that some of it can be found in practically any university library.

    (b) *Ages of rocks are based on the fossils they contain but the dating of the latter depends on the theory of evolution, hence the use of a long time scale to support evolution is a circular argument.* [70]

The quotations given to support this argument do not in fact support it, if one recognizes the difference between the relative, sequential time scale used in stratigraphy, and the quantitative radiometric ages. As we saw in section 5, radiometric dating gives an age of the earth which appears to be accurate within an error of no more than 300 million years. Since the age of the earth is determined to be more than 4 billion years, this means an error of less than ten percent. On the other hand, in classifying rocks which are less than 500 million years old, a method which may be inaccurate by 300 million years is obviously useless, and it is only in the last few decades that radiometric dating methods have become accurate enough to compete with more traditional geological methods.

Morris quotes a 1957 paper by O. H. Schindewolf as follows:

"The only chronometric scale applicable in geologic history for the stratigraphic classification of rocks and for dating geologic events is furnished by the fossils."[71]

His footnote gives the wrong volume and page numbers for the quotation, but more serious is the omission of a crucial word: the actual phrase used was ". . . for dating events *exactly* . . ."[72] Schindewolf points out that fossils provide a good *relative* time scale, whereas "absolute age determinations are by far too inaccurate for stratigraphical purposes." (A meter stick might be perfectly adequate to determine your height but useless in finding out if your right eye is bigger than your left eye.)

Another quotation used by Morris is from Everden *et al.* (1964):

> "Vertebrate paleontologists have relied upon 'stage-of-evolution' as the criterion for determining the chronologic relationships of faunas."[73]

Here the word "have" is a tip-off that the sentence has been taken out of context. Morris fails to mention that the purpose of the paper is to report new results with potassium-argon dating that provide independent confirmation of the time scale based on evolution. The text immediately following the above-quoted sentence reads:

> "Before the establishment of physical dates, evolutionary progression was the best method for dating fossiliferous strata. The physical dates presented in this paper (Table 6) demonstrate that temporal position of genera and species of fossil mammals in their accepted phylogenies is accurate at Mammal-Age degree of refinement."

The conclusion is summarized in the abstract:

> "The K/Ar ages and the Mammal Age designations are in essentially perfect agreement, thus substantiating the usefulness of the K/Ar technique throughout the Tertiary and supporting the conclusion that the defined Mammal Ages have true evolutionary significance."[74]

Tracking down the contexts of such creationist quotations may be left as an exercise for the student, provided he has been made aware of the different purposes for which geological dating is done, so that quantitative ages are not confused with relative ages. It should be pointed out that while fossils are sometimes used for the latter purpose, an examination of the geological literature before the publication of Darwin's *Origin of Species* in 1859 shows that the relative ages of most of the geological periods and epochs were established by 1840 *without* reliance on any evolutionary assumptions.[75] Conversely, as indicated in the previous sections, rocks can indeed be given quantitative ages by radioactive methods without relying on fossils at all. In fact the kinds of rocks most commonly dated radiometrically are igneous and metamorphic rocks that have no fossils at all. Two other obvious examples (which should be well known) are meteorites and the moon rocks, which have been dated very accurately even though they contain no fossils.

These two criticisms (a) and (b) are not even mentioned in the recent book by Slusher (1981), perhaps because they are so obviously phony.

(c) *"Uranium minerals always exist in open systems, not closed."*[76]

If some uranium or lead enters or leaves the system, the calculated age would be wrong.

The only evidence presented by Morris (1974) that terrestrial uranium minerals are not closed systems is a quotation from Henry Faul's book, which does indeed state that *some* uranium minerals are unreliable for age estimates.[77] But Faul also states:

> "Rigorously closed systems probably do not exist in nature, but surprisingly many minerals and rocks satisfy the requirement well enough to be useful for nuclear age determination."[78]

As mentioned earlier, the methods of estimating ages are self-checking; if some of the particular minerals selected for analysis were not closed systems, then the isotopic abundances would not fall on a straight line (the "isochron"). Our confidence in the present value of the age of the earth rests on the fact that a large body of self-consistent data does exist, and the fact that some minerals were not closed systems because of their geological history does not invalidate the ages determined from other minerals.

(d) *Lead 206 may be converted into lead 207 by free neutron capture, thus invalidating the estimation of ages from the lead 206/lead 207 ratio.*[79]

This argument was suggested by creationist Melvin Cook (1966), on the basis of what he himself calls "circumstantial evidence."[80] In a particular uranium ore from Katanga, lead 208 was found, but there was no thorium which could have produced it (see equation 13). Moreover, according to Cook, no lead 204 was found, "the absence of Pb^{204} implies that there is no original lead in this ore; apparently all of the lead is radiogenic."[81] Hence it must have been generated by neutrons acting on lead 207,

$$Pb^{207} + n \rightarrow Pb^{208},$$

and if such neutrons were present they would also have converted some of the lead 206 to lead 207.

This argument has been completely refuted by Dalrymple (1981),[82] who points out that the statement "no Pb^{204} was found" is simply incorrect; Cook neglected to read the original source of his data, a paper by Nier (1939),[83] who did *not* say that lead 204 was absent in his Katanga samples but rather that he did not measure it.

The other example given by Cook, and repeated by Slusher and Morris, is based on the Martin Lake ore of the Canadian Shield: "Here Pb^{206}/Pb^{207} is given as 90.4/9.1 = 9.9 and the ore is assigned an age of about 1650 My. There was found an average of 0.53% Pb^{208} in this ore but enough thorium (Th/U = 0.0002) to account for less than 1% of this Pb^{208}. Apparently, practically all of it, therefore, came from the (n, Υ) transmutation of Pb^{207}."[84]

Again, if we go back to the original source of the data, we find it does not support the argument. The figures quoted for the lead isotope abundances must be corrected for the amount of primeval lead present in order to get the radiogenic component, and when this is done there is no significant amount of lead 208 left.[85]

The neutron-transformation argument relies on the assumption that enough neutrons were present at some time in the past to produce the reactions to a significant extent. Cook himself admits that the present-day neutron flux is a million times too small to do this (Morris and Slusher both omit this rather significant point in quoting Cook). The only evidence that it might have been larger in the past comes from misinterpreted data on lead isotopes. If it had indeed been large enough to produce the alleged effect on lead isotope abundances, a much greater effect would have been observed in other isotopes; for example, gadolinium 157 has a much larger cross section for neutron absorption than lead 206 or lead 207.

Incidentally, the "expert" quoted by the creationist, Cook, does not himself claim that the earth is young; he suggests that the age of the Universe may even be infinite.[86]

(e) *The uranium decay rates may well be variable.*
I will consider separately the suggested reasons for variation.

(e-1) *Cosmic radiation produced by a supernova could have "reset the atomic clocks" and thereby invalidated the carbon 14, potassium-argon, and uranium-lead dating measurements.* [87]

This objection is attributed to a column on "Scientific Speculation" by Frederick Jueneman (1972), who presents no evidence that such effects could be significant but merely says it might happen if Dudley's "neutrino sea" theory (see below) were true.[88] The last sentence of

Jueneman's piece suggests that he doesn't really believe it himself, but Morris did not include that sentence in his quotation.

A recent paper by G. Robert Brakenridge of the University of Arizona suggests that a supernova explosion sometime between 8,400 and 11,300 years ago might indeed have affected the accuracy of the carbon 14 method—but in a direction opposite to that claimed by the creationists. With the estimated energy of 10^{49} ergs, radiocarbon dates would now appear 220 years too young; with the maximum possible energy of 10^{50} ergs, they would be 400 years too young.[89] In response to my inquiry about the possible effects of a supernova on estimates of the age of the earth, Brakenridge explained that he was inferring effects from a "hard photon flash" and that he could "see no way that hard photons could affect uranium decay within crystalline igneous rocks."[90]

Perrin proposed more than 50 years ago that radioactive decay may be due to the absorption of cosmic rays. This hypothesis was tested by L. R. Maxwell (1928),[91] who took a polonium source 1150 feet below the earth's surface to the bottom of a mine in New Jersey; he found that there was no change in activity within the limits of accuracy of his experiment (about one percent). Since most cosmic rays would be blocked at this depth, Maxwell's result is evidence against any suggestion that the rate of radioactive decay can be significantly influenced by cosmic rays.

> (e-2) *Changes in cosmic radiation produced by reversal of the earth's magnetic field could change the uranium decay rate.*[92]

No evidence is presented for this hypothesis, so it does not seem worthwhile to try to refute it. However, it should be noted that according to Barnes' theory, which is the one primarily relied upon by the creationists to support their own 10,000 year time scale (see below), there have *not* been any reversals of the earth's magnetic field.

> (e-3) *Decay rates can be changed by pressure, temperature, or chemical state.*[93]

Slusher first attempted to support this claim by citing a review article by Indiana physicist G. T. Emery (1972).[94] None of the experimental results mentioned by Emery, however, indicated a change of more than four percent. For the decays actually used to estimate the age of the earth there is no evidence for a change of more than one percent, which means that the earth is still at least four billion years old (cf. equation 21). In a recent letter, Emery stated that no significant changes had been reported since he wrote the review article, and further, that "from what is known about decay rate changes due to chemical and physical effects, there is no reason to doubt the accuracy of current radioactive dating results for the age of the earth."[95]

In his 1981 critique of radiometric dating, Slusher relies heavily on a book by H. C. Dudley (1976).[96] Dudley also cites Emery's review article along with some earlier papers, but he does not claim that any change in the decay rates of the uranium, thorium, rubidium or potassium isotopes used to estimate the age of the earth has ever been observed. Thus Slusher's criticism rests not on any direct experimental evidence, but only on Dudley's general theory of radioactive decay, which therefore deserves our consideration.

Dudley states that he rejects the modern physical theories that have "won almost unquestioning acceptance by the scientific community," and sees a need "to construct more complete conceptual *mechanistic* models of both the macrocosmos and the microcosmos, à la Rutherford and Michelson."[97] In particular he rejects Einstein's theory of relativity and wants to revive the classical "ether" under the modern-sounding name "neutrino sea." He also identifies this ether with the "subquantic medium" of L. de Broglie and J. P. Vigier (1963).[98] Such a medium would allow one to reduce the apparent randomness of quantum effects such as radioactive decay to a determinism on the level of "hidden variables." As Dudley recognizes, if this subquantic ether really does exist, then both relativity theory and quantum mechanics are wrong.[99]

It is completely consistent with the philosophy of creationism to reject the randomness of natural processes.[100] But neither Slusher nor Morris inform the reader that their critique of radiometric dating methods involves the rejection of the two most spectacularly successful theories of modern physics, relativity and quantum mechanics. Nor do they seem to realize that the "hidden variables" postulate used by Dudley has been thoroughly disproved by recent experiments which reconfirm the extraordinary accuracy of quantum mechanics even in its most counter-intuitive predictions.[101] To attack the theory of radioactive decay by abandoning quantum mechanics seems almost suicidal; one can only suppose that the creationists know nothing about modern atomic physics (in spite of their "qualifications") or that they hope no one will notice how absurd their position is. Dudley himself rejects the conclusion that Slusher has drawn from his neutrino-sea theory. He points out that "since laboratory findings indicate that the maximum observed alterations of *any* decay rate, *to date* is $\sim 10\%$, usually less, the figure of 4.5 billion years for the earth's age seems to be a good ball park figure."[102]

Aside from their reliance on Dudley's ether theory, the creationists are prepared to throw out relativity theory in another area of science by developing an alternative theory of gravity. They hope that "consideration of the radiation of energy through gravitational waves might help to prove the youth of the universe, by setting an upper limit on the age of double stars, planetary systems, etc."[103] Another approach that involves rejecting relativity theory is to adopt the discredited hypothesis of Walther Ritz, that the speed of light depends on the motion of the body emitting it, in order to obtain a very young age for the entire Universe.[104]

(e-4) *Evidence for past variations of radioactive decay constants comes from variations in the sizes of radioactive halos in minerals.* [105]

According to the Geiger-Nuttall law, there is a direct relation between the decay constant and the average distance travelled by the particle before it is absorbed by matter. If a speck of radioactive substance is enclosed in a crystal, the alpha particles emitted by the substance will be absorbed at a particular distance from the speck, producing a "halo" due to change in the crystal structure there. Variations in the radii of the halos indicate that the decay constant has changed.

This claim was originally made by J. Joly in the 1920s but was eventually abandoned when more was learned about the radioactive decay series. The decay from uranium to lead involves the successive emission of alpha particles with different decay constants and therefore several different halos should be formed. Joly himself eventually conceded that the age of the earth is "much more than a thousand million years,"[106] abandoning his earlier claim that variations in the decay constant implied a much shorter age. A thorough discussion of the early work may be found in the article by Arthur Holmes.[107]

Slusher cites only two other alleged discoveries of variation in halo size, both in creationist publications, while ignoring the large amount of research showing that halo radius is generally constant.[108] He gives special emphasis to the work of Robert V. Gentry on polonium halos.[109] Although Gentry appears to support the creationists' short time scale,[110] he has published research on radioactive halos in journals such as *Science* and *Nature* as well as in the creationist magazines. His approach appears to be considerably more objective than that of other creationists, for example he admits his own error "in inventing new α-activity to account for some ... ghost rings" in halos.[111] He claims to have shown that "uncertainties in radius measurements alone preclude establishing the stability of λ for $_{238}U$ to more than

35%."[112] Such a large inaccuracy in the uranium-lead dating method, if confirmed by other workers, would certainly be quite disturbing to geophysicists; but it would not be evidence for the creationist model since it would still leave the minimum age of the earth at three billion years.

Since Gentry is the only creationist who seems to have substantial qualifications in the field of radioactivity and his views may therefore carry some weight, it is necessary to mention a controversy in which he has been involved. In 1976 Gentry and six other scientists claimed to have found evidence for superheavy elements in microscopic crystalline monazite inclusions showing giant halo formation in biotite mica, using proton-induced x-ray emission. These were halos generally considered to involve uranium and thorium and are therefore directly related to the type of evidence mentioned above. They claimed that element number 126 is *now* present in the inclusions. The paper was rushed into print without going through the usual review process and created a considerable amount of excitement among nuclear physicists.[113] But it was quickly found that other researchers could not confirm the result, and Gentry's group eventually had to retract their claim when re-analysis of the sample which had shown the strongest evidence for superheavy elements, with much more sensitive techniques, indicated that no superheavy elements could be detected.[114] One of the groups which disconfirmed the results of Gentry and his colleagues pointed out that even if the observations had been valid, their theoretical interpretation was wrong and the atomic number of the superheavy elements would have had to be at least 250 rather than 126; they concluded by asking "why did we do all those experiments?"[115]

Since it is generally believed that superheavy elements would have fairly short half-lives, evidence that they are now present in measurable quantities would have provided a sensational new argument for a young earth, or at least for some recent catastrophic event that could have deposited them on the earth's surface. No one has suggested that Gentry intentionally falsified or misinterpreted his experiment on halos in order to support creationism. Other scientists, much more famous than Gentry, have made much bigger mistakes without being accused of fraud. I simply point out that one should not draw earth-shaking conclusions from any single isolated experiment that has not been replicated independently by other scientists.

(e-5) *Experiments have shown that the decay rates of cesium 133 and iron 57 vary, hence there may be similar variations in other radioactive decay rates.*[116]

These are both *stable* isotopes so there is no decay rate to be changed. This statement merely reveals Slusher's ignorance of nuclear physics. (Gamma decay of an excited state of iron 57 has been studied, but this has nothing to do with the kinds of decays used in radiometric dating.)

(f) *The potassium-argon dating method is invalid because rocks may contain excess argon from the atmosphere or from neighboring minerals, or from the magma from which the rocks were derived, thus appearing to be older than they really are.*[117]

For example, rocks formed in 1801 near Hualalei, Hawaii, gave potassium/argon ages ranging from 160 million years to 3 billion years.

This and other examples have been discussed by Dalrymple (1981) who points out that some rocks are suitable for this dating method while others, such as the xenoliths studied in the 1801 Hawaiian lava, are not. The point is that there are enough ways of cross-checking the results to eliminate the kinds of rocks that are unreliable because they tend to contain excess argon, without invalidating the results obtained from other rocks.

Professor John J. Naughton of the University of Hawaii, one of the scientists whose paper was quoted by the creationists, points out that his conclusions have been misinterpreted. Morris claimed that in the paper by Funkhouser and Naughton "the reason given for the anomalously high ages was the incorporation of environmental argon at the time of lava flow," whereas they had stated specifically that "such gases represent a portion of the environment in the magma chamber," which is a peculiar and importantly different physical situation from what can occur in a lava flow. Naughton writes,

"I have felt that these articles only give strength to the applications of K-Ar dating as a geochronological method, in that they point out instances (ultrabasics and basalts erupted in the very deep ocean) where it is unsuitable to apply the K-Ar method. To say that because the K-Ar method should not be applied to such rocks, therefore it cannot be applied to *any* rocks is as logical as saying that because some plants cannot be used for human food, therefore all plants should not be used for this purpose. . . . We have continued to use K-Ar dating in our researches out here with caution, but without any hesitation as to its reliability when properly used."[118]

Morris states that "Since there is no way at all to distinguish argon 40 as formed by unknown processes in primeval times and now dispersed around the world, from radiogenic argon 40, it seems clear that potassium-argon ages are meaningless in so far as *true* ages are concerned."[119] But in fact the proportion of atmospheric argon incorporated into the rock can be easily estimated by measuring the amount of argon 36 present and using the known isotopic composition of atmospheric argon to make this correction.[120] Neither Morris nor Slusher even mentions this point, so I must assume that they are unfamiliar with one of the most elementary parts of the procedure they are criticizing.

(g) *The existence of primordial polonium 218 halos in minerals indicates that the earth was not formed gradually over a long period of time but was created in a few hours "by Fiat nearly 6 millenia ago."*[121]

According to Gentry, the halos he has observed in certain minerals were produced by the decay of primordial polonium 218, an isotope with a half-life of only three minutes. If his interpretation were correct, it would imply that the earth was created in a few minutes. However, there are alternative explanations for the halos which he attributes to primordial polonium.[122] In particular, Hashemi-Nezhad *et al.* (1979) showed experimentally that the diffusion of lead in mica can be rapid enough to explain the anomalous polonium halos.[123] According to one of the experimenters in this group, "the haloes are inconsistent with creation less than tens to hundreds of millions of years ago unless one invents two easily observable but unobserved lead isomers of quite improbable characteristics."[124]

Gentry himself does not claim that his results lead directly to a value for the age of the earth, but he argues that the uranium 238/lead 206 ratios found in coalified wood from the Colorado Plateau could be explained by an infiltration of uranium a few thousand years ago.[125] To accept his view that the infiltration event was associated with the creation of the earth would require discarding theories based on a large amount of data from many areas of science, in order to explain a single isolated type of observation. It does not seem sensible to throw out well-established principles of science without having alternative principles that could explain at least most of the same observations, and no such alternative exists.[126] This is a good illustration of the fact that no scientific theory can or must explain *all* observations, and that a theory which is able to give a satisfactory explanation of most observations will not be replaced unless a better one is available. Gentry's postulate of recent creation of the earth is contradicted by so many other facts that it has gained no support from other scientists who are familiar with this field.

CREATIONIST ESTIMATE OF THE AGE OF THE EARTH

Morris (1977) lists 70 estimates of the age of the earth, ranging from 100 years to 500 million years. As he points out, all are based on the "uniformitarian" assumption that a process has gone on at the same rate in the past that we now observe, and therefore none is reliable.

These are processes such as the influx of various elements to the ocean, cooling of the earth by heat efflux, and erosion of sediments from continents. What he fails to note is that we have definite reasons, in the present state of scientific knowledge, to suppose that these rates *did* vary in the past, whereas we have definite reason to suppose the rate of radioactive decay did *not* vary. Arthur Holmes carefully explained in 1936 why these methods could not give accurate estimates of the age of the earth, but some creationists continue to use them.[127] Wonderly has shown that the best nonradiometric techniques clearly refute "young-earth" creationism.[128]

The one method which the creationists seem to consider especially reliable, which also happens to be almost the only one that gives an age anywhere near 10,000 years, is the decay of the earth's magnetic field. Morris says this method demonstrates "that the earth almost certainly was created less than 10,000 years ago."[129]

Thomas G. Barnes, who proposed the magnetic field method for estimating the age of the earth, was appointed Dean of the Graduate School of the Institute for Creation Research in 1981, following his retirement from the physics faculty at the University of Texas at El Paso. He was chairman of the committee responsible for the development of the Creation Research Society's biology textbook (Moore and Slusher, 1970).

According to Barnes, "in 1883 Sir Horace Lamb proved theoretically that the earth's magnetic field could be due to an original event (creation) from which it has been decaying ever since."[130] This is not a correct description of Lamb's 1883 paper, which dealt only with electric currents and did not mention geomagnetism at all; Barnes assumes that the same mathematical equations apply to magnetism.

Barnes uses data assembled by Keith L. McDonald and Robert H. Gunst (1967) on the intensity of the earth's magnetic field from 1835 to 1965, but picks out only the "dipole component" for analysis.[131] He fits the data to an exponential curve,

$$M \text{ (magnetic moment)} = M_0 \, e^{-t/T},$$

where M_0 is the value of M in 1835, namely 8.558×10^{-22} amp meter^2. The time constant T is evaluated by substituting the value of M in 1965, when $t = 130$ years later, namely 8.017×10^{-22} amp meter^2. This gives

$$(130/T) = \ln (8.558/8.017) = \ln 1.0675 = 0.0653$$

$$T = 1990$$

The half-life for this exponential decay would be $(.693)(1990) = 1380$ years (rounded off by Barnes to 1400 years).

Do the data actually fit this exponential formula? Barnes gives no evidence that they do; in fact he does not even bother to present a plot showing the experimental points in relation to his theoretical curve. When one does construct such a plot (Fig. 1) it becomes immediately obvious that the fit is not very good, and that a straight line (shown as a dotted line) is equally good considering the scatter of the observational points. Indeed that is what McDonald and Gunst themselves stated: "Since the time of Gauss' measurements the earth's dipole moment has decreased, sensibly linearly, at approximately the rate of 5% per hundred years."[132] Thus empirically there is no justification for extrapolating with an exponential curve.

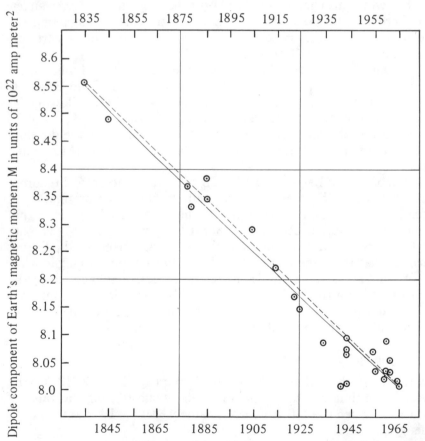

Figure 1. Decay of the Earth's magnetic field (data quoted by Barnes, 1973 from McDonald and Gunst, 1967. The dashed line is a straight line drawn between the first and last points; the solid line is an exponential curve fitted by Barnes to the first and last points).

In order to have a theoretical justification for his exponential decay curve, Barnes cannot rely on Lamb (who did not propose a physical theory of the earth's magnetism in the paper cited). He asserts that the earth's magnetic field is produced by currents that dissipate energy through Joule heating, and that there is no energy source to restore the energy dissipated by heating.

It appears that Barnes is making the same kind of mistake made by Lord Kelvin in the nineteenth century when he assumed that the earth was cooling by irreversible dissipation of energy with no source to replace the lost energy. But Kelvin made that assumption because of ignorance—radioactivity had not yet been discovered. Barnes makes a similar assumption even though it is flatly contradicted both by the source of his own data, McDonald and Gunst, and by almost every geophysicist at the present time.

To begin with, McDonald and Gunst state explicitly that "the magnetic dipole field is being driven destructively to smaller values by fluid motions which transform its magnetic energy into that of the near neighboring modes rather than expend it more directly as Joule heat."[133] In other words, the energy is being transferred from the dipole field to the quadrupole and higher moments rather than being dissipated as heat. This implies that the dipole field could not have been very much greater in the past; its value is limited by the total magnetic energy, which does not change very rapidly.

The other reason Barnes' extrapolation is completely illegitimate is that we now have very good evidence for complete reversals of the earth's magnetic field in the past; over a period of thousands or millions of years, the field has not been continually decreasing but has been fluctuating. The first definite evidence was found in lava flows by Bernard Brunhes and Pierre David at the beginning of the twentieth century; a time scale for polarity reversals extending back to 3.5 million years was estimated by Cox, Doell and Dalrymple in the 1960s. This work is closely associated with the development of the new theories of sea-floor spreading and plate tectonics.[134]

The theoretical basis for magnetic field reversals is Walter Elasasser's dynamo theory, based on fluid motions in the earth's core.[135] The dynamo theory assumes an energy source to keep the fluid moving. It is not yet established what the main source of energy is, but there are various possibilities such as radioactive heating, growth of the inner core, differential rotation of the core and mantle, etc.[136] In any case there is no justification for Barnes' assumption that there is *no* energy source.

Barnes rejects the possibility that the earth's magnetic field has reversed itself, but his basis for doing so is simply obsolete. He cites

Cowling's 1934 theorem which shows "that it is not possible for fluid motions to generate a magnetic field with axial symmetry (such as the dipole field of the earth)."[137] However, recent work shows that Cowling's theorem does not forbid a model with axially symmetric fluid motions generating a field with lower symmetry[138] and indeed the earth's field does not have a pure dipole character, a fact which Barnes finds it convenient to ignore.

The most striking example of Barnes' use of obsolete evidence is his long quotation from a 1962 book by J. A. Jacobs on difficulties with the reversal hypothesis.[139] In the same section of the later edition of this book Jacobs states that "the evidence seems compelling" that such reversals have occurred.[140]

In his most recent paper on the earth's magnetic field, Barnes repeats most of the same quotation about the need for caution in concluding that there have been reversals, citing Jacobs' 1962 book but omitting its date and ignoring the fact that Jacobs changed his position in the 1975 edition.[141] In fact, the principal creationist "expert" on geomagnetism writes as if the "revolution in the earth sciences" of the last two decades had never happened; he quotes A. A. and Howard Meyerhoff, two diehard opponents of plate tectonics, as if their "refutation" of it had been successful.

Barnes refuses to accept the overwhelming evidence for magnetic reversals, not because he can find anything wrong with the evidence itself, but simply on the grounds that there is not yet a satisfactory theoretical explanation of why the reversals occur at a particular time. For this objection he quotes an article by Carrigan and Gubbins (1979), despite the fact that these authors express no doubt whatsoever that reversals have actually occurred.[142]

When it comes to his own theory, Barnes conveniently forgets about any need for a satisfactory theoretical explanation. He derives his short time scale by extrapolating the supposed exponential decay of the earth's magnetic field into the past, with no justification whatever. This leads to a value of 18,000 gauss for the magnetic field in 20,000 B.C., which is "stronger than the field between the pole pieces of the most powerful magnets. It is not very plausible that the core of the earth could have stayed together with the Joule heat that would have been associated with the currents producing such a strong field. [Hence] the origin of the earth's magnetic moment is much less than 20,000 years ago."[143]

If any backwards extrapolation of these data were justified (which of course it is not), it would have to be linear rather than exponential. In order to find a field as large as 18,000 gauss, we would then have to make the dipole moment about 60,000 times as great as it is now, which means we would have to go back more than 100 million years. In other words, an *empirical* analysis of the magnetic data, ignoring all theories, would immediately refute the creationists' short time scale. It is only by using a completely fallacious theory that Barnes can arrive at the conclusion that the earth is less than 20,000 years old.

Even the linear extrapolation breaks down almost immediately. There is some reason to believe that the dipole field reached a maximum around 1800, and that it was smaller in 1600 than in 1800.[144] Other recent work also suggests that the dipole field has fluctuated on a fairly short time scale.[145] K. P. Games, who studied sun-dried adobe bricks in Egypt for the period 3000–0 BC, found "at least three maxima in the field at approximately 2600, 1350 and 200 BC, and two minima at about 1900 and 1000 BC . . . these results are consistent with the changes in the archaeomagnetic field found by other authors."[146] Thus there is absolutely no basis for the creationist claim that the decay of the earth's magnetic field puts a limit of 10,000 years on the age of the earth.[147]

CONCLUSION

The evidence is now overwhelming that the earth is several billion years old; many different determinations by radiometric dating methods give an age close to 4.5 billion years. The criticisms of radiometric dating by creationists have no scientific basis, and can be justified only by arbitrarily rejecting well-established results of modern physical science. Their argument for a young (10,000 year old) earth from decay of the earth's magnetic field is completely refuted by empirical data and is incompatible with all currently-accepted principles of geomagnetism.

This conclusion reinforces my earlier argument that creationism does not deserve "equal time" with evolution in public school science classes.[148]

ACKNOWLEDGEMENTS

I am grateful to the following persons for information, suggestions, and/or useful criticism on an earlier draft of this paper: Louis Brown, G. Brent Dalrymple, H. C. Dudley, J. H. Fremlin, Robert V. Gentry, David James, John O'Keefe, J. J. Naughton, James Shea and Dan Wonderly. None of these people is responsible for the content and opinions expressed in this article.

REFERENCES

1. See Francis C. Haber, *The Age of the World: Moses to Darwin* (Baltimore, MD: Johns Hopkins Press, 1959); Claude C. Albritton, *The Abyss of Time: Changing Conceptions of the Earth's Antiquity after the Sixteenth Century* (San Francisco: Freeman and Cooper, 1980); Dennis R. Dean, "The Age of the Earth Controversy: Beginnings to Hutton," *Annals of Science* 38 (1981): 435–56. On creationism, see R. H. Abelson, "Creationism and the Age of the Earth," Science 215 (1982): 119.

2. John N. Moore and Harold S. Slusher, *Biology: A Search for Order in Complexity* (Grand Rapids, MI: Zondervan, 1970), p. 416.

3. See Henry M. Morris, ed. *Scientific Creationism,* Public School ed. (Prepared by the Technical Staff and consultants of the Institute for Creation Research) (San Diego: Creation-Life Publishers, 1974), pp. 176–77.

4. Morris, 1974, p. 136.

5. Morris, 1974, p. iv.

6. Robert T. Ingram, "The Theological Necessity of a Young Universe," *Creation Society Research Quarterly* 12 (1975): 32–33.

7. See Dan Wonderly, *God's Time-Records in Ancient Sediments: Evidences of Long Time Spans in Earth's History* (Flint, MI: Crystal Press, 1977); a discussion among creationists on this topic is reported by Marvin L. Lubenow, "Does a Proper Interpretation of Scripture Require a Recent Creation?" *ICR Impact Series* No. 65 and 66 (1978).

8. Included in this volume. Ark. Stat. Ann., § 80-1663, et seq. (1981 Supp.).

9. Morris, 1974, p. 158.

10. D. T. Gish and R. B. Bliss, "Summary of Scientific Evidence for Creation (Parts IV–VII)." Included in this volume.

11. Henry M. Morris, *The Scientific Case for Creation* (San Diego, CA: Creation-Life Publishers, 1977), p. 79.

12. Henry M. Morris, *The Remarkable Birth of Planet Earth* (Minneapolis, MN: Dimension Books, 1972), p. 94.

13. See below, pp. 39–40.

14. H. Suess, "Radiocarbon Geophysics," *Endeavor* 4 (1980): 113–17.

15. Morris, 1974, pp. 97–130.

16. Ark. Stat. Ann. . . . Included in this volume.

17. Ark. Stat. Ann. . . . Included in this volume.

18. Miles F. Harris *et al, Investigating the Earth,* 2d ed. (Boston: Houghton Mifflin Co., 1973).

19. Marcel de Serres, *De la Création de la Terre et des corps céleste, ou Examen de cette question: l'oeuvre de la création est-elle aussi complète pour l'universe qu'elle paraît l'être pour la terre?* (Paris: Lagny Frères, 1843).

20. Morris, 1972, pp. 61–62.

21. See Martin Gardner, *Fads and Fallacies in the Name of Science,* 2d ed. (New York: Dover Publications, 1957), pp. 124–27; Haber, 1959, pp. 246–50.

22. Joe Burchfield, *Lord Kelvin and the Age of the Earth* (New York: Science History Publishers, 1975); S. G. Brush, *The Temperature of History: Phases of Science and Culture in the Nineteenth Century* (New York: Burt Franklin and Company, 1978),

ch. 3. On Kelvin see also S. P. Thompson, *The Life of Lord Kelvin* (1910: reprinted by Chelsea Publishing Co., New York, 1976).

23. Henry Faul, "A History of Geologic Time," *American Scientist* 66 (1978): 159–65.

24. See Loren Eiseley, *Darwin's Century* (New York: Doubleday, 1958), p. 245. see also C. Darwin, *The Origin of Species. Variorum Text.*, ed. M. Peckham Philadelphia: University of Pennsylvania Press, 1959 for changes in later editions.

25. Lord Kelvin, "On the Secular Cooling of the Earth," *Transactions, Royal Society of Edinburgh* 23 (1862): 157–70.

26. Lord Kelvin, "On the Age of the Sun's Heat," *Macmillan's Magazine* 5 (1862): 388–93.

27. Lord Kelvin, "The Age of the Earth as an Abode Fitted for Life," *Annual Report of the Board of Regents of the Smithsonian Institution,* July, 1897, pp. 337–57 (reprinted from Victoria Institute Transactions).

28. Lord Kelvin, *Popular Lectures and Addresses* (London: Macmillan, 1894), pp. 2, 89–90, 204–05.

29. T. H. Huxley, *Science and the Hebrew Tradition* (New York: Appleton, 1894), p. 134.

30. Pierre Curie and Albert Laborde, "Sur la chaleur dégagée spontanément par les sels de radium," *Comptes Rendus Hebdomadaires des Séances de l'Academie des Sciences* 136 (1903): 673–75.

31. F. Himstedt, "Über die radioaktive Emanation der Wasser—und Ölquellen," *Physikalische Zeitschrift* 5 (1904): 210–13.

32. C. H. Liebenow, "Notiz über die Radiummenge der Erde," *Physikalische Zeitschrift* 5 (1904): 625–26.

33. See Ernest Rutherford, *Radio-activity,* 2d ed. (Cambridge: Cambridge University Press, 1905), p. 485; Ernest Rutherford, *Radioactive Transformations* (New York: Scribner's Sons, 1906).

34. R. J. Strutt, "On the Radio-active Minerals," *Proceedings, Royal Society of London* A76 (1905): 88–101; B. B. Boltwood, "On the Ultimate Disintegration Products of the Radioactive Elements, Part II. The Disintegration of Uranium," *American Journal of Science* 23 (1907): 77–80, 86–88.

35. Burchfield, 1975, ch. 6; Lawrence Badash, "Rutherford, Boltwood, and the Age of the Earth: The Origin of Radioactive Dating Techniques," *Proceedings, American Philosophical Society* 112 (1968): 157–69; L. Badash, ed. *Rutherford and Boltwood: Letters on Radioactivity* (New Haven: Yale University Press, 1969); L. Badash, *Radioactivity in America: Growth and Decay of a Science* (Baltimore, MD: Johns Hopkins University Press, 1979).

36. For a history of this discovery, see S. G. Brush, "Discovery of the Earth's Core," *American Journal of Physics* 48 (1980): 705–24.

37. See A. Hallam, *A Revolution in the Earth Sciences, from Continental Drift to Plate Tectonics* (New York: Oxford University Press, 1973), p. 75; H. Takeuchi, S. Uyeda, and H. Kanamori, *Debate about the Earth* (revised edition), *Approach to Geophysics through Analysis of Continental Drift* (San Francisco: Freeman, Cooper, and Company, 1970), pp. 195, 229, 234.

38. See T. C. Chamberlin, "On Lord Kelvin's Address on the Age of the Earth as an Abode Fitted for Life," *Science,* n.s. 9 (1899), vol. 9, pp. 889–901, vol. 10, pp.

11–18; S. G. Brush, "A Geologist among Astronomers: The Rise and Fall of the Chamberlin-Moulton Cosmogony," *Journal for the History of Astronomy* 9 (1978): 1–41, 77–104; G. W. Wetherill, "Formation of the Terrestrial Planets," *Annual Review of Astronomy and Astrophysics* 18 (1980): 77–113; G. W. Wetherill, "The Formation of the Earth from Planetesimals," *Scientific American* 244 (6) (1981): 163–74.

39. G. T. Emery, "Perturbations of Nuclear Decay Rates," *Annual Review of Nuclear Science* 22 (1972): 165–202; Philip Hopke, "Extranuclear Effects of Nuclear Decay Rates," *Journal of Chemical Education* 51 (1974): 517–19.

40. George Gamow, "Zur Quantentheorie des Atomkerne," *Zeitschrift für Physik* 51 (1928): 204–12; Ronald Gurney and Edward Condon, "Wave Mechanics and Radioactive Disintegration," *Nature* 122 (1928): 439.

41. See L. F. Curtiss, "Probability Fluctuations in the Rate of Emission of Alpha Particles," *Journal of Research of the National Bureau of Standards* A4 (1930): 595–99; L. F. Curtiss, "Fluctuations of the Rate of Emission of Alpha Particles for Weak Sources and Large Solid Angles," *Journal of Research of the National Bureau of Standards* A8 (1932): 339–46; J. Berkson, "Examination of Randomness of Alpha Particle Emissions," in *Research Papers in Statistics,* ed. by F. N. David (New York: Wiley, 1966), pp. 37–54.

42. See S. Weinberg, "The Decay of the Protons," *Scientific American* 244 (6) (1981): 64–75.

43. For details, see C. T. Harper, ed. *Geochronology: Radiometric Dating of Rocks and Minerals* (Stroudsburg, PA: Dowden, Hutchinson and Ross, 1973), pp. 68–96.

44. Alfred O. Nier, R. W. Thompson, and B. F. Murphey, "The Isotopic Constitution of Lead and the Measurement of Geologic Time. III," *Physical Review,* ser. 2, 60 (1941): 112–16; reprinted in Harper, 1973, pp. 110–14. (See note 43.)

45. Arthur Holmes, "An Estimate of the Age of the Earth," *Nature* 157 (1946): 680–84 (reprinted in Harper, 1973, pp. 124–28); F. G. Houtermans, "Die Isotopenhäufigkeiten in natürlichen Blei und das Alter des Urans," *Naturwissenschaften* 33 (1946): 185–86, 219 (English translation in Harper, 1973, pp. 129–31. See note 43.)

46. Holmes, 1946 (1973); Houtermans, 1946 (1973).

47. See Harper, 1973, pp. 127–30.

48. See R. A. Alpher and R. C. Herman, "The Primeval Lead Isotopic Abundances and the Age of the Earth's Crust," *Physical Review* series 2, 84 (1951): 1111–14; A. P. Vinogradov, I. K. Zadorozhny, and S. I. Zhukov, "The Isotopic Composition of Leads and the Age of the Earth," *Report of the Committee on the Measurement of Geologic Time,* 1952–53 (Washington, DC: National Research Council, 1954), pp. 181–87, translated from Doklady Akademii Nauk SSSR, vol. 87: 1107–10; G. V. Voitkevich, "Concerning the Age of the Earth," *Report of the Committee on the Measurement of Geologic Time,* 1951–52 (Washington, DC: National Research Council, 1953), pp. 122–28, translated from Doklady Akademii Nauk SSSR, vol. 77: 471–64.

49. See, for example, S. C. Curran, "The Determination of Geological Age by Means of Radioactivity," *Quarterly Reviews, Chemical Society of London* 7 (1953): 1–18.

50. See C. Patterson, H. Brown, G. Tilton, and M. Inghram, "Concentration of Uranium and Lead and the Isotopic Composition of Lead in Meteoritic Materials," *Physical Review,* series 2, 92 (1953): 1234–35.

51. C. Patterson, "The Isotopic Composition of Meteoric, Basaltic and Oceanic Leads, and the Age of the Earth," in *Proceedings of the Conference on Nuclear Processes*

in Geologic Settings (Williams Bay, WI: September 21–23, 1953), pp. 36–40; F. G. Houtermans, "Determination of the Age of the Earth from the Isotopic Composition of Lead," *Nuovo Cimento,* series 9, vol. 10 (1953): 1623–33.

52. C. C. Patterson, G. R. Tilton, and M. Inghram, "Age of the Earth," *Science* 121 (1955): 74.

53. Patterson, et al, 1955, p. 74.

54. C. Patterson, "Age of Meteorites and the Earth," *Geochimica et Cosmochimica Acta* 10 (1956): 230–37; reprinted in Harper, 1973, pp. 132–39.

55. See G. Manhes, C. J. Allegre, B. Dupre, and B. Hamelin, "Lead-lead Systematics, the 'Age of the Earth' and the Chemical Evolution of our Planet in a New Representation Space," *Earth and Planetary Science Letters* 44 (1979): 91–104.

56. F. Tera, "Reassessment of the 'Age of the Earth'," *Carnegie Institution of Washington, Yearbook no. 79,* (1980): 524–31.

57. Bruce R. Doe, *Lead Isotopes* (New York: Springer Verlag, 1970); E. I. Hamilton, *Applied Geochronology* (New York: Academic Press, 1965); E. I. Hamilton and R. M. Farquhar, eds., *Radiometric Dating for Geologists* (New York: Interscience, 1968), chapter by Kanasewich; E. Jager and J. C. Hunziker, eds. *Lectures in Isotope Geology* (New York: Springer-Verlag, 1979), chapters by Koppel & Grunefelder and Gebauer & Grunefelder; Frank D. Stacey, *Physics of the Earth,* 2d ed. (New York: Wiley, 1977).

58. See Harper, 1973, pp. 216–41.

59. See G. Faure, *Principles of Isotope Geology* (New York: John Wiley and Sons, 1977); and C. J. Allegre, J. L. Birck, S. Fourcade, and M. P. Semet, "Rubidium-87/Strontium-87 Age of Juvinas Basaltic Achondrite and Early Igneous Activity in the Solar System," *Science* 187 (1975): 436–38.

60. See Harper, 1973, pp. 306–42.

61. See, for example, the papers by Everden and Everden and by Cox and Dalrymple in Harper, 1973, pp. 343–74; see also J. F. Everden, D. C. Savage, G. H. Curtis, and G. T. James, "Potassium-Argon Dates and the Cenozoic Mammalian Chronology of North America," *American Journal of Science* 262 (1964): 146–98.

62. E. M. D. Symbalisty and D. N. Schramm, "Neucleocosmochronology," *Reports on Progress in Physics* 44 (1981): 293–328.

63. See A. E. Mussett, "The Age of the Earth and Meteorites," *Geophysics* 1 (1970): 65–73. See also N. H. Gale and A. E. Mussett, "Episodic Uranium-Lead Models and the Interpretation of Variations in the Isotopic Composition of Lead in Rocks," *Reviews of Geophysics and Space Physics,* 11 (1973): 37–86.

64. Mussett, 1970, pp. 69–70, 72.

65. D. B. McIntyre, "Precision and Resolution in Geochronometry," in *The Fabric of Geology,* ed. by C. C. Albritton (San Francisco: Freeman, Cooper and Company, 1963), pp. 112–34.

66. Morris, *Scientific Creationism,* 1974; Harold Slusher, *Critique of Radiometric Dating,* 2d ed. (San Diego: Institute for Creation Research, 1981).

67. G. Brent Dalrymple, "Radiometric Dating, Geologic Time, and the Age of the Earth: A Reply to 'Scientific Creationism'," (1981, preprint).

68. Morris, 1974, p. 133, emphasis in original.

69. James D. Dana, *Manual of Geology,* 3d ed. (New York: Ivison, Blakeman, Taylor, and Company, 1880), p. 590.

70. Morris, 1974, pp. 134–36.

71. Morris, 1974, p. 95.

72. O. H. Schindewolf, "Comments on Some Stratigraphic Terms," *American Journal of Science* 255 (1957): 395, my italics.

73. Everden, et al, p. 166.

74. Everden, et al, p. 146.

75. See J. M. Weller, *Stratigraphic Principles and Practice* (New York: Harper, 1960), p. 36; cf. D. M. Raup, "Evolution and the Fossil Record," *Science* 213 (1981): 289.

76. Morris, 1974, p. 140.

77. Henry Faul, *Ages of Rocks, Planets, and Stars* (New York: McGraw-Hill, 1966), p. 61.

78. Faul, 1966, p. 18.

79. Morris, 1974, p. 141; Slusher, 1981, p. 34.

80. Melvin A. Cook, *Prehistory and Earth Models* (London: Max Parrish, 1966), pp. 53–72.

81. Cook, 1966, p. 54.

82. Dalrymple, 1981.

83. See A. O. Nier, "Some Reminiscences of Isotopes, Geochronology, and Mass Spectrometry," *Annual Review of Earth and Planetary Science* 9 (1981): 1–17.

84. Cook, 1966, p. 55.

85. See C. B. Collins, R. M. Farquhar, and R. D. Russell, "Isotopic Constitution of Radiogenic Leads and the Measurement of Geologic Time," *Bulletin, Geological Society of America* 65 (1954): table 1, samples 23–24.

86. Cook, 1966, pp. xiii, xiv.

87. Morris, 1974, pp. 142–43.

88. F. B. Jueneman, "Scientific Speculation . . . Will the Real Monster Please Stand Up," *Industrial Research* 14 (9) (1972): 15.

89. G. R. Brakenridge, "Terrestrial Paleo-environmental Effects of a Late Quaternary-age Supernova," *Icarus* 46 (1981): 81–93.

90. G. R. Brakenridge, "Letter to S. G. Brush," 1981.

91. L. R. Maxwell, "Cosmic Radiation and Radioactive Disintegration," *Nature* 122 (1928): 997.

92. Morris, 1974, p. 142.

93. Slusher, 1981, p. 20.

94. G. T. Emery, "Perturbations of Nuclear Decay Rates," *Annual Review of Nuclear Science* 22 (1972): 165–202.

95. G. T. Emery, "Letter to S. G. Brush," 1980; see also P. K. Hopke, "Extranuclear Effects of Nuclear Decay Rates," *Journal of Chemical Education* 51 (1974): 517–19.

96. H. C. Dudley, *The Morality of Nuclear Planning??* (Glassboro, NJ: Kronos Press, 1976).

97. Dudley, 1976, p. 4.

98. L. de Broglie and J. P. Vigier, *Theory of Elementary Particles* (Elsevier Publishing Company, 1963).

99. Dudley, 1976, p. 24.

100. See Morris, 1974, pp. 15–16, 22, 33, 59.

101. See J. F. Clauser and A. Shimony, "Bell's Theorem: Experimental Tests and Implications," *Reports on Progress in Physics* 41 (1978): 1181–1927; B. d'Espagnat, "The Quantum Theory and Reality," *Scientific American* 241 (9) (1979): 158–81. "The Quantum Theory and Reality," *Scientific American* 241 (9) (1979): 158–81.

102. H. C. Dudley, "Letter to S. G. Brush," 1981.

103. T. G. Barnes and R. J. Upham, Jr., "Another Theory of Gravitation: An Alternative to Einstein's General Theory of Relativity," *Creation Research Society Quarterly* 12 (4) (1976): editorial note on p. 197.

104. See H. S. Slusher, *Age of the Cosmos* (San Diego, CA: Institute for Creation Research, 1980).

105. Slusher, 1981, pp. 25–26.

106. John Joly, "The Theory of Thermal Cycles (A Reply to Dr. Lotze)," *Gerlands Beitrage Zur Geophysik* 20 (1928): 288–92; (editor's note: no citation given for July, 1972).

107. A. Holmes, "Radioactivity and Geological Time," in *Physics of the Earth—IV: The Age of the Earth,* Bulletin of the National Research Council, (Washington, DC, 1931), pp. 124–459.)

108. See G. H. Henderson et. al., "A Quantitative Study of Pleochroic Haloes," *Proceedings, Royal Society of London* (1934–39): vol. A145, pp. 563–81, 582–91, vol. A158, pp. 199–211, vol. A173, pp. 238–49, 250–62; K. Rankama, *Isotope Geology* (New York: McGraw-Hill, 1954).

109. R. V. Gentry, "Cosmology and Earth's Invisible Realm," *Medical Opinion and Review,* 3 (10) (1967): 64–79; R. V. Gentry, "Radioactive Halos," *Annual Review of Nuclear Science* 23 (1973): 347–62; R. V. Gentry et. al., "Evidence for Primordial Superheavy Elements," *Physical Review Letters* 37 (1976): 11–15; R. V. Gentry et al, "Radiohalos in Coalified Wood: New Evidence Relating to the Time of Uranium Introduction and Coalification," *Science* 194 (1976): 315–18; R. V. Gentry, "Time: Measured Responses," *EOS, Transactions of the American Geophysical Union,* May 29, 1979, p. 474.

110. See Raphael Kazmann, "It's about Time: 4.5 Billion Years Ago," *Geotimes* 23 (9) (1978): 18–20.

111. Gentry, 1973, p. 352.

112. Gentry, 1973, p. 353.

113. Gentry, et al, "Evidence for Primordial Superheavy Elements," 1976.

114. See G. B. Lubkin, "Mica Giant Halos Suggest Natural Superheavy Elements," *Physics Today* 29 (8) (1976): 17–20; G. B. Lubkin, "Evidence for Superheavies in Mica Looks Weaker," *Physics Today,* 30 (1) (1977): 17–20; C. J. Sparks, S. Raman, E. Ricci, R. V. Gentry, and M. O. Kraus, "Evidence against Superheavy Elements in Giant Halo Inclusions Re-examined with Synchroton Radiation," *Physical Review Letters* 40 (1978): 507–11.

115. H. J. Rose and D. Sinclair, "A Search for High-Energy Alpha Particles from Superheavy Elements," *Journal of Physics* G 5 (1979): 781–96.

116. Slusher, 1981, pp. 22, 49.

117. Slusher, 1981, p. 39; Morris, 1974, pp. 145–48.

118. J. J. Naughton, "Letter to S. G. Brush," 1981.

119. Morris, 1974, p. 148.

120. See Calvin Alexander, Jr., "Radiometric Dating and the Atmospheric Argon Correction in K/Ar Dating." Included in this volume.

121. Gentry, 1979.

122. See Derek York, "Polonium Halos and Geochronology," *EOS,* August 14, 1979, p. 617.

123. S. R. Hashemi-Nezhad, J. H. Fremlin, and S. A. Durrani, "Polonium Haloes in Mica," *Nature* 278 (1979): 333–35.

124. Fremlin (1981) [editor's note: no citation given.]

125. Gentry, et al., 1976; and telephone conversation, 16 September, 1981.

126. See Paul Damon, "Time: Measured Responses," *EOS,* May 29, 1979, p. 474; York, 1979.

127. A. Holmes, "A Reply to 'The Earth is Not old'," *Report of the Committee on the Measurement of Geologic Time* (Washington, DC: National Research Council, 1936), pp. 45–48.

128. Dan Wonderly, *God's Time-records in Ancient Sediments: Evidences of Long Time Spans in Earth's History* (Flint, MI: Crystal Press, 1977).

129. Morris, 1977, p. 79.

130. T. G. Barnes, *Origin and Destiny of the Earth's Magnetic Field* (San Diego, CA: Creation-Life Publishers, 1973), p. viii; see Horace Lamb, "On Electrical Motions in a Spherical Conductor," *Philosophical Transactions, Royal Society of London* 174 (1883): 519–49.

131. K. L. McDonald and R. H. Gunst, "Recent Trends in Earth's Magnetic Field," *Journal of Geophysical Research* 73 (1968): 2057–67.

132. Quoted by Barnes, 1973, p. 34.

133. McDonald and Gunst, 1968, p. 2057.

134. See Takeuchi et al, 1970; Hallam, 1973; J. A. Jacobs, *The Earth's Core* (New York: Academic Press, 1975).

135. W. M. Elsasser, "Induction Effects in Terrestrial Magnetism," *Physical Review,* series 2 (1946–47): vol. 69, pp. 106–16, vol. 70, pp. 202–12; see also Jacobs, 1975, ch. 4; or Stacey, 1977, chs. 5, 6.

136. See C. R. Carrigan and D. Gubbins, "The Source of the Earth's Magnetic Field," *Scientific American* 240 (2) (1979): 118–30.

137. Barnes, 1973, pp. 44–45.

138. See Jacobs, 1975, pp. 128–31.

139. Barnes, 1973, pp. 46–47.

140. Jacobs, 1975, p. 140.

141. T. G. Barnes, "Depletion of the Earth's Magnetic Field," *ICR Impact Series* no. 100 (1981): iv.

142. Carrigan and Gubbins, 1979.

143. Barnes, 1973, p. 38.

144. See Takesi Yukutake, "Spherical Harmonic Analysis of the Earth's Magnetic Field for the 17th and 18th Centuries," *Journal of Geomagnetism and Geoelectricity* 23 (1971): 11–31. See also T. Yuketake, "Review of the Geomagnetic Secular Variations on the Historical Time Scale," *Physics of the Earth and Planetary Interiors* 20 (1979): 83–95.

145. See S. I. Braginsky, "Oscillation Spectrum of the Hydromagnetic Dynamo of the Earth," *Geomagnetism and Aeronomy* 10 (1970): 172–81; papers by J. C. Cain and others in R. M. Fisher, M. Fuller, V. A. Schmidt, and P. J. Wasilewski, *Proceedings of the Takesi Nagata Conference, Magnetic Fields: Past and Present* (Greenbelt, MY: Goddard Space Flight Center, June 3–4, 1974).

146. K. P. Games, "The Magnitude of the Archaeomagnetic Field in Egypt between 3000 and O. B. C.," *Geophysical Journal of the Royal Astronomical Society* 63 (1980): 49.

147. Morris, 1974, p. 158.

148. See S. G. Brush, "Creationism/Evolution: The Case Against 'Equal Time'," *The Science Teacher* 48 (1981): 29–33.

Radiometric Dating and the Atmospheric Argon Correction in K/Ar Dating

by E. Calvin Alexander, Jr.

K/Ar chronologists are not fools, knaves, nor rabid Evolutionists documenting our preconceived notions. We are conscientious scientists who check and recheck our assumptions and stand ready to modify those assumptions when presented with scientific evidence that the assumptions are wrong.

Radiometric dating is under attack by Creationists since the expanse of geologic time documented by radiometric dating techniques is one of the many pieces of scientific evidence which is completely incompatible with any concept of a young earth. The attack is completely baseless in fact. Creationists apparently hope that if they make enough nonsensical noise the unwary observer might become worried 'that where there is smoke there must be a fire' and think that a real scientific controversy exists. In this case, the smoke indicates only the presence of 'hot air' and no such controversy exists. The 'critiques' of radiometric dating produced by various Creationists contain so many errors of fact and logic as to defy rational comment. The purpose of this article is not to rebut the 'critiques' but is rather to reiterate the basic principles of radiometric dating and to comment on one of the Creationists' most recent 'revelations' on radiometric dating.

First, radiometric dating is neither magical or infallible nor was it developed purely to provide a timescale long enough to satisfy evolutionists. Radiometric dating is rather a tool which follows logically from, and is an integral part of, modern science and technology. It was initially developed by physicists and chemists with no evolutionary axes to grind. For a local example, the basic development of mass spectrometers (which permit modern geochronology to be done) as well as much of the pioneering work in modern geochronology was done in the 1930s and 1940s, by Prof. Al Nier of the Physics Department at the University of Minnesota. Radiometric dating is, as Creationists loudly point out as though it were proof that the techniques don't work, based on assumptions. Any simple radiometric age determination assumes: (1) that the initial state of the system can be determined, (2) that the system has been closed between the event being dated and the present, and (3) that the fundamental physical constants of nature have not changed over the time interval in question. Major classes of geologically interesting samples unfortunately fail to meet assumptions (1) and/or (2) — and are therefore undatable by direct radiometric methods. The assertions of Henry Morris and other Creationists notwithstanding, however, there do exist significant numbers of geological samples for which *these assumptions are certainly possible, are eminently reasonable, are completely testable, and are provable approximations of reality.* Such samples can be and are being correctly dated by radiometric techniques. The recent text *Principles of Isotope Geology* by G. Faure is an excellent, readable, up-to-date reference for anyone interested in geochronology.[1]

Four radiometric 'clocks' are in common use in geochronology and several additional 'clocks' are in limited use. All four are based on the decay of naturally occurring radioactive isotopes. One clock is based on the rate of disappearance of ^{14}C. The other three 'clocks' are based on the accumulation of ^{40}Ar, ^{87}Sr, ^{206}Pb, ^{207}Pb and ^{208}Pb from the decay of ^{40}K, ^{87}Rb, ^{238}U, ^{235}U and ^{232}Th respectively. More K/Ar age determinations are made than any other accumulation 'clock' and the K/Ar clock is currently under attack by Creationists. As a geochronologist working with the K/Ar system I am qualified to address these attacks.

Much of the Creationists' current indignation over K/Ar ages concerns the common way the nonradiogenic ^{40}Ar component is calculated in a K/Ar age determination. When the argon is extracted from a sample in the laboratory, it is in general found to consist not only of the ^{40}Ar produced by the decay of ^{40}K in the sample since that sample formed (the radiogenic ^{40}Ar) but also contains varying amounts of nonradiogenic ^{40}Ar. This nonradiogenic component of ^{40}Ar is accompanied by two other stable argon isotopes, ^{36}Ar and ^{38}Ar, and is termed 'contamination' or 'trapped' argon. This nonradiogenic argon must be subtracted from the total measured ^{40}Ar before a valid age can be calculated. For old, potassium rich minerals the nonradiogenic component is often very small to vanishing. For young and/or potassium poor minerals, the nonradiogenic component can be significant and its presence is the current barrier to the use of K/Ar techniques for such materials.

In most routine K/Ar work the nonradiogenic component is indeed a 'contamination' component. The atmosphere contains almost 1% by volume argon and small amounts of that argon are absorbed onto the surfaces of the minerals and/or are present in the experimental apparatus itself. Since ^{36}Ar is not produced by the radioactive decay of ^{40}K it provides a convenient measure of the 'atmospheric contamination' present in each analysis. The normal operative procedure is to multiply the ^{36}Ar content by the atmospheric ^{40}Ar/^{36}Ar ratio (295.6) and to subtract the resulting 'atmospheric contamination' ^{40}Ar from the measured ^{40}Ar to yield the radiogenic ^{40}Ar. This procedure is not an assumption, since it can be and is checked in most competent measurements. Replicate analyses of the same rock or mineral separate normally yield varying (often by large factors) ratios of radiogenic to contamination ^{40}Ar. The variation is entirely in the amount of contamination ^{40}Ar, and the radiogenic component for a given rock or mineral reproduces nicely in replicate analyses. That reproducibility demonstrates the validity of the procedure particularly when as is often the case, Rb/Sr and/or U,Th/Pb techniques yield concordant ages for the samples.

There is no *a priori* reason that a rock or mineral can not include a truly 'trapped' Ar component when it forms. Such a nonradiogenic 'trapped' component can have essentially any ^{40}Ar/^{36}Ar ratio. Indeed a number of cases are known where such a situation can be demonstrated. In some of these situations, by analyzing more than one cogenetic sample and by using isochron diagrams, it is possible to determine both the age of the samples and the 'initial' or 'trapped' ^{40}Ar/^{36}Ar composition. In other situations, the samples are just undatable. When one goes to the trouble to apply isochron techniques to K/Ar data from common rocks and/or minerals, one usually obtains an atmospheric ^{40}Ar/^{36}Ar ratio for the trapped component. It has been proven time and time again that in most cases the nonradiogenic argon in common igneous rocks and minerals is modern atmospheric contamination.

The atmospheric ^{40}Ar/^{36}Ar ratio should be changing (dare I say evolving?) with time since each isotope comes from a separate source —^{40}Ar from the decay of ^{40}K and ^{36}Ar from primordial, etc. sources. Plausible models can be constructed in which the terrestrial atmospheric ^{40}Ar/^{36}Ar ratio has increased, decreased, or remained essentially constant over geologic time, depending on the assumptions made about the history of the earth. The atmospheric ^{40}Ar/^{36}Ar ratio should be a unique record of the composition and history of the earth's atmosphere.

The recent discoveries that extraterrestrial bodies contain argon with widely varying ^{40}Ar/^{36}Ar ratios simply reinforce this view. Lunar soils yield trapped $^{40}Ar/^{36}Ar$ ratios ranging from ~ 0.1 to ~ 10. The Viking space craft proved that the Martian atmosphere has a $^{40}Ar/^{36}Ar$ ratio of ~ 3000. Some individual meteorites contain trapped $^{40}Ar/^{36}Ar$ values as low as 10^{-3}. The recent Pioneer spacecraft seem to have found a $^{40}Ar/^{36}Ar$ ratio of ~ 1 in the Venusian atmosphere. When measuring the K/Ar age of a meteoritic or lunar sample one cannot (and does not) use a trapped $^{40}Ar/^{36}Ar$ ratio equal to the terrestrial atmosphere value. If and when Martian or Venusian samples are returned to earth for dating purposes, those chronologists (the lucky stiffs!) making the analyses will not use the terrestrial atmospheric $^{40}Ar/^{36}Ar$ value to calculate the nonradiogenic ^{40}Ar component.

The trapped, common, or nonradiogenic components in the isotope geology of Sr and Pb are known to change through time. Those changing records provide very useful boundary conditions on the thermal and chemical history of the earth. I have personally searched, without success, for evidence of any analogous change in the atmospheric $^{40}Ar/^{36}Ar$ ratio through time.[2] The basic *observation* (not assumption) remains. Most of the nonradiogenic argon present in the analyses of terrestrial samples is indeed correctly measured by the standard calculation.

K/Ar chronologists are not fools, knaves, nor rabid Evolutionists documenting our preconceived notions. We are conscientious scientists who check and recheck our assumptions and stand ready to modify those assumptions when presented with scientific evidence that the assumptions are wrong. Until such evidence is presented, however, the data seems overwhelming that the earth is about 4.5 billion years old and the universe on the order of 10 billion years older than the earth.

REFERENCES

1. G. Faure, *Principles of Isotope Geology* (New York: John Wiley and Sons, 1977).

2. E. C. Alexander, Jr., "^{40}Ar-^{39}Ar Studies of Precambrian Cherts: An Unsuccessful Attempt to Measure the Time Evolution of the Atmospheric $^{40}Ar/^{36}Ar$ Ratio," *Precambrian Research* 2 (1975): 329–44.

The Association of "Human" and Fossil Footprints

by Robert E. Sloan

The documented cases of "human" footprints associated with pre-Pliocene fossils are all hoaxes. In all likelihood, the hoaxes were not created by the creationists, but it clearly suits their purpose to use them.

One of the standard arguments that Creationists use against evolution and the geological sciences is the presence of "human footprints" along with ancient fossils. Their stated conclusion is that the fossil record is misinterpreted by geologists and that these footprints prove geologists wrong. The documented cases of "human" footprints associated with pre-Pliocene fossils are all hoaxes. In all likelihood, the hoaxes were not created by the Creationists, but it clearly suits their purpose to use them.

At least some of these "footprints" were apparently carved by pre-Columbian American Indians and are gross cartoons of human feet, no two identical or even similar in shape. Albert G. Ingalls, writing in *Scientific American,* for January 1940, quotes ethnologist David I. Bushell of the Smithsonian Institution as saying they were unquestionably carvings made by Indians. Bushell suggested that the human foot was a symbol associated with a watering place. Nine very crude "footprints," no two alike or capable of being made by the same type of feet are illustrated in that article.

The footprints usually cited by Creationists and the subject of a movie they widely circulate are those in the 120 million year Early Cretaceous Glen Rose Limestone on the Paluxy River near the town of Glen Rose, Somervell County, Texas. These were first widely reported by Dr. Roland T. Bird of the Department of Vertebrate Paleontology of the American Museum of Natural History in New York, in an article in that museum's popular magazine *Natural History* for May

1939. Bird's article was on the well known and important occurrences of dinosaur footprints and trackways that occur in the Glen Rose Limestone. The dinosaur footprints were made in several feet of sea water by wading animals. He correctly identified the "human footprints" as hoaxes in the article on the basis of size and anatomical impossibility. Nonetheless the joking fun he made of these "footprints" has lead Creationists to quote him out of context as saying "perfect in every detail" about them.

Dr. Richard P. Aulie of the Department of Natural Science at Loyola University of Chicago wrote an excellent but unfavorable critical review of a Creation Research Society high school biology text in the *American Biology Teacher,* April and May 1972. It was later reprinted with revision in the *Journal of the American Scientific Affiliation* for March, June, September and December 1975. The revised review is a very scholarly article well worth reading for the history of the artificially prolonged Evolution-Creation controversy. Dr. Aulie requested opinions from paleontologists of national stature with decades of experience working on the particular rock units in the areas of two of these groups of "footprints."

Dr. Keith Young of the Department of Geology of the University of Texas at Austin spent his professional career working on the marine fossils of the Early Cretaceous of Texas. On the subject of the footprints, Dr. Young's letter to Dr. Aulie dated May 21, 1971 reads in part:

> I have visited the Glen Rose locality several times. Never have I been shown, nor have I seen, human footprints in association with dinosaur tracks, although I've seen many dinosaur tracks.
>
> The thing that bothers me most are the alleged "human" tracks, since they:
>
> a. show no left or right;
> b. show no pressure points in the mud as the result of walking or other movement; the dinosaur tracks in this area, on the other hand, illustrate the flow of mud under the foot as the weight of the animal shifted during movement.
> c. There is no narrowing of the "human" footprints at the instep as in man, even when man wears sandals. Even when sandals are the same width throughout, there is still a narrowing at the instep with man because there is less weight on the inside of the foot.
> d. The "tracks" of the men are chiseled cleanly, whereas those of even the smaller dinosaurs were made in the soft mud and the soft mud rolled back into the track after the foot left, and deformed the track. For this reason it is obvious that the dinosaur tracks and the "human" tracks were made under greatly different conditions.

My only conclusion has been, and is now, without even considering the anomalous size of the tracks and geological conditions, that the "human tracks" are a hoax, and a very poorly done hoax at that. This hoax does not appear to have been created by the authors of the book, but since these men have an overwhelming drive to believe the Old Testament, they have swallowed the hoax hook, line, and sinker.

Dr. Richard A. Robison, then Professor of Geology, University of Utah, now at the University of Kansas and directing the monumental Treatise of Invertebrate Paleontology, wrote Dr. Aulie as follows on June 1, 1971:

I am familiar with the "human-like sandal print" in Cambrian rocks from western Utah that you ask about. Much of my professional research (about 15 years) has been centered on the fossil assemblages and rocks found in the area where the "footprint" was discovered. In my opinion, the report is preposterous. The supposed "footprint" is the result of an unusual fracture pattern rather than a primary structure preserved in the rock. That type of fracturing is characteristic of certain sedimentary layers in the area, numerous odd-shaped forms can be found. Attention to detail and a little common sense show them to be of nonorganic origin. For example, the question might be asked, "If one footprint was discovered, wouldn't one expect to find others of the same size and kind on the same bedding surface?" More than one "footprint" has been reported from the locality, but none are of the same size or shape, and I doubt that more than one has been found at any one horizon. Furthermore, the "footprints" occur in a formation with brachiopods and echinoderms, which can only exist in a marine environment. Paleoecologic evidence indicates deposition in deep water (at least below wave base), and paleogeographic evidence indicates deposition at least 200 miles from shore. This would seem to be a strange habitat for prehistoric man!"

If there were any further truth to the association of true human or humanoid footprints with pre-Pliocene fossils, it would make the scientific reputation of the paleontologist who could demonstrate it. But the footprint would have to be demonstrably impressed in the original wet sediment with appropriate sedimentary deformational features, would have to be anatomically accurate, preferably excavated as part of a continuing trackway to show strides, and preferably covered by unaltered and undisturbed sediments of the next younger bed of the same geologic age. None of these criteria are fulfilled by the examples cited in the Creationist literature. They all appear to be hoaxes.

ADDENDUM

In 1980, after the article above was originally published, Creation Life Publishers of San Diego printed a book by John D. Morris, *Tracking Those Incredible Dinosaurs . . . and the People who Knew Them.* This is the most scholarly and most interesting of all the Creationist footprint literature. Morris discusses in great detail all the footprints both real and fake in the Glen Rose Limestone of the Paluxy River near Glen Rose Texas. Every surviving proposed human footprint is discussed case by case in detail. He dismisses every single case as dinosaur footprints or tail drags, carvings by known or unknown hoaxers, as undiagnostic or unidentifiable. Nonetheless he remains convinced that humans were living and making footprints during the Early Cretaceous! He does this mainly on the basis of testimony of untrained witnesses 40 or more years after they saw them. How he can disprove his own case and then maintain that it is still viable boggles the mind.

Since this article was originally published, two separate field parties working in East Africa under the direction of Mary Leakey and of Kay Behrensmeyer have found fossil footprints of fossil hominids in Pliocene and early Pleistocene rocks. These footprints are much less detailed than the Texas hoaxes, but satisfy the criteria I listed above. They have enhanced the scientific reputations of their finders. They have also been found in the same sediments as bones of the hominids that made them. One reason they have been accepted is that they are no more detailed than the mode of deposition of the sediments would permit. Footprints in soft soupy river mud are not very detailed whether they are hippopotamus or hominid. They were also found as part of complete trails, and of course were buried in just younger sediments than the footprints themselves. The older of these footprints are of the species of hominid known as *Australopithecus afarensis*, of which the well-known "Lucy" is an example. Older hominid footprints will probably be found in Africa, but probably not older than late Miocene in age.

Part 5
The Legal Issues

Why Creationism Should Not Be Taught as Science: The Legal Issues

by Frederick Edwords

> ... It would seem the creationists have been rather clumsy in sticking to their new tactic of secularizing creationism. But, even if they had managed to carry off such a plan with any efficiency, their position would still fall short of legal acceptability.

The legal objections to placing Special Creation doctrines in the science classroom form what, quite frankly, can only be called an air-tight case. For once one understands the history of what Biblical creationists have been trying to do, once one grasps the full significance of their new tactic, and once one is aware of the nature of their latest legal moves, no choice is left but to acknowledge that the creationist's aims can never be legal under our present constitution. Let us, then, explore the history, tactics, and legal efforts of the creationist movement so as to better understand why it has never won a constitutional battle.

A HISTORY OF THE LEGAL CONFLICT

Large scale challenges by creationists to the teaching of evolution have occurred on three significant occasions in the last century and a half. The first was after the publication of Darwin's *Origin of Species,* the second was at the time of the Scopes trial, and the third is taking place today. On each occasion, creationists have attacked those in the scientific and educational community desiring to teach or promote evolution.

Looking back on the first battle, Andrew White, in his 1896 book, *A History of the Warfare of Science with Theology in Christendom,* recalled that, "Darwin's *Origin of Species* had come into the theological world like a plough into an ant hill. Everywhere those thus rudely awakened from their old comfort and repose had swarmed forth angry and confused. Reviews, sermons, books light and heavy, came flying at the new thinker from all sides."

Specifically, one English clergyman, who was vice president of a Protestant institute to combat "dangerous" science, had denounced Darwinism as "an attempt to dethrone God." Another creationist, Whewell, succeeded in preventing a copy of the *Origin* from being placed in the Trinity College Library. Rougemont had called for a crusade against evolution in Switzerland. And a similar crusade had almost taken place among the scientific community in America until Asa Gray, the foremost American botanist, won it over in a series of stunning public debates at Harvard that defeated the anti-evolution movement for a time.[1]

But a dozen years later it flared up again with Darwin's *Descent of Man.* In England, Gladstone condemned it. In America the Reverend Dr. Hodge of Princeton declared that Christians "have a right to protest against the arraying of probabilities against the clear evidence of the scriptures."

However, the problem of the teaching of evolution in the *public* schools was not yet an issue. In those days the issue was the teaching of science in *any* form to children. Huxley had his hands full in England just trying to lay to rest the old classical and theological education so as to make room for such "liberal" studies as science, geography, history, grammar, composition, drawing, and physical education.

This meant that it wasn't until the decade of the Scopes trial that teaching children about evolution became an issue. And it became an issue largely because its teaching had finally become frequent enough to alarm the conservative American religious community. So, once again the anti-evolutionists formed their battle lines, thereby setting off the second great conflict.

Between 1922 and 1929, forty-six pieces of legislation aimed at preventing the teaching of evolution were introduced. Of these, only three were passed, all of which were later declared unconstitutional.

Writing in 1927 in the *Bulletin of the American Association of University Professors,* S. J. Holmes said that "the worst feature of the situation is not so much the intellectual backwardness revealed by the passage of these statutes as the spirit of religious intolerance and disregard of intellectual liberty which prompted their enactment."[2]

Many feel it was this sentiment, becoming widely held, that brought an end to the legislative attacks by fundamentalists. Yet nothing could be further from the truth.

The only real reason the attacks came to an end was because evolutionists made a compromising retreat. As Mayer points out, "In most American schoolbooks the word *evolution* simply disappeared." Many times this was done as a mere camouflage maneuver, evolution still being taught under different names like "change through time" or "heredity." But at other times it was done in an apparent recognition of defeat.[3]

As Bette Chambers noted when president of the American Humanist Association, "Years ago we were made painfully aware that this intricate and beautiful principle of modern biology is taught almost nowhere without extensive apologetics or having first been filtered through a sieve of nervous religious disclaimers." She was describing the case of her own daughter who, in 1965, had come home angrily from junior high school after seeing a Moody Bible Institute nature film in her science class. "*Must* I believe that the spider makes the web perfectly the first first time she tries because God has 'programmed' her brain like a computer?" she cried.[4]

So, even though the legislative track record of creationists was poor, they had an impressive long-term success in convincing teachers and publishers to soft-pedal evolution.[5] That is, they managed to set up an environment where evolution was "selected out" of text books by a "slow and gradual process" which went almost unnoticed. No wonder only one piece of legislation attempting to prohibit the teaching of evolution was introduced in the 33 years between 1930 and 1963.

But this couldn't go on forever, not with evolutionary science developing by leaps and bounds. Sooner or later the scientific and academic community would have to wake up to the fact that only a shadow of evolution, if any at all, was being presented in the public schools. And to help bring about this awakening, biologist Hermann J. Muller, on the centennial of Darwin's *Origin,* wrote an article entitled "One Hundred Years Without Darwin Are Enough."

When the drive finally got underway to bring evolution back into the classroom, it seemed the public would be receptive. Russian advances in the space race had parents and school boards calling for more science education. So, in 1964, biology textbooks sponsored by the National Science Foundation went into use with government funding. These textbooks reintroduced evolution and, as a consequence, also reintroduced the creation/evolution controversy.

This time, however, the religious conservatives were not so blunt as to reject all science, or even to reject evolution alone. The new ploy was to appeal to "fairness," and thereby demand "equal time" for creationism. As a result, since 1964 pieces of legislation relating to the "equal time" idea have been proposed in the legislatures of most of the

50 states. At present, more than 20 states have policies allowing local school districts to include creationism as an alternative.

And it hasn't stopped there. With the creationists gaining momentum and putting forth ever more sophisticated legal arguments (they at first wanted equal time for Genesis, but now usually seek it for "creation science"), they have burst forth in a new wave that is beginning to blanket the nation.

So far, two "equal time" bills have passed in state legislatures and been signed into law, one of which has already been declared unconstitutional by a Federal District Court judge.[6] The real threat, however, has come from the creationist influence on individual school boards to either allow or require their "two-model" teaching program. A large number of local school boards in a variety of states have been persuaded that equal time for creationism is both fair and legal.

THE NEW TACTIC

Obviously the creationists have learned a lot in their long struggle to unseat evolution. Trial and error has shown them what doesn't work: Anti-science doesn't, efforts to ban evolution don't, and purely religious invective is also a losing proposition. The idea of being open-minded, religiously neutral, and scientific has gained such wide credence (or at least lip-service) that creationists can't successfully oppose it, no matter how much they might like to.

So, their new tactic is to declare creationism scientific, then join in with the majority and espouse the virtues of the times in their own name. In this way they can pose as latter-day Galileos being persecuted by "orthodox" science. They can become the champions for fairness fighting against the "dogmatic" evolutionists who have hauled them into the "Scopes trial in reverse."[7] In fact, they can even declare themselves Jeffersonian fighters for church-state separation against "the religion of evolutionary humanism" in the public schools, as well as revolutionaries for progress bringing new truths into play against "the establishment."

How have the creationists accomplished this? With one simple sentence. Dr. Henry Morris of the Institute for Creation Research probably deserves the credit for it. In his debates he simply says, "Creation is just as much a science as is evolution, and evolution is just as much a religion as is creation."

Such a statement serves three purposes at once. First, it declares creationism to be an alternate scientific theory to evolution. Second, it criticizes evolution for being a belief held only on faith. And third, it confuses school boards and legislatures.

To back up this statement, Morris throws in a variety of scientific-sounding arguments and legalistic appeals for "equal time" and "church-state separation." The effect of this on his average audience is one of producing doubt. And in the face of such doubt, these people begin to think, "Since I can't tell who is right, it's only reasonable to let both views be taught." And so it happens: through clever word manipulation and appeals to "equal opportunity," the creationists win the day.

When objections are raised, however, the first one is invariably that creationism is derived from the Bible, that the Bible is a religious book, that it is unconstitutional to mandate teaching sectarian religion in the public school science curriculum, and therefore creationism should not be introduced.

The creationists, however, have a ready answer. The two-model approach, they declare, "is *not* the introduction of the Bible or Bible stories about creation into the science books or classrooms. It is the fair and balanced presentation of the evidence and arguments both pro and con relative to both models of origins. . . ."[8]

In addition to this ready answer, they also have ready-made textbooks. Probably the most famous is Dr. Morris' *Scientific Creationism,* put out by his Institute for Creation Research. The preface states *"Scientific Creationism* . . . deals with all the important aspects of the creation/evolution question from a strictly scientific point of view, attempting to evaluate the physical evidence from the relevant scientific fields without reference to the Bible or other religious literature."[9]

However, in spite of this nice-sounding opener, this textbook is nothing more than a polemical attack on the evidences for evolution, with almost no statement of the case for creationism or the nature of the creation model. Such a ploy is necessary since, outside the Bible, there *is* no creation model. This is readily proved by Dr. Morris' revealing statement, "The Bible account of creation can be taught in the public schools if only the scientific aspects of creationism are taught, keeping the Bible and religion out of it altogether." This seems to mean that Biblical ideas suddenly become scientific once one hides the fact that the Bible is the source.

So, hide the Bible they do. For example, in another anti-evolution book entitled *Evolution: The Fossils Say No!,* Dr. Morris' colleague, Dr. Gish, writes, "By creation we mean that bringing into being of the basic kinds of plants and animals by the process of sudden, or fiat, creation described in the first two chapters of Genesis."[10] This seems plain enough. But Dr. Gish wanted his book used in the public schools. So, what did he do? He wrote a revision of it that left out this reference to his ultimate authority.

Perhaps he had learned something from the recent experiences of John N. Moore and Harold Slusher, two other creationists. They co-edited the controversial high school science textbook *Biology, A Search for Order in Complexity*. Although this book was purported to be objective, scientific, and non-sectarian, an Indiana Superior Court found it riddled with religious references such as: ". . . the second law (increasing entropy) is essentially a confirmation of the universal law of decay and death postulated in accordance with the biblical version of the creation model." ". . . most fossil material was laid down by the flood in Noah's time." ". . . the most reasonable explanation for the actual facts of biology as they are known scientifically is that of biblical creationism."[11]

The court's verdict, issued by Judge Michael T. Dugan II, was probably the most embarrassing judicial exposé of modern-day creationism handed down from the bench in the 1970s. The Court declared,

> Clearly, the purpose of *A Search for Order in Complexity* is the promotion and inclusion of fundamentalist Christian doctrine in the public schools. The publishers, themselves, admit that this text is designed to find its way into the public schools to *stress* Biblical Creationism. . . . The question is whether a text obviously designed to present *only* the view of Biblical Creationism in a favorable light is constitutionally acceptable in the public schools of Indiana. Two hundred years of constitutional government demand that the answer be *no*. The asserted object of the text to present a balanced or neutral argument is a sham that breaches that 'wall of separation' between church and state voiced by Thomas Jefferson. Any doubt of the text's fairness is dispelled by the demand for 'correct' Christian answers demanded by the *Teacher's Guide*. The prospect of biology teachers and students alike, forced to answer and respond to continued demand for 'correct' fundamentalist Christian doctrines, has no place in the public schools.[12]

As one watches creationists, one can see that they learn their lessons very well. *Scientific Creationism*, though it mentions a world-wide flood that occurred less than 10,000 years ago, the "survivors" of which "emerged" "near the site of Mount Ararat," and though it refers to a miraculous origin of languages "near Babylon" ". . . where tradition indicates the confusion of languages took place," it never mentions the Bible.

At least it never mentions the Bible in the "Public School Edition." The "General Edition," however, is quite another story. It is "essentially identical with the public school edition, except for the addition of a comprehensive chapter which places the scientific evidence in its proper Biblical and theological context," says the Foreword.[13] This version is for the *Christian* schools.

In their public debates, the creationists are even more careful to avoid stating their creation model. They invariably start out by saying that what they're talking about has nothing to do with Genesis. After that, the rest of their material is evolutionary criticism. If their opponents try to bring up the Bible, they counter-attack by declaring they came to talk about *science,* not religion. They further add that they have a right to their religious faith and should not have to hear criticism of it during a discussion of the scientific issues.

This approach seems to do well for them most of the time. But constant demands by evolutionists for creationists to explicitly state their model has lately forced them to formulate a secularized version of what they really believe. This version, contrived by attorney Wendell R. Bird, was published for all the world to see in the December 1978 issue of *Acts & Facts,* put out by the Institute for Creation Research.[14] [See Figure 1.]

Clearly, Bird felt it was important to carefully define the differences between the Biblical creation and *scientific* creation models. It was and is his view that a sharp and consistent distinction can be made. *Acts & Facts* declared, "The scientific creation model is based on scientific evidence, and the Biblical creation model is based on Genesis and other Biblical revelations. Mixing presentation of the scientific creation model and supporting scientific evidence with references to the Bible, Genesis, Adam, Noah, or the Ark will cause scientific creationism to be barred from the public schools."[15]

It would seem by all this that the differences between the two models must be quite radical. Are they? You can find out for yourself by comparing them side-by-side as is done in Figure 1. No doubt you'll notice that the actual differences between the "scientific" and Biblical creation models are quite small, in some places only amounting to a change of two or three words.

As I pointed out rather bluntly to Dr. Kofahl, a leading creationist, during a recent debate in which we both participated, "The differences between the Biblical model and the science model are so minor, so minute, that nobody is kidding anyone and nobody is being fooled. Once you hear the creationist model laid out, you're going to recognize it immediately as a Biblical model unless you were born in Borneo somewhere and never heard of the Bible."

In response, Dr. Kofahl argued that this wasn't the creationism he was interested in, and that he had no desire to bring Dr. Morris'

FIGURE 1.

The Two Creation Models of Wendell R. Bird
As Taken From the December 1978 Issue of *Acts & Facts.*

Scientific Creation Model:

I. Special creation of the universe and earth (by a Creator), on the basis of scientific evidence.

II. Application of the entropy law to produce deterioration in the earth and life, on the basis of scientific evidence.

III. Special creation of life (by a Creator), on the basis of scientific evidence.

IV. Fixity of original plant and animal kinds, on the basis of scientific evidence.

V. Distinct ancestry of man and apes, on the basis of scientific evidence.

VI. Explanation of much of the earth's geology by a worldwide deluge, on the basis of scientific evidence.

VII. Relatively recent origin of the earth and living kinds (in comparison with several billion years), on the basis of scientific evidence.

Biblical Creation Model:

I. Divine creation of the heaven, stars, and earth by God, on the basis of Genesis.

II. Application of the curse, pronounced by God after Adam's fall, to produce deterioration in the earth and life, on the basis of Genesis.

III. Divine creation of plant and animal life, Adam the first man, and Eve from Adam's side by God, on the basis of Genesis.

IV. Fixity of original plant and animal kinds, determined by God, on the basis of Genesis.

V. Distinct ancestry of Adam and apes, on the basis of Genesis.

VI. Explanation of the earth's geology by a world-wide flood in which only Noah, his family, and animal pairs were preserved in an ark, on the basis of Genesis.

VII. Approximately six thousand year time span since creation of the earth, life, and Adam, on the basis of Genesis.

Scientific Creationism into the classroom. What he wanted to see was "the evolution model criticized on the basis of the scientific evidence."

Well, there we have it again. They don't want to talk about their own model. They only want to critique evolution. This is the only *real* way they can avoid the problem of bringing in the Bible, and they know it.

And yet, when they get careless, they *do* bring it in. The Creation-Science Research Center, of which Dr. Kofahl is a representative, publishes the *Science and Creation Series,* which is a set of graded public school textbooks. This set was examined by Richard M. Lemmon for the California State Board of Education in 1975. In his report he drew attention to a number of religious references in the series.

He wrote, "In the 'Handbook for Teachers', page 75, it is stated that 'It is known that the nation of Israel began about 3700 years ago with the patriarch Jacob.' ... In 'The World of Long Ago, 3T', page

29, it is stated that 'The Bible also records a great flood, one that covered the highest mountains.' . . . In 'Man and His World, 7T', page 11, we find that a French explorer 'found timber which he believes came from the Ark of Noah.' . . . In 'Beginning of the World, 7T', page 5, there is a reference to 'one eternal personal God as the Creator of all things (as in Genesis).' . . . In the 'Handbook for Teachers', page 77, we find a statement about '. . . great world catastrophes occurring after the creation, including especially the great flood recorded in the book of Genesis . . .'"[16]

On the basis of these and other discoveries, Lemmon argued that "The entire purpose of these books is to use science classes to indoctrinate students in a particularly narrow brand of religious sectarianism. That sectarianism ignores most of the world's great religions; its promulgation in the public schools would violate the Education Code, Article 2, Section 9014, and the California State Constitution, Article IX, Section 8."[17]

THE NEW LEGAL MOVES

From the foregoing, it would seem the creationists have been rather clumsy in sticking to their new tactic of secularizing creationism. But, even if they had managed to carry off such a plan with any efficiency, their position would still fall short of legal acceptability.

There are a number of reasons for this; but to understand them correctly, it will be necessary to first reveal what the creationists are trying to do with their new "scientific creationism" now that they have formulated its rhetoric.

In a September 1977 letter of appeal for contributions, Dr. Morris wrote, "As you know, one of our main purposes here at ICR has been to reach the schools and colleges of our nation with the message of creation, so that young people would know there is a valid alternative to the evolutionary humanism that dominates our society today." In October he added, "We especially appreciate the splendid efforts of so many of you to accomplish the goal of getting creation into your own local schools and colleges."

Nell Segraves, Administrative Assistant of the Creation-Science Research Center, and one of the founders of the modern creationist movement, stated in a recent letter to Frank Mortyn of San Diego Mesa College that, "we are advocating the introduction into science textbooks and classrooms of scientific data which support the alternative explanation of origins, namely, intelligent purposeful design and special creation. In other words, we are calling for a reform in the teaching of science."[18]

Segraves authored the Center's "action Manual," a guide for implementing Creation-science curricula in the public schools, the legal rationale for teaching it, and guides for evaluating textbooks. In a recent debate she declared, "We feel that we are entitled to at least 50 percent of the public education system for our point of view."

Her reasoning is simple and straightforward: The Scopes trial, in showing the illegality of banning evolution, also showed the illegality of banning *any* theory of origins. Since creationism is such a theory, then by the logic of the Scopes trial, it cannot be banned either.

When confronted with the argument that creationism has definite religious overtones, she responds by claiming that the same is true for evolution. It is "the religion of secular humanism" in the public schools. This means that any school that teaches evolution without balancing it with special creation is operating contrary to the religious neutrality requirement of the U.S. Constitution. It is setting up a state religion in the science classroom.

These are her arguments; and on the basis of these, the Creation-Science Research Center is seeking to cut off millions of dollars in federal funds that come into California. Since the state-supported schools don't teach both theories of origins as science, it is claimed the schools are religiously biased and therefore undeserving of the monies.

Late in 1980 CSRC sued the state of California for setting up textbook guidelines that left out special creation.[19] CSRC's goal was to prevent the guidelines from going into effect. Although their suit failed to accomplish its goal, the judge did rule that the State Board of Education's 1973 policy on avoiding dogmatism in the teaching of origins be sent to all school districts and science teachers in the state as well as to textbook publishers, and that it be included in future editions of the guidelines.

More blatant efforts to require the teaching of creationism were made by other organizations, particularly Citizens for Fairness in Education, a South Carolina group headed by Paul Ellwanger. Such thrusts began to make national headlines early in 1980 when the joint houses of the Georgia state legislature came very close to passing a bill that would have required equal time for creationism any time the issue of origins came up.[20]

In the same month, the Florida House Education Committee voted 7 to 6 for a similar bill. An editorial in the St. Petersburg *Times* declared:

> This bill would not prohibit the teaching of evolution, at least not in
> so many words. But any school that undertook to acknowledge the

theory of evolution—whether in class or merely on its library shelves —would have to give 'balanced treatment' to what is called 'the theory of scientific creationism.'

And what is that? The bill defines it with a lot of gibberish and mumbo-jumbo, all of which boils down to this: The biblical account of creation can be proven literally, with scientific 'evidence.' . . .

IN PRACTICE, the bill would simply end the teaching of evolution—and perhaps all science—because few teachers and school boards would consent to teach the alternative theories the bill espouses.[21]

And this may be something creationists would like to see. The April 1979 *Acts & Facts* stated: "We are not trying to exclude evolution from public schools, unless creation is also excluded."[22] Nell Segraves put it more plainly in debate: "It's totally unnecessary to bring origins into a science discussion. Textbooks today can give good science without discussing philosophy of origins at all." Dr. Kofahl, in the same debate, then immediately added, "We would really be satisfied to see the subject of origins removed entirely from public school science . . . Let's *forget* about origins. Let's put *all* origins discussions into the philosophy department."

In Medford, Oregon, it seems creationists easily got their wish. When a young student of "scientific creationism" started stumping for equal time, the Medford School Superintendent, Richard Langton, declared the following:

Evolution is not taught in any of the schools of District 549C [Medford]; neither is creation for that matter. Down through the years, educators have learned that this is such a controversial subject that it is far better not to deal with it at all than to try to deal with it, even on a fair basis, pointing out the claims of both sides. At appropriate levels, where it is understood, we do teach simple genetics, but we in no way get into the question of the evolution of man.[23]

We can now see the entire creationist legal program in all its glory. First they stump for equal time on the grounds that creationism is an alternate scientific view. When that fails, they argue for equal time on the grounds that creationism is an excluded *religion.* When that fails, they say that *neither* should be taught because both are *philosophies.* And by the time that fails, the school officials are so intimidated they begin to wish they had never even heard of evolution.

Still, however, the creationists have one more legal gambit up their sleeves. Nell Segraves probably deserves all the credit for it. Her argument runs thusly:

The atheists have won a number of significant court cases that have resulted in the removal from the public schools of everything offensive to their atheistic viewpoint. They have gotten rid of prayers, religious references in text books, religious displays, etc. Women's rightists have also had much success in removing things that offend them, such as sexist language in textbooks. Well, now it's time for Christian fundamentalists to use these *same* court decisions in *their* favor—that is, to remove everything offensive to the *Christian fundamentalist* viewpoint. ". . . *we* now are on the outside demanding equal treatment and equal recognition for our point of view under the First and Fourteenth Amendments of the U.S. Constitution and the Civil Rights Act of 1964," she argues.

Such an interpretation of the relevant court decisions has far-reaching implications, and the Creation-Science Research Center reaches most of them. They aren't satisfied with calling only *evolution* "offensive," but go on to add sex education to the list. They further object to the teaching in history classes of the theory that human societies evolved from tribe to village to cradle of civilization. (They believe that man was civilized when he came off the Ark.)

In the general public sector they use the same arguments to condemn rehabilitation of criminals, abortion, government grants to Planned Parenthood, and research grants to behaviorists. In their January, 1980 *Creation-Science Report* they made their position very plain: "As theists and creationists, possessing equal rights and privileges under the Constitution and Federal Civil Rights legislation, we can set forth creationist position papers on any and *all* problems affecting public morals or health, domestic or foreign policy, whenever government funding is required."[24] This is why seeking cutoffs of funds is one of their major tactics.

One can only ask, in the face of this line of reasoning, where it will stop. Obviously, there is *no* view taught in our schools that at least *somebody* won't find offensive to their religion or value system. The teaching of physical science in any form is offensive to mystics who hold that matter is an illusion. If the school nurse talks about health, she had better not mention medicine or vaccinations, or it will offend the followers of Christian Science. Teaching English is bound to be an offense to those who uphold the sacred languages of Hebrew or Sanskrit. Any geography or astronomy which declares the world to be round will create problems in the homes of religious children who were raised by Bible-believing flat earthers.

So, we must ask the practical and legal question: how far must the schools go to avoid offending someone's religion, and how far must they go in giving balanced presentations of all viewpoints every time an "offensive" issue is raised? Furthermore, what state and federal programs will have to be cut off because someone comes up with a religious reason for not liking them? Would we have any government programs or modern education left?

Two creationist women I met during a lecture in Seattle had a simple solution. Get rid of public schools altogether. Let parents choose what kind of schooling they want their children to have. In fact, let them opt for no schooling at all, if they so desire.

THE LEGAL CASE AGAINST CREATIONISM

Since 1980, efforts have been stepped up to pass "equal time" bills in a number of state legislatures. Some of the creationists promoting such action probably think they can win, that the law is on their side. But many others know better, like Senator Hugh Carter, who, in speaking for Georgia's 1980 creation bill, declared cynically from the floor of the State Senate, ". . . look at all the good we can do between now and the time it is declared unconstitutional."[25]

Those on both sides who have really looked into the matter can see hopeless flaws in the legal case for creationism. Right off the bat it starts out with a basic contradiction. First the creationists try to define science so narrowly that it leaves out evolution. This renders evolution a *religion,* right along with creation. Then they try to so broadly define the science *curriculum* that it allows both "religions" to be taught in a scientific context. Putting it another way, creationists demand equal time for creation on *religious* grounds, so they can get it into the schools, and then demand equal time on *science* grounds, so they can get science instructors to teach it! No case this absurd can be tried for long without trying the patience of everyone.

In the new legal battles, creationists will often deny they are trying to replay the Scopes trial. They don't want to ban evolution, they just want to make sure it won't be taught without creation having a place too. But the idea that evolution is OK only if creation is included is really two ideas in one. First, it is the idea that when evolution is taught, creation is *mandated.* Second, it is the idea that if creation is *not* taught, evolution is *banned.* The two must be dealt with separately. Let's begin with the second.

The banning of evolution on religious grounds has the unenviable legal status of being totally unconstitutional. In the case of *Epperson* v. *Arkansas* in 1968, the U.S. Supreme Court held that no religious group had the right to blot out any public school teaching just because it was "deemed to conflict with a particular religious doctrine." For to do so would be to, in effect, establish a religion, or at least a religion's prohibitions, in the public sector. This is contrary to the First Amendment of the U.S. Constitution which reads in part: "Congress shall make no law respecting an establishment of religion or prohibiting the free exercise thereof; . . ."[26]

And it doesn't seem to matter if the anti-evolution law is stridently religious, or is vague on the matter, it is unconstitutional all the same. For example, the Tennessee law which John Scopes was charged with breaking made it unlawful "to teach any theory that denies the story of Divine Creation of man as taught in the Bible and to teach that man has descended from a lower order of animals." But the Arkansas law challenged in *Epperson* v. *Arkansas* was less explicit. Both were, in effect, declared unconstitutional by the *Epperson* decision. The Court declared in *Epperson* that it was "clear that fundamentalist sectarian conviction was and is the law's reason for existence." It was noted that "Arkansas did not seek to excise from the curricula of its schools and universities all discussion of the origin of man. The law's effort was confined to an attempt to blot out a particular theory because of its supposed conflict with the Biblical account, literally read." Recent legal moves, though more camouflaged than ever, seem to come to the same thing. The creationists are trying to remove evolution on religious grounds.

It would seem strange, in the light of the *Epperson* decision, that creationists wouldn't move to do what the Court seemed to allow, that is, remove *all* teachings of origins. But I doubt if that is their first preference. They would probably prefer to find a way to teach special creation (or, more correctly, Biblical fundamentalism). And it isn't likely they would be satisfied to have it taught in comparative religion classes either. Why? Because the science classes will continue to teach things creationists regard as persuasive in the "wrong" direction, things that would be devastating to their belief system if true. So, they want to get religion into the science classes also. When they can't ban evolution and teach creation, they usually strive to *require* creation and *neutralize* evolution. Teaching neither, then, is hardly satisfactory to them. This is probably why they don't really push for that except as a footnote to their bills and lawsuits, as an afterthought in their debates. (The aforementioned announcement by the Medford, Oregon School Superintendent that evolution was not being taught did not put an end to the creationist movement there.) So, let's look into this idea of *requiring* creationism.

Obviously, if banning a teaching in the schools on religious grounds constitutes the establishment of religion in the public sector, then it is all the more true that *requiring a religious doctrine* in the schools is to do the same thing! Yet creationists somehow think they can do better with this idea than with the previous one.

True, appeals for "equal time," "fair play," and "academic freedom" are more persuasive with the public. But it isn't the public who decides constitutionality. That operates according to a basic *principle,* one that is to be unchanging, for the most part.

But even if the doctrine being required wasn't religious, it would still be questionable. As Professor Richard D. Alexander noted in the February 1978 *American Biology Teacher,* "If evolutionists were attempting to require that evolution be taught it would be no less pernicious. . . . when anyone attempts to establish laws or rules requiring that certain theories be taught or not be taught, he or she invites us to take a step toward totalitarianism. Whether a law is to prevent the teaching of a theory or to require it is immaterial. It does not matter if equal time is being demanded or something called 'reasonable' time, because there can be no reasonable time in such a law."[27]

In the past when a scientific view was mandated by government, it resulted in disaster and a stiffling of progress. One particular example occurred 40 years ago in Russia. A man named Lysenko temporarily established that Lamarckian evolution was true science and that Darwin was wrong. This resulted in, first, a mandating of Lamarckianism. But following shortly on its heels was a banning of Darwinism. It took decades for Russia to recover from this legal action and catch up to the modern world in the realm of science.[28]

In the 1980 Georgia battle, Julian Bond, a black State Senator, expressed the point in this way. "Thirteen years ago, I sponsored a bill that called for the teaching of black history in the public schools. Everybody said, 'It's a fine idea, but we can't legislate the curriculum.' What will we tell the large body of non-Christian children who sit in Georgia's classrooms and are taught the creation theory?"

In the Georgia State Legislature, Representative Bill McKinney argued much the same way. He noted that if government was now going to enter the business of curriculum design, it should demand equal time for black history. After all, "There are more black folks in this country now than there are scientific creationists."

The *Epperson* decision, while dealing with a law banning evolution, had something to say about requiring creation as well. The Court declared, "There is and can be no doubt that the First Amendment does not permit the State to require that teaching and learning must be tailored to the principles or prohibitions of any religious sect or dogma. . . . the State may not adopt programs or practices in its public schools or colleges which 'aid or oppose' any religion. . . . This prohibition is absolute. It forbids alike the preference of a religious doctrine or the prohibition of theory which is deemed antagonistic to a particular dogma."

In *Zorach* v. *Clauson* in 1951, Justice Douglas wrote the majority opinion, saying: "Government may not finance religious groups nor undertake religious instruction nor use secular institutions to force one or some religion on any person." The comment about *financing* religious groups is instructive, because the teaching of creation would require the use of creationist textbooks and learning materials. Since only religious creationists offer them, then to make such purchases could easily amount to the financing of religion by government.

In California, religious ideas may be discussed in the schools, provided they "do not constitute instruction in religious principles or aid to any religious sect, church, creed, or sectarian purpose. . . ."[29] The teaching of special creation would do at least one of these things. Therefore creationism, even the "scientific" version, is a religious doctrine. This was brought out most completely in the recent case of *McLean* v. *Arkansas,* where U.S. District Judge William Overton overturned a 1981 "equal time" creationism law in Arkansas, declaring the law to be "simply and purely an effort to introduce the biblical version of creation into the public school curricula." Overton noted that the "equal time" or "two model" idea was "simply a contrived dualism" having no scientific basis or "legitimate educational purpose." He declared the law unconstitutional because it lacked a secular legislative purpose, because its principal or primary effect was the advancement of religion, and because the law, if carried out, would foster an excessive government entanglement with religion.

As for the question of whether evolution is *also* a religion, Evelle Younger, Attorney General for California, had this to say to the Creation-Science Research Center in 1975:

> The "neutrality requirements" of the First Amendment are not violated by the inclusion in textbooks by the State Board of Education of a scientific treatment of evolution. The degree to which a scientific subject should be made more or less "dogmatic" does not involve considerations of "religion." Such considerations, in the exercise of the Board's sound discretion, turn upon the degree of scientific certainty supporting a subject presented in a textbook. Action by the State Board of Education or local boards of education to modify a scientific theory may be judicially proscribed if it can be demonstrated that it is an attempt to modify such theory because of its supposed conflict with religion.[30]

The issue Younger was commenting upon was the Creation-Science Research Center's efforts to have evolution taught in a less "dogmatic" way in California schools and textbooks. His arguments indicate that not only can evolution not be banned or "balanced," but it also cannot be modified (at least not unless the scientific facts, as determined by the State Board of Education, warrant such modification independent of religious criteria).

As for Nell Segraves' argument that prayers in school were removed because they were an "offense" to atheism: this is nothing but more creationist revisionist history. The reason prayers were removed was because they constituted an establishment of sectarian religion in the public sector.[31] This means Mrs. Segraves can gain no legal advantage by claiming evolution is a religious "offense" to creationism.

And if she tries to point to civil rights legislation that bars "offenses" to blacks, women, etc., her argument will still miss the point. The civil rights laws ban disparaging remarks, not courses of study. Therefore, if blacks are depicted as lazy, women as emotional, or Christians as bigoted, *then* legal action will be taken. But no one can, under these laws, either ban courses or *require "equal time"* for black studies, women's studies, or creationism.

It is true, however, that in the case of *West Virginia* v. *Barnette,* Justice Murphy wrote in his concurring opinion: "Official compulsion to affirm what is contrary to one's religious beliefs is the antithesis of freedom of worship. . . ." But this only applied to the compelling of unconscionable statements. Evolution, as normally taught, does not require the student's allegiance. Only his or her understanding of the objectively presented concepts is sought. Therefore, the teaching of evolution is neither a threat to nor an imposition on the religious freedom of any child. Students are always free to disagree with any theory they learn.

In the case of *Wisconsin* v. *Yoder,* the Court granted Amish parents the right to take their children out of the public schools after eighth grade, provided those children were participating in the "long established program of informal vocational education" that the Amish taught. The Court declared that "the values of parental direction of the religious upbringing and education of their children in their early and formative years have a high place in our society." A similar right of parents to send their children to private, religious schools was upheld in *Pierce* v. *Society of Sisters,* so long as the children were prepared "for additional obligation" in society.

All these cases, then, seem to offer a solution to parents like Mrs. Segraves. If they are "offended" by evolution, they can send their children to private religious schools, or, as in the case of sex education, have them released from the class when the subjects at issue are being taught.

Regarding this solution, creationist lawyer Wendell Bird points out the apparent unfairness of requiring an individual to make a choice between his or her faith and a public benefit. He or she has a right to both. Free education and free exercise of religion need not be mutually exclusive.[32] Bird also criticizes the released time plan, citing the case of atheists who were not satisfied with merely having the right to leave the classroom during school prayers. Creationists, too, who in *their* situation might desire to leave evolution studies "would probably be prevented by pressure from fellow students, respect for teacher opinions, and need for other course material missed."[33]

What Bird fails to see, however, is that in the school prayer cases, religion was being introduced into the school curriculum. That would have been unacceptable even if *nobody* had protested, since it is unconstitutional to teach sectarian religion in the public schools. But in the case of the teaching of evolution, there is no unconstitutional teaching of religion. Only *science* is being taught, and that is acceptable. The problem, then, is not in the schools. So the only solution is to allow the offended students the right to remove themselves.

There may be something to Bird's argument that evolution serves to undermine faith in a literal interpretation of the Bible and is therefore a burden on a fundamentalist's rights to free exercise of religion. But this is hardly sufficient to justify forcing *all the rest* of the pupils to study creationist religious doctrines or to go without learning about evolution and thereby receive an inferior education. It would seem there is a state interest in teaching this material, and teaching it exclusively. This is made clear by the fact that evolution is the one great unifying principle of all science. Students cannot be adequately prepared for scientific careers if they are left in the dark about its existence. And if it is "balanced" with a non-scientific theory, then they will get an inaccurate picture of science and be misled into believing there is a significant split of opinion among scientists on the issue, when there is not.

CREATIONIST GUERRILLA WARFARE

Unlike the Creation-Science Research Center and other similar organizations, the Institute for Creation Research does not engage in law suits or legislation, at least not directly. In the January–February 1973 *Acts & Facts,* Dr. Morris wrote that "no recommendation is made for political or legal pressure to *force* the teaching of creationism in the schools. Some well-meaning people have tried this, and it may serve the purpose of generating publicity for the creationist movement. In general, however, such pressures are self-defeating. . . . The hatchet job accomplished on the fundamentalists by the news media and the educational establishment following the Scopes trial in 1925 is a type of what could happen, in the unlikely event that favorable legislation or court decisions could be obtained by this route."[34]

The clear admission that creationism doesn't have a legal case is even more explicitly stated by Morris in a December 1974 article. He wrote: "Even if a favorable statute or court decision is obtained, it would probably be declared unconstitutional, especially if the legislation or injunction refers to the *Bible* account of creation."[35]

Since Dr. Morris and ICR, then, clearly recognize the legal shakiness of their two-model position, what is their plan for getting creationism into the schools?

Well, they outline it in detail in a number of issues of *Acts & Facts*. Here are its salient points:

I. Parents should—
    1. Buy and read ICR creationist books, both religious and scientific.
    2. Teach their children and those of other parents about creationism, and encourage them to bring the issue up in the classroom.
    3. Talk to the school teachers about it, and if they aren't receptive, go to the principal or superintendent.
    4. Convince local school boards that the two-model approach is legal, non-religious, and in no way contrary to the U.S. Constitution.
    5. Purchase copies of *Biology, A Search for Order in Complexity* and *Scientific Creationism* (Public School Edition) to show to school officials. Recommend the former for students, the latter for teachers.
    6. Get permission and speak at the next state board of education meeting or meeting of the proper state curriculum authority.
    7. Get permission and speak at the next state textbook commission meeting after seeing advance copies of the textbooks and reviewing them.
    8. Petition that a resolution (not a law) be passed "permitting" or "encouraging" (not requiring) the teaching of creationism.
    9. Establish a community pressure group with an appropriate name like "Citizens for Scientific Creationism" or "Civil Rights for Creationists." Then do things like take a community census poll, raise funds to buy the school and public libraries creationist books from ICR, promote a workshop on creationism for teachers or a seminar for the general public, sponsor debates using ICR experts, and/or work up a lot of media publicity in local and school papers, etc.
    10. Donate money to ICR for further creation research.

II. School administrators should—
1. Encourage teachers to teach creationism.
2. Conduct workshops on creationism for teachers on a graduate credit basis, bringing in ICR experts.
3. Provide substitute teachers to teach creationism when the regular teacher isn't willing, or have regular specialists in the subject.
4. Have creationist materials purchased for the school(s).

III. Teachers should—
1. Introduce creationism into their own classrooms "no matter what the course subject or grade level may be. . . . whenever the textbook or course plan contains evolutionary teachings or implications." This not only includes science, but geography, history, social science, and other subjects.
2. Rent or order ICR two-model and creationist audio-visual aids.
3. Invite creationist speakers to address a school assembly.
4. Talk to fellow teachers over coffee and win them over to the two-model approach.

IV. Scientists should—
1. Stand firm to their creationist convictions when faced with the derision of their colleagues.
2. Serve as consultants and lecturers for schools and citizen groups.
3. Join the Creation Research Society.

V. Pastors should—
1. Promote *Biblical* creationism in their church and Sunday school.
2. Lead community-wide creationist movements involving the churches.
3. Talk with school administrators.
4. Promote creationism over the airwaves.

VI. Students should—
1. Give "careful, courteous, consistent Christian testimony" to the teacher in a way that is "winsome and tactful, kind and patient."
2. Raise questions and offer alternative suggestions in class discussions.
3. Bring creationism into speeches, papers, and class projects.

4. Invite the teacher and classmates to creation seminars.
5. Suggest a creation/evolution debate in the classroom.
6. Give ICR tracts and publications to the teacher and principal.
7. Answer relevant test questions with the prefacing words "Evolutionists believe that—" when an evolutionary answer is required to get a test question correct.
8. Withdraw from the course if the teacher is too hostile.

Dr. Morris has said: "Creationist teachers are in a unique position to play a critical role in this strategic conflict," and he has his strategy all worked out. He notes further that pastors "are especially capable at the arts of persuasion and instruction" and should use these to promote the cause. "Scientists and other professionals who are Christians have a peculiar trust from the Lord."[36] The aim here is obviously to bring as much pressure to bear as possible in order to "bring creation back into the public schools."

Sample resolutions for presentation before school boards and state curriculum authorities have been published for easy use in both the July–August 1975 and May 1979 issues of *Acts & Facts.*[37] They have been used widely all over the country with some success. The 1975 *Acts & Facts,* however, recommends a bit of secrecy as to the source of the legal wording, saying "it would be better not to mention ICR at all in connection with it [the resolution], so that the officials will realize that it is their own constituents who are concerned with the issue."

What this boils down to is an ICR engineered local grass roots pressure movement to sneak creationism into the schools through every back door they can find. But, failing that, they will settle for the intimidation caused in their wake, knowing full well that such intimidation tends to prevent, or water down, the teaching of evolution.

CONCLUSION

It should be clear by now that, legally, the creationists do not have a case. Any effort to ban evolution because it conflicts with a religion is an effort to bring sectarian religious prohibitions into the public schools. This is unconstitutional. On the other hand, any effort to *add creationism* to the science curriculum will amount to the teaching of sectarian *doctrines.* This too would be unconstitutional.

To get around this problem, creationists have sought to establish creationism as secular science. They have gathered data and tried to remove references to the Bible. But, because they have made little effort to work through the scientific community, to participate in the peer review of the journals, to do more than just token field research; and since they have promoted a rather dogmatic "science," the courts have exposed this effort to be a sham.

Yet even if they *had* become truly secular in their ideas, mandating inclusion of these through legislation would remain illegal and contrary to academic freedom. Even *evolution* can't be forced in this way. It is not the business of the legislature to determine what is and is not science. This task belongs to the scientific community. Therefore, only if there is a legitimate controversy among knowledgeable field workers on an issue is it proper for more than one model to be taught. Since there is no such controversy at this time, creationism is without academic grounds for inclusion (except, perhaps, as a discredited theory in the same class as Lamarckianism).

This realization has forced creationists to try another ploy: If you can't join them, beat them—that is, ban *all* discussions of origins from the science curriculum, and send them off to the philosophy department.

Of course there's no need to ban creationism. It isn't part of the curriculum. And if it's proposed that we ban evolution, we're headed for another Scopes trial. We must therefore ask creationists *why* they want it banned. If it's because it conflicts with their religion, the constitution will prohibit such a move. But if it's because evolution is *itself* supposedly a religion, they will have to prove that. And they will have to prove it using scientific means, submitting their arguments to peer review, and actually showing that evolution is untestable and non-scientific in nature.

Because of the difficulty of this endeavor, and because they cannot win in the courts, some creationist groups have given up legal action altogether and have emphasized a kind of "religious smuggling." One part of their plan involves telling school officials that the two-model approach is both constitutional and scientific, even though creationists have never won a court case or convinced a scientific symposium. Another part involves gathering pressure groups to intimidate school authorities so evolution can be pulled out, or creation brought in, through the back door. (In such cases, it should occur to school authorities to ask why pressure is necessary if creationism is scientifically sound, and why ICR has avoided the courts if their position is supposed to be constitutional.)

So far, not having a good case hasn't been fatal to the creationists. In fact they have flourished!—which shows that the problem will be with us a long time. Obviously it isn't enough for evolutionists to have the law on their side, to sit back and let the lawyers do the work. Creationists have been losing the battles and winning the war. That is, they have successfully intimidated the schools and textbook publishers into near submission. This has effectively won them the Scopes trial. With their continued persistence, and with further neglect by evolutionists, they may, through their "guerrilla warfare," succeed in their primary goal of getting Biblical fundamentalism taught in our public schools.

Meanwhile, their constant battling costs the taxpayers money and gains them the supporters they need. As a result, in time they could feel confident enough to push for a constitutional amendment that would turn the legal case around in their favor.

Because it isn't safe to neglect this threat any longer, the time has come to inform the public of the facts—and to guarantee students an adequate education. Respect for science in America is waning. The popularity of both creationism and mysticism are symptomatic of it. It's no longer possible for academics to ignore the public while advancing their scientific careers. If they try, they will soon find creationism in the schools and anti-science in the electorate.

The public never fully accepted evolution. Now that we realize this, we can work to remedy the situation. We can study the creationist arguments to learn where evolution is being misunderstood or feared. We can then tell the public *why* scientists accept evolution, instead of telling them merely *that* they do. We can improve the public relations of science in general, and thereby bring it back into respect. But, most importantly, we can update Muller's statement and boldly declare, "One hundred twenty-four years without Darwin *are* enough!"

REFERENCES

1. On the initial reaction to Darwin, see William Irvine, *Apes, Angels, and Victorians* (New York: McGraw-Hill, 1955).

2. S. J. Holmes, "Proposed Laws Against the Teaching of Evolution," in *A Compendium of Information on the Theory of Evolution and the Evolution-Creation Controversy* (Reston, VA: National Association of Biology Teachers, 1978).

3. W. V. Mayer, "Creation Concepts Should Not Be Taught in Public Schools," *Liberty* (September/October 1978): 3ff.

4. Bette Chambers, "Why a Statement Affirming Evolution," *The Humanist* (January/February 1977): 23.

5. See R. C. Cowen, "Evolution: Equal Time for God," *Technology Review* (June/July 1979): 10ff. See also Jerry Bergman, *Teaching about the Creation/Evolution Controversy* (Bloomington, IN: Phi Delta Kappa Educational Foundation, 1979); Larry Hatfield, "Educators against Darwin," *Science Digest* (Special) (Winter 1979): 94; and R. H. Utt, "A Tale of Two Theories," *Liberty* (January/February 1980): 12. On recent litigation, see Frederic S. LeClerq, "The Constitution and Creationism," in *A Compendium of Information,* 1978.

6. See William Overton, Memorandum Opinion in *Rev. Bill McLean,* et al. vs. *The Arkansas Board of Education* et al, " (United States District Court, Eastern District of Arkansas, Western Division: January 5, 1982). Included in this volume. This decision has also been reprinted in the *American Biology Teacher,* 44 (3) (1982): 172–79 and *Science* 215 (February, 1982): 934–43.

7. See D. T. Gish, "The Scopes Trial in Reverse," *The Humanist* (November/ December 1977): 50.

8. Creation-Science Research Center, *Decide: Evolution and Creation—One or Both in Public School Science?* (San Diego, CA: Creation-Science Research Center, 1980).

9. H. M. Morris, ed. *Scientific Creationism,* General and Public School editions. (San Diego, CA: Creation-Life Publishers, 1974). The quotation is from the Public School edition.

10. D. T. Gish, *Evolution: The Fossils Say No!* (San Diego, CA: Creation-Life Publishers, 1973).

11. J. N. Moore and H. Slusher, eds., *Biology: A Search for Order in Complexity* (Grand Rapids, MI: Zondervan, 1974).

12. M. T. Dugan, "Hendren v. Campbell," in *A Compendium of Information,* 1978. In the same work, see also Edward, Celebrezze, and Lively, "Daniel v. Waters."

13. Morris, *Scientific Creationism,* Public School edition, foreword.

14. W. R. Bird, "Distinction between Scientific Creationism and Biblical Creationism," *Acts and Facts, ICR Impact Series* (December 1978).

15. Bird, 1978.

16. Richard Lemmon, *Review of Science and Creation Series* (Sacramento, CA: California State Department of Education, February 28, 1975).

17. Lemmon, 1975.

18. Nell Segraves, "Letter to Frank Mortyn" (Creation-Science Research Center, April 15, 1980)

19. See John Gilmore, "Evolution vs. Creation: New Battle Brews," *San Diego Evening Tribune,* March 19, 1980, p. A-2; Ellen Goodman, "Creationists Get Some Knocks," *San Diego Evening Tribune,* April 18, 1980, p. B-11. See also M. Scott-Blair, "Creationists Seek Say in Schools," *San Diego Union,* October 14, 1979, pp. A-1.

20. See Paul Ellwanger, "Proposal to Anderson School District No. 5 Board of Trustees," *Acts and Facts, ICR Impact Series,* 67 (January 1979); Jeff Prugh, "Georgia Senate Passes Bill to Teach 'Creation by God'," *Los Angeles Times,* March 6, 1980, p. 8, part 1; Kenneth S. Saladin, "Creationism Bill Dies in Georgia Legislature," *The Humanist* (May/June 1980): 59.

21. "Bunkum in the Capital," *St. Petersburg Times,* March 13, 1980, editorial page. On the bill, see Bush, "Balanced Treatment for Scientific Creationism and Evolution Act (HB 107)," Florida House of Representatives, 1980.

22. Wendell R. Bird, "Evolution in Public Schools and Creation in Student's Homes: What Creationists Can Do," *Acts and Facts, ICR Impact Series,* nos. 69 and 70 (March and April 1979).

23. See Denise Stanley, "Kindell: 'Setting Minds Afire'," *Medford Mail Tribune,* March 19, 1978, pp. 1A.

24. *Creation Science Report* (San Diego, CA: Creation-Science Research Center, 1980).

25. See Jeff Prugh, 1980. For other quotations on the Georgia bill see this article by Prugh and also Saladin, 1980.

26. A. Fortas, "Epperson v. Arkansas," 393 U.S. 97 (1968).

27. R. D. Alexander, "Evolution, Creation, and Biology Teaching." Included in this volume.

28. See M. Gardner, *Fads and Fallacies in the Name of Science* (New York: Dover Publications, 1957), ch. on Lysenko.

29. Evelle Younger, "Opinion," Office of the Attorney General, State of California, April 2, 1975.

30. Younger, 1975.

31. See F. Swancara, *The Separation of Religion and Government* (New York: The Truth Seeking Company, 1950).

32. Wendell R. Bird, "Freedom of Religion and Science Instruction in Public Schools," *Yale Law Journal* (January 1978): 515ff.

33. Bird, *Acts and Facts*, 1979.

34. H. M. Morris, "Evolution, Creation, and the Public Schools," *Acts and Facts, ICR Impact Series*, no. 1 (January/February 1973).

35. H. M. Morris, "Introducing Creationism into the Public Schools," *Acts and Facts, ICR Impact Series*, no. 20 (December 1974).

36. *Ibid.*

37. H. M. Morris, "Resolution for Equitable Treatment of Both Creation and Evolution," *Acts and Facts, ICR Impact Series*, no. 26 (July/August, 1975); W. R. Bird, "Resolution for Balanced Presentation of Evolution and Scientific Creationism," *Acts and Facts, ICR Impact Series*, no. 71 (May 1979).

The Tennessee Anti-Evolution Act (1925)

Chapter 27, House Bill 185
Public Acts of Tennessee for 1925
Passed March 13, 1925
Approved March 21, 1925
Repealed 1967

An act prohibiting the teaching of the Evolution Theory in all the Universities, Normals and all other public schools of Tennessee, which are supported in whole or in part by the public school funds of the State, and to provide penalties for the violations thereof.

Section 1. Be it enacted by the General Assembly of the State of Tennessee, That it shall be unlawful for any teacher in any of the Universities, Normals and all other public schools of the State which are supported in whole or in part by the public school funds of the State, to teach any theory that denies the story of the Divine Creation of man as taught in the Bible, and to teach instead that man has descended from a lower order of animals.

Section 2. Be it further enacted, That any teacher found guilty of the violation of this Act, shall be guilty of a misdemeanor and upon conviction, shall be fined not less than One Hundred ($100.00) Dollars nor more than Five Hundred ($500.00) Dollars for each offense.

Section 3. Be it further enacted, That this Act take effect from and after its passage, the public welfare requiring it.

The Tennessee Creationism Act (1973)

Amendment to the Tennessee Annotated Code
Section 49-2008
April 30, 1973

Any biology textbook used for teaching in the public schools which expresses an opinion of, or relates to a theory about origins or creation of man and his world shall be prohibited from being used as a textbook in such system unless it specifically states that it is a theory as to the origin and creation of man and his world and is not represented to be a scientific fact. Any textbook so used in the public education system which expresses an opinion or relates to a theory or theories shall give in the same textbook and under the same subject commensurate attention to, and an equal amount of emphasis on, the origins and creation of man and his world as the same is recorded in other theories including, but not limited to, the Genesis account in the Bible.

Editor's Note: A court of appeals in Tennessee overruled this legislation on April 10, 1975, declaring that it violated the Establishment of Religion clause of the First Amendment and, as such, was unconstitutional (Edwards, Celebrezze, and Lively, "Daniel v. Waters," April 10, 1975).

Proposed National Bill: An Act to Protect Academic Freedom and to Prevent Federal Censorship in Scientific Inquiry Funded with Federal Tax Monies

Prepared by: Citizens Against Federal Establishment of Evolutionary Dogma (Marshall and Sandra Hall, National Coordinators, Murphy, North Carolina, n.d.)

Be it enacted by the Congress of the United States of America:

Section 1. This Act shall be known as the "Academic Freedom in Scientific Inquiry without Federal Censorship Act."

Section 2. The purposes of this Act are: (1) "To promote the Progress of Science"; (2) To protect academic freedom in the award and use of federal funds; and (3) To prevent federal censorship of minority scientific viewpoints about the origin of the universe, the earth, life, and man.

Section 3. Science-related applications of federal funds shall not prefer the evolution model (frequently referred to as the theory of evolution) or any other model of origins over the creation-science model (frequently referred to as the theory of scientific creationism). In particular:
(1) Federal funds for research that deals with the origin of the universe, the earth, life, and man shall be awarded to creationist scientist research applicants to the extent such funds are awarded to evolutionist scientist research applicants or other scientist research

applicants without censorship of scientific viewpoints about the origin of the universe, the earth, life, and man. Research applicants shall be classified as creationist scientists, evolutionist scientists, or other scientists according to their signed statements of adherence to the creation-science model as defined in Section 5, to the evolution model as defined in Section 5, or to any other scientific model of origins.

(2) Federal funds for museum exhibits that deal with the origin of the universe, the earth, life or man, or to museums that contain such exhibits, including the Smithsonian Institution, shall be disbursed to creation-science exhibits or museums to the extent such funds are disbursed to evolutionist exhibits or museums.

(3) Federal funds disbursed or awarded by the National Science Foundation that are used for research or other activities dealing with the origin of the universe, the earth, life or man shall be disbursed or awarded to creation-science research or related activities to the extent such funds are applied to evolutionist research or activities or to other research or activities dealing with origins.

(4) Federal funds for the National Park Service that are used for exhibits or signs that deal with the origin of the universe, the earth, life, or man, or to any other governmental agency, shall be used for creation-science exhibits or signs to the extent such funds are used for evolutionist exhibits or signs or for other exhibits or signs dealing with origins.

(5) Federal funds shall not be used for curriculum development or textbook writing for elementary or secondary school courses, other than military science courses, or for college courses dealing with origins, including the Biological Sciences Curriculum Study and Man: A Course of Study.

(6) No federal funds shall be used for any theistic, non-theistic, humanistic, or other religious doctrine about the origin of the universe, the earth, life, or man.

Section 4. Hearings shall be held by the U.S. House of Representatives Committee on Science and Technology and the U.S. Senate Committee on Commerce, Science, and Transportation, or their successors, on the scientific evidence for the creation-science model, for the evolution model, and for other science models of origins. The purpose of these hearings shall be to delineate the scientific evidence for these models, and to determine the existing uses of federal funds dealing with the origin of the universe, the earth, life, and man. The same number of witnesses shall be heard in favor of the creation-science model as the number in favor of the evolution model or the number in favor of all other science models of origins. Witnesses shall be classified as creationist scientists, evolutionist scientists, or other, according to their signed statements of adherence to creation-science as defined in Section 5, to evolution as defined in Section 5, or to any other science-model of origins. Witnesses must have an earned college or university degree from any institution in some field of science. No testimony shall be heard about religious beliefs but only about scientific evidence and existing federal programs.

Section 5.   As used in this Act:
(1) "Creation-science model" means the positive and negative evidences for creation and inferences from such scientific evidences based on scientific principles. The creation-science model includes belief in:
  (a) Sudden creation of the universe, energy, and life;
  (b) The insufficiency of mutation and natural selection in bringing about development of all living kinds from a single organism;
  (c) Changes only within fixed limits of originally created kinds of plants and animals;
  (d) Separate ancestry for man and apes;
  (e) Explanation of the earth's geology by catastrophism, including the occurrence of a worldwide flood; and
  (f) A relatively recent inception of the earth and living kinds.
(2) Scientific evidences of and inferences for creation include, but are not necessarily limited to:
  (a) The sudden appearance of complex living forms in the fossil record;
  (b) The harmful nature of most if not all mutations, and the tautologous nature of natural selection;
  (c) Application of the second law of thermodynamics to the transitions from greater order to lesser order, and the mathematical improbability of evolution of complex living forms;
  (d) The systematic absence of transitional forms between kinds in the alleged evolutionary chain;
  (e) The potential errors in the radiometric dating methods that are alleged to support an ancient age for the universe, the earth, life, and man; and the alternate interpretations and methods of dating the universe, the earth, life, and man to support a relatively young age.

(3) "Evolution model" means the positive and negative evidences for evolution and inferences from such scientific evidence based on scientific principles. "Evolution" includes belief in:
- (a) Emergence by naturalistic processes of the universe from disordered matter and emergence of life from nonlife;
- (b) The sufficiency of mutation and natural selection of present living kinds from simple earlier kinds;
- (c) Emergence by mutation and natural selection of present living kinds from simple earlier kinds;
- (d) Emergence of man from a common ancestor with apes;
- (e) Explanation of the earth's geology and the evolutionary sequence by uniformitarianism; and
- (f) An inception several billion years ago of the earth and somewhat later of life.

Section 6. This Act shall take effect in the federal fiscal year beginning after this Act's passage, and shall apply thereafter.

The Minnesota Creationism Bill (1978)

Senate Version
Introduced January 19, 1978
Referred to and defeated in
The Committee on Education

A bill for an act relating to education; curriculum; requiring certain theories of origins to be presented in Minnesota schools.

Be It Enacted By the Legislature of the State of Minnesota:

Section 1. It is the policy of the legislature that in Minnesota schools neither the theory of evolution nor the theory of special creation be taught exclusively in courses dealing with origins of man and the universe and that both theories receive equal consideration by students.

Section 2. For the purposes of sections 1 to 3, the following words shall have the meanings given to them in this section:
 (a) "Theory of evolution" means the belief that present processes, acting essentially as at present, suffice to explain the past history of the cosmos and its assumed evolutionary development from primeval chaos into its present form, including all life, the elements and the solar system;
 (b) "Special creation" means the belief that all matter and life was created out of nothing (ex nihilo) by divine power approximately 6,000 to 10,000 years ago in six solar days;
 (c) "Global deluge" means that there was subsequent to special creation a complete covering of the earth by water and upheaval of the earth's crust, followed by a stabilizing period.

Section 3. It is required of all schools in this state receiving public funds that the following conditions be observed in the teaching of the origins of man and the universe in any course or instruction included in the school curriculum.

(a) The theory of evolution shall not be presented as basic truth, a unifying principle of science, or the exclusive explanation of origins of life and matter;

(b) The theory of special creation shall be presented with equal emphasis both as to time and content as is given to the theory of evolution;

(c) The reading of bible passages as part of the instruction in the theory of special creation is permitted but shall not include any denominational religious belief and shall be accompanied by scientific rationale for the biblical account of creation;

(d) The two-model approach to the teaching of origins shall be applied by allowing pupils to present their belief in either the theory of evolution or special creation in discussion or written examinations with equal credit;

(e) Textbooks or other instructional materials dealing with origins shall either contain equal content and emphasis regarding the theory of evolution and the theory of special creation or they shall be used in conjunction with materials stating the opposite view; and

(f) Scientific data may not be presented in support of the theory of evolution unless the theory of the global deluge and other qualifications are attached to the scientific findings.

Section 4.   Failure to comply with the conditions of section 3 constitutes cause for withdrawal of public funds.

Section 5.   Sections 1 to 3 are effective the day following final enactment. Section 4 is effective on July 15, 1980.

The Louisiana Creationism Act (1981)

Passed Regular Session, 1981

AN ACT

To amend Part III of Chapter I of Title 17 of the Louisiana Revised Statutes of 1950 by adding thereto a new Sub-Part, to be designated as Sub-Part D-2 thereof, comprised of Sections 286.1 through 286.7, both inclusive, relative to balanced treatment of creation-science and evolution-science in public schools, to require such balanced treatment, to bar discrimination on the basis of creationist or evolutionist belief, to provide definitions and clarifications, to declare the legislative purpose, to provide relative to inservice teacher training and materials acquisition, to provide relative to curriculum development, and otherwise to provide with respect thereto.

Be it enacted by the Legislature of Louisiana:

Section 1. Sub-Part D-2 of Part III or Chapter I of Title 17 of the Louisiana Revised Statutes of 1950, comprised of Sections 286.1 through 286.7, both inclusive, is hereby enacted to read as follows:

CHAPTER I. GENERAL SCHOOL LAW

PART III. PUBLIC SCHOOLS AND SCHOOL CHILDREN

SUB-PART D-2. BALANCED TREATMENT FOR CREATION-SCIENCE AND EVOLUTION-SCIENCE IN PUBLIC SCHOOL INSTRUCTION

286.1. Short Title
This Subpart shall be known as the "Balanced Treatment for Creation-Science and Evolution-Science Act."

286.2. Purpose
This Subpart is enacted for the purposes of protecting academic freedom.

286.3. Definitions
As used in this Subpart, unless otherwise clearly indicated, these terms have the following meanings:
(1) "Balanced treatment" means providing whatever information and instruction in both creation and evolution models the classroom teacher determines is necessary and appropriate to provide insight into both theories in view of the textbooks and other instructional materials available for use in his classroom.
(2) "Creation-science" means the scientific evidences for creation and inferences from those scientific evidences.
(3) "Evolution-science" means the scientific evidences for evolution and inferences from those scientific evidences.
(4) "Public schools" means public secondary and elementary schools.

286.4 Authorization for balanced treatment; requirement for nondiscrimination
A. Commencing with the 1982-1983 school year, public schools within this state shall give balanced treatment to creation-science and to evolution-science. Balanced treatment of these two models shall be given in classroom lectures taken as a whole for each course, in textbook materials taken as a whole for each course, in library materials taken as a whole for the sciences and taken as a whole for the humanities, and in other educational programs in public schools, to the extent that such lectures, textbooks, library materials, or educational programs deal in any way with the subject of the origin of man, life, the earth, or the universe. When creation or evolution is taught, each shall be taught as a theory, rather than as proven scientific fact.
B. Public schools within this state and their personnel shall not discriminate by reducing a grade of a student or by singling out and publicly criticizing any student who demonstrates a satisfactory understanding of both evolution-science or creation-science and who accepts or rejects either model in whole or part.
C. No teacher in public elementary or secondary school or instructor in any state-supported university in Louisiana, who chooses to be a creation-scientist or to teach scientific data which points to creationism shall, for that reason, be discriminated against in any way by any school board, college board, or administrator.

286.5. Clarifications
This Subpart does not require any instruction in the subject of
origins but simply permits instruction in both scientific models (of
evolution-science and creation-science) if public schools choose to
teach either. This Subpart does not require each individual textbook
or library book to give balanced treatment to the models of
evolution-science and creation-science; it does not require any
school books to be discarded. This Subpart does not require each
individual classroom lecture in a course to give such balanced
treatment but simply permits the lectures as a whole to give
balanced treatment; it permits some lectures to present
evolution-science and other lectures to present creation-science.

286.6 Funding of inservice training and materials acquisition
Any public school that elects to present any model of origins shall
use existing teacher inservice training funds to prepare teachers of
public school courses presenting any model of origins to give
balanced treatment to the creation-science model and the
evolution-science model. Existing library acquisition funds shall be
used to purchase nonreligious library books as are necessary to give
balanced treatment to the creation-science model and the
evolution-science model.

286.7. Curriculum Development
A. Each city and parish school board shall develop and provide to
each public school classroom teacher in the system a curriculum
guide on presentation of creation-science.
B. The governor shall designate seven creation-scientists who shall
provide resource services in the development of curriculum guides
to any city or parish school board upon request. Each such
creation-scientist shall be designated from among the full-time
faculty members teaching in any college and university in
Louisiana. These creation-scientists shall serve at the pleasure of the
governor and without compensation.

 Section 2. If any provision or item of this Act or the application
thereof is held invalid, such invalidity shall not affect other provisions,
items, or applications of this Act which can be given effect without the
invalid provisions, items, or applications, and to this end the provisions
of this Act are hereby declared severable.

 Section 3. All laws or parts of laws in conflict herewith are hereby
repealed.

The Arkansas Creationism Act (1981)

Act 590
Passed 73rd General Assembly,
Regular Session, 1981

For An Act To Be Entitled

"An act to require balanced treatment of creation-science and evolution-science in public schools; to protect academic freedom by providing student choice; to ensure freedom of religious exercise; to guarantee freedom of belief and speech; to prevent establishment of religion; to prohibit religious instruction concerning origins; to bar discrimination on the basis of creationist or evolutionist belief; to provide definitions and clarifications; to declare the legislative purpose and legislative findings of fact; to provide for severability of provisions; to provide for repeal of contrary laws; and to set forth an effective date."

Be It Enacted by the General Assembly of the State of Arkansas:

Section 1. Requirement for Balanced Treatment. Public schools within this State shall give balanced treatment to creation-science and to evolution-science. Balanced treatment to these two models shall be given in classroom lectures taken as a whole for each course, in textbook materials taken as a whole for each course, in library materials taken as a whole for the sciences and taken as a whole for the humanities, and in other educational programs in public schools, to the extent that such lectures, textbooks, library materials, or educational programs deal in any way with the subject of the origin of man, life, the earth, or the universe.

Section 2. Prohibition against Religious Instruction. Treatment of either evolution-science or creation-science shall be limited to scientific evidences for each model and inferences from those scientific evidences, and must not include any religious instruction or references to religious writings.

Section 3. Requirement for Nondiscrimination. Public schools within this State, or their personnel, shall not discriminate, by reducing a grade of a student or by singling out and making public criticism, against any student who demonstrates a satisfactory understanding of both evolution-science and creation-science and who accepts or rejects either model in whole or part.

Section 4. Definitions. As used in this Act:
(a) "Creation-science" means the scientific evidences for creation and inferences from those scientific evidences. Creation-science includes the scientific evidences and related inferences that indicate: (1) Sudden creation of the universe, energy, and life from nothing; (2) The insufficiency of mutation and natural selection in bringing about development of all living kinds from a single organism; (3) Changes only within fixed limits of originally created kinds of plants and animals; (4) Separate ancestry for man and apes; (5) Explanation of the earth's geology by catastrophism, including the occurrence of a worldwide flood; and (6) A relatively recent inception of the earth and living kinds.
(b) "Evolution-science" means the scientific evidences for evolution and inferences from those scientific evidences. Evolution-science includes the scientific evidences and related inferences that indicate: (1) Emergence by naturalistic processes of the universe from disordered matter and emergence of life from nonlife; (2) The sufficiency of mutation and natural selection in bringing about development of present living kinds from simple earlier kinds; (3) Emergence by mutation and natural selection of present living kinds from simple earlier kinds; (4) Emergence of man from a common ancestor with apes; (5) Explanation of the earth's geology and the evolutionary sequence by uniformitarianism; and (6) An inception several billion years ago of the earth and somewhat later of life.
(c) "Public schools" mean public secondary and elementary schools.

Section 5. Clarifications. This Act does not require or permit instruction in any religious doctrine or materials. This Act does not require any instruction in the subject of origins, but simply requires instruction in both scientific models (of evolution-science and creation-science) if public schools choose to teach either. This Act does not require each individual textbook or library book to give balanced treatment to the models of evolution-science and creation-science; it does not require any school books to be discarded. This Act does not require each individual classroom lecture in a course to give such balanced treatment, but simply requires the lectures as a whole to give balanced treatment; it permits some lectures to present evolution-science and other lectures to present creation-science.

Section 6. Legislative Declaration of Purpose. This Legislature enacts this Act for public schools with the purpose of protecting academic freedom for students' differing values and beliefs; ensuring neutrality toward students' diverse religious convictions; ensuring freedom of religious exercise for students and their parents; guaranteeing freedom of belief and speech for students; preventing establishment of Theologically Liberal, Humanist, Nontheist, or Atheist religions; preventing discrimination against students on the basis of their personal beliefs concerning creation and evolution; and assisting students in their search for truth. This Legislature does not have the purpose of causing instruction in religious concepts or making an establishment of religion.

Section 7. Legislative Findings of Fact. This Legislature finds that:
(a) The subject of the origin of the universe, earth, life, and man is treated within many public school courses, such as biology, life science, anthropology, sociology, and often also in physics, chemistry, world history, philosophy, and social studies.
(b) Only evolution-science is presented to students in virtually all of those courses that discuss the subject of origins. Public schools generally censor creation-science and evidence contrary to evolution.
(c) Evolution-science is not an unquestionable fact of science, because evolution cannot be experimentally observed, fully verified, or logically falsified, and because evolution-science is not accepted by some scientists.
(d) Evolution-science is contrary to the religious convictions or moral values or philosophical beliefs of many students and parents, including individuals of many different religious faiths and with diverse moral values and philosophical beliefs.
(e) Public school presentation of only evolution-science without any alternative model of origins abridges the United States Constitution's protections of freedom of religious exercise and of freedom of belief and speech for students and parents because it undermines their religious convictions and moral or philosophical values, compels their

unconscionable professions of belief, and hinders religious training and moral training by parents.

(f) Public school presentation of only evolution-science furthermore abridges the Constitution's prohibition against establishment of religion, because it produces hostility toward many Theistic religions and brings preference to Theological Liberalism, Humanism, Nontheistic religions, and Atheism, in that these religious faiths generally include a religious belief in evolution.

(g) Public school instruction in only evolution-science also violates the principle of academic freedom, because it denies students a choice between scientific models and instead indoctrinates them in evolution-science alone.

(h) Presentation of only one model rather than alternative scientific models of origins is not required by any compelling interest of the State, and exemption of such students from a course or class presenting only evolution-science does not provide an adequate remedy because of teacher influence and student pressure to remain in that course or class.

(i) Attendance of those students who are at public schools is compelled by law, and school taxes from their parents and other citizens are mandated by law.

(j) Creation-science is an alternative scientific model of origins and can be presented from a strictly scientific standpoint without any religious doctrine just as evolution-science can, because there are scientists who conclude that scientific data best support creation-science and because scientific evidences and inferences have been presented for creation-science.

(k) Public school presentation of both evolution-science and creation-science would not violate the Constitution's prohibition against establishment of religion, because it would involve presentation of the scientific evidences and related inferences for each model rather than any religious instruction.

(l) Most citizens, whatever their religious beliefs about origins, favor balanced treatment in public schools of alternative scientific models of origins for better guiding students in their search for knowledge, and they favor a neutral approach toward subjects affecting the religious and moral and philosophical convictions of students.

Section 8. Short Title. This Act shall be know as the "Balanced Treatment for Creation-Science and Evolution-Science Act."

Section 9. Severability of Provisions. If any provision of this Act is held invalid, that invalidity shall not affect other provisions that can be applied in the absence of the invalidated provisions, and the provisions of this Act are declared to be severable.

Section 10. Repeal of Contrary Laws. All State laws or parts of State laws in conflict with this Act are hereby repealed.

Section 11. Effective Date. The requirements of the Act shall be met by and may be met before the beginning of the next school year if that is more than six months from the date of enactment, or otherwise one year after the beginning of the next school year, and in all subsequent school years.

The Arkansas Decision: Memorandum Opinion in *Rev. Bill McLean et al* v. *The Arkansas Board of Education et al* (January 5, 1982)

William R. Overton, U.S. District Court,
Eastern District of Arkansas,
Western Division

The application and content of First Amendment principles are not determined by public opinion polls or by a majority vote. Whether the proponents of Act 590 constitute the majority or the minority is quite irrelevant under a constitutional system of government. No group, no matter how large or small, may use the organs of government, of which the public schools are the most conspicuous and influential, to foist its religious beliefs on others.

JUDGMENT

Pursuant to the Court's Memorandum Opinion filed this date, judgment is hereby entered in favor of the plaintiffs and against the defendants. The relief prayed for is granted.

Dated this January 5, 1982.

INJUNCTION

Pursuant to the Court's Memorandum Opinion filed this date, the defendants and each of them and all their servants and employees are hereby permanently enjoined from implementing in any manner Act 590 of the Acts of Arkansas of 1981.

It is so ordered this January 5, 1982.

MEMORANDUM OPINION

Introduction

On March 19, 1981, the Governor of Arkansas signed into law Act 590 of 1981, entitled the "Balanced Treatment for Creation-Science and Evolution-Science Act." The Act is codified as Ark. Stat. Ann. §80–1663, *et seq.,* (1981 Supp.). Its essential mandate is stated in its first sentence: "Public schools within this State shall give balanced treatment to creation-science and to evolution-science." On May 27, 1981, this suit was filed[1] challenging the constitutional validity of Act 590 on three distinct grounds.

First, it is contended that Act 590 constitutes an establishment of religion prohibited by the First Amendment to the Constitution, which is made applicable to the states by the Fourteenth Amendment. Second, the plaintiffs argue the Act violates a right to academic freedom which they say is guaranteed to students and teachers by the Free Speech Clause of the First Amendment. Third, plaintiffs allege the Act is impermissibly vague and thereby violates the Due Process Clause of the Fourteenth Amendment.

The individual plaintiffs include the resident Arkansas Bishops of the United Methodist, Episcopal, Roman Catholic and African Methodist Episcopal Churches, the principal official of the Presbyterian Churches in Arkansas, other United Methodist, Southern Baptist and Presbyterian clergy, as well as several persons who sue as parents and next friends of minor children attending Arkansas public schools. One plaintiff is a high school biology teacher. All are also Arkansas taxpayers. Among the organizational plaintiffs are the American Jewish Congress, the Union of American Hebrew Congregations, the American Jewish Committee, the Arkansas Education Association, the National Association of Biology Teachers and the National Coalition for Public Education and Religious Liberty, all of which sue on behalf of members living in Arkansas.[2]

The defendants include the Arkansas Board of Education and its members, the Director of the Department of Education, and the State Textbooks and Instructional Materials Selecting Committee.[3] The Pulaski County Special School District and its Directors and Superintendent were voluntarily dismissed by the plaintiffs at the pre-trial conference held October 1, 1981.

The trial commenced December 7, 1981, and continued through December 17, 1981. This Memorandum Opinion constitutes the Court's findings of fact and conclusions of law. Further orders and judgment will be in conformity with this opinion.

I

There is no controversy over the legal standards under which the Establishment Clause portion of this case must be judged. The Supreme Court has on a number of occasions expounded on the meaning of the clause, and the pronouncements are clear. Often the issue has arisen in the context of public education, as it has here. In *Everson* v. *Board of Education,* 330 U.S. 1, 15–16(1947), Justice Black stated:

> The "establishment of religion" clause of the First Amendment means at least this: Neither a state nor the Federal Government can set up a church. Neither can pass laws which aid one religion, aid all religions, or prefer one religion over another. Neither can force nor influence a person to go to or to remain away from church against his will or force him to profess a belief or disbelief in any religion. No person can be punished for entertaining or professing religious beliefs or disbeliefs, for church-attendance or non-attendance. No tax, large or small, can be levied to support any religious activities or institutions, whatever they may be called, or whatever form they may adopt to teach or practice religion. Neither a state nor the Federal Government can, openly or secretly, participate in the affairs of any religious organizations or groups and *vice versa.* In the words of Jefferson, the clause...was intended to erect "a wall of separation between church and State."

The Establishment Clause thus enshrines two central values: voluntarism and pluralism. And it is in the area of the public schools that these values must be guarded most vigilantly.

> Designed to serve as perhaps the most powerful agency for promoting cohesion among a heterogeneous democratic people, the public school must keep scrupulously free from entanglement in the strife of sects. The preservation of the community from divisive conflicts, of Government from irreconcilable pressures by religious groups, of religion from censorship and coercion however subtly exercised, requires strict confinement of the State to instruction other than religious, leaving to the individual's church and home, indoctrination in the faith of his choice. [*McCollum* v. *Board of Education,* 333 U.S. 203, 216–217 (1948), (Opinion of Frankfurter, J., joined by Jackson, Burton and Rutledge, J. J.)]

The specific formulation of the establishment prohibition has been refined over the years, but its meaning has not varied from the principles articulated by Justice Black in *Everson*. In *Abbington School District* v. *Schempp,* 374 U.S. 203, 222 (1963), Justice Clark stated that "to withstand the strictures of the Establishment Clause there must be a secular legislative purpose and a primary effect that neither advances nor inhibits religion." The Court found it quite clear that the First Amendment does not permit a state to require the daily reading of the Bible in public schools, for "[s]urely the place of the Bible as an instrument of religion cannot be gainsaid." *Id.* at 224. Similarly, in *Engel* v. *Vitale,* 370 U.S. 421 (1962), the Court held that the First Amendment prohibited the New York Board of Regents from requiring the daily recitation of a certain prayer in the schools. With characteristic succinctness, Justice Black wrote. "Under [the First] Amendment's prohibition against governmental establishment of religion, as reinforced by the provisions of the Fourteenth Amendment, government in this country, be it state or federal, is without power to prescribe by law any particular form of prayer which is to be used as an official prayer in carrying on any program of governmentally sponsored religious activity." *Id.* at 430. Black also identified the objective at which the Establishment Clause was aimed: "Its first and most immediate purpose rested on the belief that a union of government and religion tends to destroy government and to degrade religion." *Id.* at 431.

Most recently, the Supreme Court has held that the clause prohibits a state from requiring the posting of the Ten Commandments in public school classrooms for the same reasons that officially imposed daily Bible reading is prohibited. *Stone* v. *Graham,* 449 U.S. 39 (1980). The opinion in *Stone* relies on the most recent formulation of the Establishment Clause test, that of *Lemon* v. *Kurtzman,* 403 U.S. 602, 612–613 (1971):

> First, the statute must have a secular legislative purpose; second, its principal or primary effect must be one that neither advances nor inhibits religion. . .; finally, the statute must not foster "an excessive government entanglement with religion." [*Stone* v. *Graham,* 449 U.S. at 40]

It is under this three part test that the evidence in this case must be judged. Failure on any of these grounds is fatal to the enactment.

II

The religious movement known as Fundamentalism began in nineteenth century America as part of evangelical Protestantism's response to social changes, new religious thought and Darwinism. Fundamentalists viewed these developments as attacks on the Bible and as responsible for a decline in traditional values.

The various manifestations of Fundamentalism have had a number of common characteristics,[4] but a central premise has always been a literal interpretation of the Bible and a belief in the inerrancy of the Scriptures. Following World War I, there was again a perceived decline in traditional morality, and Fundamentalism focused on evolution as responsible for the decline. One aspect of their efforts, particularly in the South, was the promotion of statutes prohibiting the teaching of evolution in public schools. In Arkansas, this resulted in the adoption of Initiated Act 1 of 1929.[5]

Between the 1920's and early 1960's, anti-evolutionary sentiment had a subtle but pervasive influence on the teaching of biology in public schools. Generally, textbooks avoided the topic of evolution and did not mention the name of Darwin. Following the launch of the Sputnik satellite by the Soviet Union in 1957, the National Science Foundation funded several programs designed to modernize the teaching of science in the nation's schools. The Biological Sciences Curriculum Study (BSCS), a nonprofit organization, was among those receiving grants for curriculum study and revision. Working with scientists and teachers, BSCS developed a series of biology texts which, although emphasizing different aspects of biology, incorporated the theory of evolution as a major theme. The success of the BSCS effort is shown by the fact that fifty percent of American school children currently use BSCS books directly and the curriculum is incorporated indirectly in virtually all biology texts. (Testimony of Mayer; Nelkin, Px1).[6]

In the early 1960's, there was again a resurgence of concern among Fundamentalists about the loss of traditional values and a fear of growing secularism in society. The Fundamentalist movement became more active and has steadily grown in numbers and political influence. There is an emphasis among current Fundamentalists on the literal interpretation of the Bible and the Book of Genesis as the sole source of knowledge about origins.

The term "scientific creationism" first gained currency around 1965 following publication of *The Genesis Flood* in 1961 by Whitcomb and Morris. There is undoubtedly some connection between the appearance of the BSCS texts emphasizing evolutionary thought and efforts by Fundamentalists to attack the theory. (Mayer)

In the 1960's and early 1970's, several Fundamentalist organizations were formed to promote the idea that the Book of Genesis was supported by scientific data. The terms "creation science" and "scientific creationism" have been adopted by these Fundamentalists as descriptive of their study of creation and the origins of man. Perhaps the leading creationist organization is the Institute for Creation Research (ICR), which is affiliated with the Christian Heritage College and supported by the Scott Memorial Baptist Church in San Diego, California. The ICR, through the Creation-Life Publishing Company, is the leading publisher of creation science material. Other creation science organizations include the Creation Science Research Center (CSRC) of San Diego and the Bible Science Association of Minneapolis, Minnesota. In 1963, the Creation Research Society (CRS) was formed from a schism in the American Scientific Affiliation (ASA). It is an organization of literal Fundamentalists[7] who have the equivalent of a master's degree in some recognized area of science. A purpose of the organization is "to reach all people with the vital message of the scientific and historic truth about creation." Nelkin, *The Science Textbook Controversies and the Politics of Equal Time,* 66. Similarly, the CSRC was formed in 1970 from a split in the CRS. Its aim has been "to reach the 63 million children of the United States with the scientific teaching of Biblical creationism." *Id.* at 69.

Among creationist writers who are recognized as authorities in the field by other creationists are Henry M. Morris, Duane Gish, G. E. Parker, Harold S. Slusher, Richard B. Bliss, John [N.] Moore, Martin E. Clark, W. L. Wysong, Robert E. Kofahl and Kelly L. Segraves. Morris is Director of ICR, Gish is Associate Director and Segraves is associated with CSRC.

Creationists view evolution as a source of society's ills, and the writings of Morris and Clark are typical expressions of that view.

> Evolution is thus not only anti-Biblical and anti-Christian, but it is utterly unscientific and impossible as well. But it has served effectively as the pseudo-scientific basis of atheism, agnosticism, socialism, fascism, and numerous other false and dangerous philosophies over the past century. [Morris and Clark, *The Bible Has the Answer,* (Px 31 and Pretrial Px 89)[8]]

Creationists have adopted the view of Fundamentalists generally that there are only two positions with respect to the origins of the earth and life: belief in the inerrancy of the Genesis story of creation and of a worldwide flood as fact, or belief in what they call evolution.

Henry Morris has stated, "It is impossible to devise a legitimate means of harmonizing the Bible with evolution." Morris, "Evolution and the Bible," *ICR Impact Series* Number 5 (undated, unpaged), quoted in Mayer, Px 8, at 3. This dualistic approach to the subject of origins permeates the creationist literature.

The creationist organizations consider the introduction of creation science into the public schools part of their ministry. The ICR has published at least two pamphlets[9] containing suggested methods for convincing school boards, administrators and teachers that creationism should be taught in public schools. The ICR has urged its proponents to encourage school officials to voluntarily add creationism to the curriculum.[10]

Citizens For Fairness In Education is an organization based in Anderson, South Carolina, formed by Paul Ellwanger, a respiratory therapist who is trained in neither law nor science. Mr. Ellwanger is of the opinion that evolution is the forerunner of many social ills, including Nazism, racism and abortion. (Ellwanger Depo. at 32–34). About 1977, Ellwanger collected several proposed legislative acts with the idea of preparing a model state act requiring the teaching of creationism as science in opposition to evolution. One of the proposals he collected was prepared by Wendell Bird, who is now a staff attorney for ICR.[11] From these various proposals, Ellwanger prepared a "model act" which calls for "balanced treatment" of "scientific creationism" and "evolution" in public schools. He circulated the proposed act to various people and organizations around the country.

Mr. Ellwanger's views on the nature of creation science are entitled to some weight since he personally drafted the model act which became Act 590. His evidentiary deposition with exhibits and unnumbered attachments (produced in response to a subpoena *duces tecum*) speaks to both the intent of the Act and the scientific merits of creation science. Mr. Ellwanger does not believe creation science is a science. In a letter to Pastor Robert E. Hays he states, "While neither evolution nor creation can qualify as a scientific theory, and since it is virtually impossible at this point to educate the whole world that evolution is not a true scientific theory, we have freely used these terms—the evolution theory and the theory of scientific creationism—in the bill's text." (Unnumbered attachment to Ellwanger Depo., at 2.) He further states in a letter to Mr. Tom Bethell. "As we examine evolution (remember, we're not making any scientific claims for creation, but we are challenging evolution's claim to be scientific). . ." (Unnumbered attachment to Ellwanger Depo. at 1.)

Ellwanger's correspondence on the subject shows an awareness that Act 590 is a religious crusade, coupled with a desire to conceal this fact. In a letter to State Senator Bill Keith of Louisiana, he says, "I view this whole battle as one between God and anti-God forces, though I know there are a large number of evolutionists who believe in God." And further, ". . .it behooves Satan to do all he can to thwart our efforts and confuse the issue at every turn." Yet Ellwanger suggests to Senator

Keith, "If you have a clear choice between having grassroots leaders of this statewide bill promotion effort to be ministerial or non-ministerial, be sure to opt for the non-ministerial. It does the bill effort no good to have ministers out there in the public forum and the adversary will surely pick at this point.. . . . Ministerial persons can accomplish a tremendous amount of work from behind the scenes, encouraging their congregations to take the organizational and P.R. initiatives. And they can lead their churches in storming Heaven with prayers for help against so tenacious an adversary." (Unnumbered attachment to Ellwanger Depo. at 1.)

Ellwanger shows a remarkable degree of political candor, if not finesse, in a letter to State Senator Joseph Carlucci of Florida:

> 2. It would be very wise, if not actually essential, that all of us who are engaged in this legislative effort be careful not to present our position and our work in a religious framework. For example, in written communications that might somehow be shared with those other persons whom we may be trying to convince, it would be well to exclude our own personal testimony and/or witness for Christ, but rather, if we are so moved, to give that testimony on a separate attached note. (Unnumbered attachment to Ellwanger Depo. at 1.)

The same tenor is reflected in a letter by Ellwanger to Mary Ann Miller, a member of FLAG (Family, Life, America under God) who lobbied the Arkansas Legislature in favor of Act 590:

> . . .we'd like to suggest that you and your co-workers be very cautious about mixing creation-science with creation-religion. . . Please urge your co-workers not to allow themselves to get sucked into the "religion" trap of mixing the two together, for such mixing does incalculable harm to the legislative thrust. It could even bring public opinion to bear adversely upon the higher courts that will eventually have to pass judgment on the constitutionality of this new law. (Ex. 1 to Miller Depo.)

Perhaps most interesting, however, is Mr. Ellwanger's testimony in his deposition as to his strategy for having the model act implemented:

> Q. You're trying to play on other people's religious motives.
> A. I'm trying to play on their emotions, love, hate, their likes, dislikes, because I don't know any other way to involve, to get humans to become involved in human endeavors. I see emotions as being a healthy and legitimate means of getting people's feelings into action, and. . .I believe that the predominance of population in America that represents the greatest potential for taking some kind of action in this area is a Christian community. I see

the Jewish community as far less potential in taking action. . .but I've seen a lot of interest among Christians and I feel, why not exploit that to get the bill going if that's what it takes. (Ellwanger Depo. at 146–147.)

Mr. Ellwanger's ultimate purpose is revealed in the closing of his letter to Mr. Tom Bethell: "Perhaps all this is old hat to you, Tom, and if so, I'd appreciate your telling me so and perhaps where you've heard it before—the idea of killing evolution instead of playing these debating games that we've been playing for nigh over a decade already." (Unnumbered attachment to Ellwanger Depo. at 3.)

It was out of this milieu that Act 590 emerged. The Reverend W. A. Blount, a Biblical literalist who is pastor of a church in the Little Rock area and was, in February, 1981, chairman of the Greater Little Rock Evangelical Fellowship, was among those who received a copy of the model act from Ellwanger.[12]

At Reverend Blount's request, the Evangelical Fellowship unanimously adopted a resolution to seek introduction of Ellwanger's act in the Arkansas Legislature. A committee composed of two ministers, Curtis Thomas and W. A. Young, was appointed to implement the resolution. Thomas obtained from Ellwanger a revised copy of the model act which he transmitted to Carl Hunt, a business associate of Senator James L. Holsted, with the request that Hunt prevail upon Holsted to introduce the act.

Holsted, a self-described "born again" Christian Fundamentalist, introduced the act in the Arkansas Senate. He did not consult the State Department of Education, scientists, science educators or the Arkansas Attorney General.[13] The Act was not referred to any Senate committee for hearing and was passed after only a few minutes' discussion on the Senate floor. In the House of Representatives, the bill was referred to the Education Committee which conducted a perfunctory fifteen minute hearing. No scientist testified at the hearing, nor was any representative from the State Department of Education called to testify.

Ellwanger's model act was enacted into law in Arkansas as Act 590 without amendment or modification other than minor typographical changes. The legislative "findings of fact" in Ellwanger's act and Act 590 are identical, although no meaningful fact-finding process was employed by the General Assembly.

Ellwanger's efforts in preparation of the model act and campaign for its adoption in the states were motivated by his opposition to the theory of evolution and his desire to see the Biblical version of creation taught in the public schools. There is no evidence that the pastors, Blount, Thomas, Young or The Greater Little Rock Evangelical Fellowship were motivated by anything other than their religious convictions when proposing its adoption or during their lobbying efforts in its behalf. Senator Holsted's sponsorship and lobbying efforts in behalf of the Act were motivated solely by his religious beliefs and desire to see the Biblical version of creation taught in the public schools.[14]

The State of Arkansas, like a number of states whose citizens have relatively homogeneous religious beliefs, has a long history of official opposition to evolution which is motivated by adherence to Fundamentalist beliefs in the inerrancy of the Book of Genesis. This history is documented in Justice Fortas' opinion in *Epperson* v. *Arkansas,* 393 U.S. 97 (1968), which struck down Initiated Act 1 of 1929, Ark. Stat. Ann. §§80–1627–1628, prohibiting the teaching of the theory of evolution. To this same tradition may be attributed Initiated Act 1 of 1930, Ark. Stat. Ann. §80–1606 (Repl. 1980), requiring "the reverent daily reading of a portion of the English Bible" in every public school classroom in the State.[15]

It is true, as defendants argue, that courts should look to legislative statements of a statute's purpose in Establishment Clause cases and accord such pronouncements great deference. See, e.g., *Committee for Public Education & Religious Liberty* v. *Nyquist,* 413 U.S. 756, 773 (1973) and *McGowan* v. *Maryland,* 366 U.S. 420, 445 (1961). Defendants also correctly state the principle that remarks by the sponsor or author of a bill are not considered controlling in analyzing legislative intent. See, e.g., *United States* v. *Emmons,* 410 U.S. 396 (1973) and *Chrysler Corp.* v. *Brown,* 441 U.S. 281 (1979).

Courts are not bound, however, by legislative statements of purpose or legislative disclaimers, *Stone* v. *Graham,* 449 U.S. 39 (1980): *Abbington School Dist.* v. *Schempp,* 374 U.S. 203 (1963). In determining the legislative purpose of a statute, courts may consider evidence of the historical context of the Act, *Epperson* v. *Arkansas,* 393 U.S. 97 (1968), the specific sequence of events leading up to passage of the Act, departures from normal procedural sequences, substantive departures from the normal. *Village of Arlington Heights* v. *Metropolitan Housing Corp.,* 429 U.S. 252 (1977), and contemporaneous statements of the legislative sponsor, *Fed. Energy Admin.* v. *Algonquin SNG, INC.,* 426 U.S. 548, 564 (1976).

The unusual circumstances surrounding the passage of Act 590, as well as the substantive law of the First Amendment, warrant an inquiry into the stated legislative purposes. The author of the Act had publicly proclaimed the sectarian purpose of the proposal. The Arkansas residents who sought legislative sponsorship of the bill did so for a purely sectarian purpose. These circumstances alone may not be particularly persuasive, but when considered with the publicly announced motives of the legislative sponsor made contemporaneously with the legislative process; the lack of any legislative investigation, debate or consultation with any educators or scientists; the unprecedented intrusion in school curriculum;[16] and official history of the State of Arkansas on the subject, it is obvious that the statement of purposes has little, if any, support in fact. The State failed to produce any evidence which would warrant an inference or conclusion that at any point in the process anyone considered the legitimate educational value of the Act. It was simply and purely an effort to introduce the Biblical version of creation into the public school curricula. The only inference which can be drawn from these circumstances is that the Act was passed with the specific purpose by the General Assembly of advancing religion. The Act therefore fails the first prong of the three-pronged test, that of secular legislative purpose, as articulated in *Lemon* v. *Kurtzman, supra,* and *Stone* v. *Graham, supra.*

III

If the defendants are correct and the Court is limited to an examination of the language of the Act, the evidence is overwhelming that both the purpose and effect of Act 590 is the advancement of religion in the public schools.

Section 4 of the Act provides:

Definitions, as used in this Act:

(a) "Creation-science" means the scientific evidences for creation and inferences from those scientific evidences. Creation-science includes the scientific evidences and related inferences that indicate: (1) Sudden creation of the universe, energy, and life from nothing; (2) The insufficiency of mutation and natural selection in bringing about development of all living kinds from a single organism; (3) Changes only within fixed limits of originally created kinds of plants and animals; (4) Separate ancestry for man and apes; (5) Explanation of the earth's geology by catastrophism, including the occurrence of a worldwide flood; and (6) A relatively recent inception of the earth and living kinds.

(b) "Evolution-science" means the scientific evidences for evolution and inferences from those scientific evidences. Evolution-science includes the scientific evidences and related inferences that indicate: (1) Emergence by naturalistic processes of the universe from disordered matter and emergence of life from nonlife; (2) The sufficiency of mutation and natural selection in bringing about development of present living kinds from simple earlier kinds; (3) Emergence by mutation and natural selection of present living kinds from simple earlier kinds; (4) Emergence of man from a common ancestor with apes; (5) Explanation of the earth's geology and the evolutionary sequence by uniformitarianism; and (6) An inception several billion years ago of the earth and somewhat later of life.

(c) "Public schools" mean public secondary and elementary schools.

The evidence establishes that the definition of "creation science" contained in 4(a) has as its unmentioned reference the first 11 chapters of the Book of Genesis. Among the many creation epics in human history, the account of sudden creation from nothing, or *creatio ex nihilo,* and subsequent destruction of the world by flood is unique to Genesis. The concepts of 4(a) are the literal Fundamentalists' views of Genesis. Section 4(a) is unquestionably a statement of religion, with the exception of 4(a)(2) which is a negative thrust aimed at what the creationists understand to be the theory of evolution.[17]

Both the concepts and wording of Section 4(a) convey an inescapable religiosity. Section 4(a)(1) describes "sudden creation of the universe, energy and life from nothing." Every theologian who testified, including defense witnesses, expressed the opinion that the statement referred to a supernatural creation which was performed by God.

Defendants argue that: (1) the fact that 4(a) conveys ideas similar to the literal interpretation of Genesis does not make it conclusively a statement of religion; (2) that reference to a creation from nothing is not necessarily a religious concept since the Act only suggests a creator who has power, intelligence and a sense of design and not necessarily the attributes of love, compassion and justice;[18] and (3) that simply teaching about the concept of a creator is not a religious exercise unless the student is required to make a commitment to the concept of a creator.

The evidence fully answers these arguments. The ideas of 4(a)(1) are not merely similar to the literal interpretation of Genesis; they are identical and parallel to no other story of creation.[19]

The argument that creation from nothing in 4(a)(1) does not involve a supernatural deity has no evidentiary or rational support. To the contrary, "creation out of nothing" is a concept unique to Western religions. In traditional Western religious thought, the conception of a creator of the world is a conception of God. Indeed, creation of the world "out of nothing" is the ultimate religious statement because God is the only actor. As Dr. Langdon Gilkey noted, the Act refers to one who has the power to bring all the universe into existence from nothing. The only "one" who has this power is God.[20]

The leading creationist writers, Morris and Gish, acknowledge that the idea of creation described in 4(a)(1) is the concept of creation by God and make no pretense to the contrary.[21] The idea of sudden creation from nothing, or *creatio ex nihilo,* is an inherently religious concept. (Vawter, Gilkey, Geisler, Ayala, Blount, Hicks.)

The argument advanced by defendants' witness, Dr. Norman Geisler, that teaching the existence of God is not religious unless the teaching seeks a commitment, is contrary to common understanding and contradicts settled case law. *Stone* v. *Graham,* 449 U.S. 39 (1980); *Abbington School District* v. *Schempp,* 374 U.S. 203 (1963).

The facts that creation science is inspired by the Book of Genesis and that Section 4(a) is consistent with a literal interpretation of Genesis leave no doubt that a major effect of the Act is the advancement of particular religious beliefs. The legal impact of this conclusion will be discussed further at the conclusion of the Court's evaluation of the scientific merit of creation science.

IV(A)

The approach to teaching "creation science" and "evolution science" found in Act 590 is identical to the two-model approach espoused by the Institute for Creation Research and is taken almost verbatim from ICR writings. It is an extension of Fundamentalists' view that one must either accept the literal interpretation of Genesis or else believe in the godless system of evolution.

The two model approach of the creationists is simply a contrived dualism [22] which has no scientific factual basis or legitimate educational purpose. It assumes only two explanations for the origins of life and existence of man, plants and animals: It was either the work of a creator or it was not. Application of these two models, according to Creationists, and the defendants, dictates that all scientific evidence which fails to support the theory of evolution is necessarily scientific evidence in support of creationism and is, therefore, creation science "evidence" in support of Section 4(a).

IV(B)

The emphasis on origins as an aspect of the theory of evolution is peculiar to creationist literature. Although the subject of origins of life is within the province of biology, the scientific community does not consider origins of life a part of evolutionary theory. The theory of evolution assumes the existence of life and is directed to an explanation of *how* life evolved. Evolution does not presuppose the absence of a

creator of God and the plain inference conveyed by Section 4 is erroneous.[23]

As a statement of the theory of evolution, Section 4(b) is simply
a hodgepodge of limited assertions, many of which are factually inaccurate.

For example, although 4(b)(2) asserts, as a tenet of evolutionary
theory, "the sufficiency of mutation and natural selection in bringing
about the existence of present living kinds from simple earlier kinds,"
Drs. Ayala and Gould both stated that biologists know that these two
processes do not account for all significant evolutionary change. They
testified to such phenomena as recombination, the founder effect, genetic drift and the theory of punctuated equilibrium, which are believed
to play important evolutionary roles. Section 4(b) omits any reference
to these. Moreover, 4(b) utilizes the term "kinds" which all scientists
said is not a word of science and has no fixed meaning. Additionally,
the Act presents both evolution and creation science as "package
deals." Thus, evidence critical of some aspect of what the creationists
define as evolution is taken as support for a theory which includes a
worldwide flood and a relatively young earth.[24]

IV(C)

In addition to the fallacious pedagogy of the two model approach,
Section 4(a) lacks legitimate educational value because "creation
science" as defined in that section is simply not science. Several witnesses suggested definitions of science. A descriptive definition was said
to be that science is what is "accepted by the scientific community" and
is "what scientists do." The obvious implication of this description is
that, in a free society, knowledge does not require the imprimatur of
legislation in order to become science.

More precisely, the essential characteristics of science are:

(1) It is guided by natural law;

(2) It has to be explanatory by reference to natural law;

(3) It is testable against the empirical world;

(4) Its conclusions are tentative, i.e., are not necessarily the final
word; and

(5) It is falsifiable. (Ruse and other science witnesses).

Creation science as described in Section 4(a) fails to meet these
essential characteristics. First, the section revolves around 4(a)(1)
which asserts a sudden creation "from nothing." Such a concept is not
science because it depends upon a supernatural intervention which is

not guided by natural law. It is not explanatory by reference to natural law, is not testable and is not falsifiable.[25]

If the unifying idea of supernatural creation by God is removed from Section 4, the remaining parts of the section explain nothing and are meaningless assertions.

Section 4(a)(2), relating to the "insufficiency of mutation and natural selection in bringing about development of all living kinds from a single organism," is an incomplete negative generalization directed at the theory of evolution.

Section 4(a)(3) which describes "changes only within fixed limits of originally created kinds of plants and animals" fails to conform to the essential characteristics of science for several reasons. First, there is no scientific definition of "kinds" and none of the witnesses was able to point to any scientific authority which recognized the term or knew how many "kinds" existed. One defense witness suggested there may be 100 to 10,000 different "kinds." Another believes there were "about 10,000, give or take a few thousand." Second, the assertion appears to be an effort to establish outer limits of changes within species. There is no scientific explanation for these limits which is guided by natural law and the limitations, whatever they are, cannot be explained by natural law.

The statement in 4(a)(4) of "separate ancestry of man and apes" is a bald assertion. It explains nothing and refers to no scientific fact or theory.[26]

Section 4(a)(5) refers to "explanation of the earth's geology by catastrophism, including the occurrence of a worldwide flood." This assertion completely fails as science. The Act is referring to the Noachian flood described in the Book of Genesis.[27] The creationist writers concede that *any* kind of Genesis Flood depends upon supernatural intervention. A worldwide flood as an explanation of the world's geology is not the product of natural law, nor can its occurrence be explained by natural law.

Section 4(a)(6) equally fails to meet the standards of science. "Relatively recent inception" has no scientific meaning. It can only be given meaning by reference to creationist writings which place the age at between 6,000 and 20,000 years because of the genealogy of the Old Testament. See, e.g., Px 78, Gish (6,000 to 10,000); Px 87, Segraves (6,000 to 20,000). Such a reasoning process is not the product of natural law; not explainable by natural law; nor is it tentative.

Creation science, as defined in Section 4(a), not only fails to follow the canons defining scientific theory, it also fails to fit the more general descriptions of "what scientists think" and "what scientists do." The scientific community consists of individuals and groups, nationally and internationally, who work independently in such varied fields as biology, paleontology, geology and astronomy. Their work is published and subject to review and testing by their peers. The journals for publication are both numerous and varied. There is, however, not one recognized scientific journal which has published an article espousing the creation science theory described in Section 4(a). Some of the State's witnesses suggested that the scientific community was "close-minded" on the subject of creationism and that explained the lack of acceptance of the creation science arguments. Yet no witness produced a scientific article for which publication had been refused. Perhaps some members of the scientific community are resistant to new ideas. It is, however, inconceivable that such a loose knit group of independent thinkers in all the varied fields of science could, or would, so effectively censor new scientific thought.

The creationists have difficulty maintaining among their ranks consistency in the claim that creationism is science. The author of Act 590, Ellwanger, said that neither evolution nor creationism was science. He thinks both are religion. Duane Gish recently responded to an article in *Discover* critical of creationism by stating:

> Stephen Jay Gould states that creationists claim creation is a scientific theory. This is a false accusation. Creationists have repeatedly stated that neither creation nor evolution is a scientific theory (and each is equally religious). (Gish, letter to editor of *Discover,* July, 1981, App. 30 to Plaintiffs' Pretrial Brief)

The methodology employed by creationists is another factor which is indicative that their work is not science. A scientific theory must be tentative and always subject to revision or abandonment in light of facts that are inconsistent with, or falsify, the theory. A theory that is by its own terms dogmatic, absolutist and never subject to revision is not a scientific theory.

The creationists' methods do not take data, weigh it against the opposing scientific data, and thereafter reach the conclusions stated in Section 4(a). Instead, they take the literal wording of the Book of Genesis and attempt to find scientific support for it. The method is best explained in the language of Morris in his book (Px 31) *Studies in The Bible and Science* at page 114:

> . . .it is. . .quite impossible to determine anything about Creation through a study of present processes, because present processes are not creative in character. If man wishes to know anything about Creation (the time of Creation, the duration of Creation, the order of Creation, the methods of Creation, or anything else) his sole source of true information is that of divine revelation. God was there

when it happened. We were not there.... Therefore, we are completely limited to what God has seen fit to tell us, and this information is in His written Word. This is our textbook on the science of Creation!

The Creation Research Society employs the same unscientific approach to the issue of creationism. Its applicants for membership must subscribe to the belief that the Book of Genesis is "historically and scientifically true in all of the original autographs."[28] The Court would never criticize or discredit any person's testimony based on his or her religious beliefs. While anybody is free to approach a scientific inquiry in any fashion they choose, they cannot properly describe the methodology used as scientific, if they start with a conclusion and refuse to change it regardless of the evidence developed during the course of the investigation.

IV(D)

In efforts to establish "evidence" in support of creation science, the defendants relied upon the same false premise as the two model approach contained in Section 4, i.e., all evidence which criticized evolutionary theory was proof in support of creation science. For example, the defendants established that the mathematical probability of a chance chemical combination resulting in life from non-life is so remote that such an occurrence is almost beyond imagination. Those mathematical facts, the defendants argue, are scientific evidences that life was the product of a creator. While the statistical figures may be impressive evidence against the theory of chance chemical combinations as an explanation of origins, it requires a leap of faith to interpret those figures so as to support a complex doctrine which includes a sudden creation from nothing, a worldwide flood, separate ancestry of man and apes, and a young earth.

The defendants' argument would be more persuasive if, in fact, there were only two theories or ideas about the origins of life and the world. That there are a number of theories was acknowledged by the State's witnesses. Dr. Wickramasinghe and Dr. Geisler. Dr. Wickramasinghe testified at length in support of a theory that life on earth was "seeded" by comets which delivered genetic material and perhaps organisms to the earth's surface from interstellar dust far outside the solar system. The "seeding" theory further hypothesizes that the earth remains under the continuing influence of genetic material from space which continues to affect life. While Wickramasinghe's theory[29] about the origins of life on earth has not received general acceptance within the scientific community, he has, at least, used scientific methodology to produce a theory of origins which meets the essential characteristics of science.

The Court is at a loss to understand why Dr. Wickramasinghe was called in behalf of the defendants. Perhaps it was because he was generally critical of the theory of evolution and the scientific community, a tactic consistent with the strategy of the defense. Unfortunately for the defense, he demonstrated that the simplistic approach of the two model analysis of the origins of life is false. Furthermore, he corroborated the plaintiffs' witnesses by concluding that "no rational scientist" would believe the earth's geology could be explained by reference to a worldwide flood or that the earth was less than one million years old.

The proof in support of creation science consisted almost entirely of efforts to discredit the theory of evolution through a rehash of data and theories which have been before the scientific community for decades. The arguments asserted by creationists are not based upon new scientific evidence or laboratory data which has been ignored by the scientific community.

Robert Gentry's discovery of radioactive polonium haloes in granite and coalified woods is, perhaps, the most recent scientific work which the creationists use as argument for a "relatively recent inception" of the earth and a "worldwide flood." The existence of polonium haloes in granite and coalified wood is thought to be inconsistent with radiometric dating methods based upon constant radioactive decay rates. Mr. Gentry's findings were published almost ten years ago and have been the subject of some discussion in the scientific community. The discoveries have not, however, led to the formulation of any scientific hypothesis or theory which would explain a relatively recent inception of the earth or a worldwide flood. Gentry's discovery has been treated as a minor mystery which will eventually be explained. It may deserve further investigation, but the National Science Foundation has not deemed it to be of sufficient import to support further funding.

The testimony of Marianne Wilson was persuasive evidence that creation science is not science. Ms. Wilson is in charge of the science curriculum for Pulaski County Special School District, the largest school district in the State of Arkansas. Prior to the passage of Act 590, Larry Fisher, a science teacher in the District, using materials from the ICR, convinced the School Board that it should voluntarily adopt creation science as part of its science curriculum. The District Superintendent assigned Ms. Wilson the job of producing a creation science curriculum guide. Ms. Wilson's testimony about the project was particularly convincing because she obviously approached the assignment with an open mind and no preconceived notions about the subject. She had not heard of creation science until about a year ago and did not know its meaning before she began her research.

Ms. Wilson worked with a committee of science teachers appointed from the District. They reviewed practically all of the creationist literature. Ms. Wilson and the committee members reached the unanimous conclusion that creationism is not science; it is religion. They so reported to the Board. The Board ignored the recommendation and insisted that a curriculum guide be prepared.

In researching the subject, Ms. Wilson sought the assistance of Mr. Fisher who initiated the Board action and asked professors in the science departments of the University of Arkansas at Little Rock and the University of Central Arkansas[30] for reference material and assistance, and attended a workshop conducted at Central Baptist College by Dr. Richard Bliss of the ICR staff. Act 590 became law during the course of her work so she used Section 4(a) as a format for her curriculum guide.

Ms. Wilson found all available creationists' materials unacceptable because they were permeated with religious references and reliance upon religious beliefs.

It is easy to understand why Ms. Wilson and other educators find the creationists' textbook material and teaching guides unacceptable. The materials misstate the theory of evolution in the same fashion as Section 4(b) of the Act, with emphasis on the alternative mutually exclusive nature of creationism and evolution. Students are constantly encouraged to compare and make a choice between the two models, and the material is not presented in an accurate manner.

A typical example is *Origins* (Px 76) by Richard B. Bliss, Director of Curriculum Development of the ICR. The presentation begins with a chart describing "preconceived ideas about origins" which suggests that some people believe that evolution is atheistic. Concepts of evolution, such as "adaptive radiation" are erroneously presented. At page 11, figure 1.6, of the text, a chart purports to illustrate this "very important" part of the evolution model. The chart conveys the idea that such diverse mammals as a whale, bear, bat and monkey all evolved from a shrew through the process of adaptive radiation. Such a suggestion is, of course, a totally erroneous and misleading application of the theory. Even more objectionable, especially when viewed in light of the emphasis on asking the student to elect one of the models, is the chart presentation at page 17, figure 1.6. That chart purports to illustrate the evolutionists' belief that man evolved from bacteria to fish to reptile to mammals and, thereafter, into man. The illustration indicates, however, that the mammal from which man evolved was *a rat*.

Biology, A Search For Order in Complexity[31] is a high school biology text typical of creationists' materials. The following quotations are illustrative:

> Flowers and roots do not have a mind to have purpose of their own; therefore, this planning must have been done for them by the Creator. (at page 12)
>
> The exquisite beauty of color and shape in flowers exceeds the skill of poet, artist, and king. Jesus said (from Matthew's gospel), "Consider the lilies of the field, how they grow; they toil not, neither do they spin. . ." (Px 129 at page 363)

The "public school edition" texts written by creationists simply omit Biblical references but the content and message remain the same. For example, *Evolution—The Fossils Say No!*[32] contains the following:

> Creation. By creation we mean the bringing into being by a supernatural Creator of the basic kinds of plants and animals by the process of sudden, or fiat, creation.
>
> We do not know how the Creator created, what processes He used, *for He used processes which are not now operating anywhere in the natural universe.* This is why we refer to creation as Special Creation. We cannot discover by scientific investigation anything about the creative processes used by the Creator. (page 40)

Gish's book also portrays the large majority of evolutionists as "materialistic atheists or agnostics."

Scientific Creationism (Public School Edition) by Morris, is another text reviewed by Ms. Wilson's committee and rejected as unacceptable. The following quotes illustrate the purpose and theme of the text:

> Foreword
>
> Parents and youth leaders today, and even many scientists and educators, have become concerned about the prevalence and influence of evolutionary philosophy in modern curriculum. Not only is this system inimical to orthodox Christianity and Judaism, but also, as many are convinced, to a healthy society and true science as well. (at page iii)
>
> The rationalist of course finds the concept of special creation insufferably naive, even "incredible." Such a judgment, however, is warranted only if one categorically dismisses the existence of an omnipotent God. (at page 17)

Without using creationist literature, Ms. Wilson was unable to locate one genuinely scientific article or work which supported Section 4(a). In order to comply with the mandate of the Board she used such materials as an article from *Readers Digest* about "atomic clocks" which inferentially suggested that the earth was less than 4½ billion years old. She was unable to locate any substantive teaching material for some parts of Section 4 such as the worldwide flood. The curriculum guide which she prepared cannot be taught and has no educational value as science. The defendants did not produce any text or writing in response to this evidence which they claimed was usable in the public school classroom.[33]

The conclusion that creation science has no scientific merit or educational value as science has legal significance in light of the Court's previous conclusion that creation science has, as one major effect, the advancement of religion. The second part of the three-pronged test for establishment reaches only those statutes having as their *primary* effect the advancement of religion. Secondary effects which advance religion are not constitutionally fatal. Since creation science is not science, the conclusion is inescapable that the *only* real effect of Act 590 is the advancement of religion. The Act therefore fails both the first and second portions of the test in *Lemon* v. *Kurtzman,* 403 U.S. 602 (1971).

IV(E)

Act 590 mandates "balanced treatment" for creation science and evolution science. The Act prohibits instruction in any religious doctrine or references to religious writings. The Act is self-contradictory and compliance is impossible unless the public schools elect to forego significant portions of subjects such as biology, world history, geology, zoology, botany, psychology, anthropology, sociology, philosophy, physics and chemistry. Presently, the concepts of evolutionary theory as described in 4(b) permeate the public school textbooks. There is no way teachers can teach the Genesis account of creation in a secular manner.

The State Department of Education, through its textbook selection committee, school boards and school administrators will be required to constantly monitor materials to avoid using religious references. The school boards, administrators and teachers face an impossible task. How is the teacher to respond to questions about a creation suddenly and out of nothing? How will a teacher explain the occurrence of a worldwide flood? How will a teacher explain the concept of a relatively recent age of the earth? The answer is obvious because the only source of this information is ultimately contained in the Book of Genesis.

References to the pervasive nature of religious concepts in creation science texts amply demonstrate why State entanglement with religion is inevitable under Act 590. Involvement of the State in screening texts for impermissible religious references will require State officials to make delicate religious judgments. The need to monitor classroom discussion in order to uphold the Act's prohibition against religious instruction will necessarily involve administrators in questions concerning religion. These continuing involvements of State officials in questions and issues of religion create an excessive and prohibited entanglement with religion. *Brandon* v. *Board of Education,* 487 F.Supp 1219, 1230 (N.D.N.Y.), *aff'd.,* 635 F.2d 971 (2nd Cir. 1980).

These conclusions are dispositive of the case and there is no need to reach legal conclusions with respect to the remaining issues. The plaintiffs raised two other issues questioning the constitutionality of the Act and, insofar as the factual findings relevant to these issues are not covered in the preceding discussion, the Court will address these issues. Additionally, the defendants raised two other issues which warrant discussion.

V(A)

First, plaintiff teachers argue the Act is unconstitutionally vague to the extent that they cannot comply with its mandate of "balanced" treatment without jeopardizing their employment. The argument centers around the lack of a precise definition in the Act for the word "balanced." Several witnesses expressed opinions that the word has such meanings as equal time, equal weight, or equal legitimacy. Although the Act could have been more explicit, "balanced" is a word subject to ordinary understanding. The proof is not convincing that a teacher using a reasonably acceptable understanding of the word and making a good faith effort to comply with the Act will be in jeopardy of termination. Other portions of the Act are arguably vague, such as the "relatively recent" inception of the earth and life. The evidence establishes, however, that relatively recent means from 6,000 to 20,000 years, as commonly understood in creation science literature. The meaning of this phrase, like Section 4(a) generally, is, for purposes of the Establishment Clause, all too clear.

V(B)

The plaintiffs' other argument revolves around the alleged infringement by the defendants upon the academic freedom of teachers and students. It is contended this unprecedented intrusion in the curriculum by the State prohibits teachers from teaching what they believe should be taught or requires them to teach that which they do not believe is proper. The evidence reflects that traditionally the State Department of Education, local school boards and administration officials exercise little, if any, influence upon the subject matter taught by classroom teachers. Teachers have been given freedom to teach and emphasize those portions of subjects the individual teacher considered important. The limits to this discretion have generally been derived from the approval of textbooks by the State Department and preparation of curriculum guides by the school districts.

Several witnesses testified that academic freedom for the teacher means, in substance, that the individual teacher should be permitted unlimited discretion subject only to the bounds of professional ethics. The Court is not prepared to adopt such a broad view of academic freedom in the public schools.

In any event, if Act 590 is implemented, many teachers will be required to teach material in support of creation science which they do not consider academically sound. Many teachers will simply forego teaching subjects which might trigger the "balanced treatment" aspects of Act 590 even though they think the subjects are important to a proper presentation of a course.

Implementation of Act 590 will have serious and untoward consequences for students, particularly those planning to attend college. Evolution is the cornerstone of modern biology, and many courses in public schools contain subject matter relating to such varied topics as the age of the earth, geology and relationships among living things. Any student who is deprived of instruction as to the prevailing scientific thought on these topics will be denied a significant part of science education. Such a deprivation through the high school level would undoubtedly have an impact upon the quality of education in the State's colleges and universities, especially including the pre-professional and professional programs in the health sciences.

V(C)

The defendants argue in their brief that evolution is, in effect, a religion, and that by teaching a religion which is contrary to some students' religious views, the State is infringing upon the student's free exercise rights under the First Amendment. Mr. Ellwanger's legislative findings, which were adopted as a finding of fact by the Arkansas Legislature in Act 590, provides:

> Evolution-science is contrary to the religious convictions or moral values or philosophical beliefs of many students and parents, including individuals of many different religious faiths and with diverse moral and philosophical beliefs, Act 590, §7(d).

The defendants argue that the teaching of evolution alone presents both a free exercise problem and an establishment problem which can only be redressed by giving balanced treatment to creation science, which is admittedly consistent with some religious beliefs. This argument appears to have its genesis in a student note written by Mr. Wendell Bird, "Freedom of Religion and Science Instruction in Public Schools," 87 Yale L.J. 515 (1978). The argument has no legal merit.

If creation science is, in fact, science and not religion, as the defendants claim, it is difficult to see how the teaching of such a science could "neutralize" the religious nature of evolution.

Assuming for the purposes of argument, however, that evolution is a religion or religious tenet, the remedy is to stop the teaching of evolution, not establish another religion in opposition to it. Yet it is clearly established in the case law, and perhaps also in common sense, that evolution is not a religion and that teaching evolution does not violate the Establishment Clause, *Epperson* v. *Arkansas, supra, Willoughby* v. *Stever,* No. 15574–75 (D.D.C. May 18, 1973); *aff'd.* 504 F.2d 271 (D.C. Cir. 1974), *cert. denied,* 420 U.S. 924 (1975); *Wright* v. *Houston Indep. School Dist.,* 366 F.Supp. 1208 (S.D. Tex. 1978), *aff'd.* 486 F.2d 137 (5th Cir. 1973), *cert. denied* 417 U.S. 969 (1974).

V(D)

The defendants presented Dr. Larry Parker, a specialist in devising curricula for public schools. He testified that the public school's curriculum should reflect the subjects the public wants taught in schools. The witness said that polls indicated a significant majority of the American public thought creation science should be taught if evolution was taught. The point of this testimony was never placed in a legal context. No doubt a sizeable majority of Americans believe in the concept of a Creator or, at least, are not opposed to the concept and see nothing wrong with teaching school children about the idea.

The application and content of First Amendment principles are not determined by public opinion polls or by a majority vote. Whether the proponents of Act 590 constitute the majority or the minority is quite irrelevant under a constitutional system of government. No group, no matter how large or small, may use the organs of government, of which the public schools are the most conspicuous and influential, to foist its religious beliefs on others.

The Court closes this opinion with a thought expressed eloquently by the great Justice Frankfurter:

> We renew our conviction that "we have staked the very existence of our country on the faith that complete separation between the state

and religion is best for the state and best for religion." *Everson* v. *Board of Education,* 330 U.S. at 59. If nowhere else, in the relation between Church and State, "good fences make good neighbors." [*McCollum* v. *Board of Education,* 333 U.S. 203, 232 (1948)]

An injunction will be entered permanently prohibiting enforcement of Act 590.

It is ordered this January 5, 1982.

—William R. Overton *in the U.S. District Court, Eastern District of Arkansas, Western Division*

REFERENCES

1. The complaint is based on 42 U.S.C. §1983, which provides a remedy against any person who, acting under color of state law, deprives another of any right, privilege or immunity guaranteed by the United States Constitution or federal law. This Court's jurisdiction arises under 28 U.S.C. §§1331, 1343(3) and 1343(4). The power to issue declaratory judgments is expressed in 28 U.S.C. §§2201 and 2202.

2. The facts necessary to establish the plaintiffs' standing to sue are contained in the joint stipulation of facts, which is hereby adopted and incorporated herein by reference. There is no doubt that the case is ripe for adjudication.

3. The State of Arkansas was dismissed as a defendant because of its immunity from suit under the Eleventh Amendment. *Hans* v. *Louisiana,* 134 U.S. 1 (1890).

4. The authorities differ as to generalizations which may be made about Fundamentalism. For example, Dr. Geisler testified to the widely held view that there are five beliefs characteristic of all Fundamentalist movements, in addition, of course, to the inerrancy of Scripture: (1) belief in the virgin birth of Christ, (2) belief in the deity of Christ, (3) belief in the substitutional atonement of Christ, (4) belief in the second coming of Christ, and (5) belief in the physical resurrection of all departed souls. Dr. Marsden, however, testified that this generalization, which has been common in religious scholarship, is now thought to be historical error. There is no doubt, however, that all Fundamentalists take the Scriptures as inerrant and probably most take them as literally true.

5. Initiated Act 1 of 1929, Ark. Stat. Ann. §80–1627 *et seq.,* which prohibited the teaching of evolution in Arkansas schools, is discussed *infra* at text accompanying note 26.

6. Subsequent references to the testimony will be made by the last name of the witness only. References to documentary exhibits will be by the name of the author and the exhibit number.

7. Applicants for membership in the CRS must subscribe to the following statement of belief: "(1) The Bible is the written Word of God, and because we believe it to be inspired thruout (sic), all of its assertions are historically and scientifically true in all of the original autographs. To the student of nature, this means that the account of origins in Genesis is a factual presentation of simple historical truths. (2) All basic types of living things, including man, were made by direct creative acts of God during Creation Week as described in Genesis. Whatever biological changes have occurred since Creation have accomplished only changes within the original created kinds. (3) The great Flood described in Genesis, commonly referred to as the Noachian Deluge,

was an historical event, worldwide in its extent and effect. (4) Finally, we are an organization of Christian men of science, who accept Jesus Christ as our Lord and Savior. The account of the special creation of Adam and Eve as one man and one woman, and their subsequent Fall into sin, is the basis for our belief in the necessity of a Savior for all mankind. Therefore, salvation can come only thru (sic) accepting Jesus Christ as our Savior." (Px 115)

8. Because of the voluminous nature of the documentary exhibits, the parties were directed by pre-trial order to submit their proposed exhibits for the Court's convenience prior to trial. The numbers assigned to the pre-trial submissions do not correspond with those assigned to the same documents at trial and, in some instances, the pre-trial submissions are more complete.

9. Px 130, Morris, *Introducing Scientific Creationism Into the Public Schools* (1975), and Bird, "Resolution for Balanced Presentation of Evolution and Scientific Creationism." *ICR Impact Series* No. 71, App. 14 to Plaintiffs' Pretrial Brief.

10. The creationists often show candor in their proselytization. Henry Morris has stated, "Even if a favorable statute or court decision is obtained, it will probably be declared unconstitutional, especially if the legislation or injunction refers to the Bible account of creation." In the same vein he notes, "The only effective way to get creationism taught properly is to have it taught by teachers who are both willing and able to do it. Since most teachers now are neither willing nor able, they must first be both persuaded and instructed themselves." Px 130, Morris, *Introducing Scientific Creationism Into the Public Schools* (1975) (unpaged).

11. Mr. Bird sought to participate in this litigation by representing a number of individuals who wanted to intervene as defendants. The application for intervention was denied by this Court. *McLean* v. *Arkansas,* ———F.Supp.———, (E.D. Ark. 1989), aff'd. *per curiam,* Slip Op. No. 81–2023 (8th Cir. Oct. 16, 1981).

12. The model act had been revised to insert "creation science" in lieu of creationism because Ellwanger had the impression people thought creationism was too religious a term. (Ellwanger Depo. at 79.)

13. The original model act had been introduced in the South Carolina Legislature, but had died without action after the South Carolina Attorney General had opined that the act was unconstitutional.

14. Specifically, Senator Holsted testified that he holds to a literal interpretation of the Bible: that the bill was compatible with his religious beliefs: that the bill does favor the position of literalists: that his religious convictions were a factor in his sponsorship of the bill: and that he stated publicly to the *Arkansas Gazette* (although not on the floor of the Senate) contemporaneously with the legislative debate that the bill does presuppose the existence of a divine creator. There is no doubt that Senator Holsted knew he was sponsoring the teaching of a religious doctrine. His view was that the bill did not violate the First Amendment because, as he saw it, it did not favor one denomination over another.

15. This statute is, of course, clearly unconstitutional under the Supreme Court's decision in *Abbington School Dist.* v. *Schempp,* 374 U.S. 203 (1963).

16. The joint stipulation of facts establishes that the following areas are the only *information* specifically required by statute to be taught in all Arkansas schools: (1) the effects of alcohol and narcotics on the human body, (2) conservation of national resources, (3) Bird Week, (4) Fire Prevention, and (5) Flag etiquette. Additionally, certain specific courses, such as American history and Arkansas history, must be completed by each student before graduation from high school.

17. Paul Ellwanger stated in his deposition that he did not know why Section 4(a)(2) (insufficiency of mutation and natural selection) was included as an evidence supporting creation science. He indicated that he was not a scientist, "but these are the postulates that have been laid down by creation scientists." Ellwanger Depo. at 136.

18. Although defendants must make some effort to cast the concept of creation in non-religious terms, this effort surely causes discomfort to some of the Act's more theologically sophisticated supporters. The concept of a creator God distinct from the God of love and mercy is closely similar to the Marcion and Gnostic heresies, among the deadliest to threaten the early Christian church. These heresies had much to do with development and adoption of the Apostle's Creed as the official creedal statement of the Roman Catholic Church in the West. (Gilkey.)

19. The parallels between Section 4(a) and Genesis are quite specific: (1) "sudden creation from nothing" is taken from Genesis 1:1–10 (Vawter, Gilkey); (2) destruction of the world by a flood of divine origin is a notion peculiar to Judeo-Christian tradition and is based on Chapters 7 and 8 of Genesis (Vawter); (3) the term "kinds" has no fixed scientific meaning, but appears repeatedly in Genesis (all scientific witnesses); (4) "relatively recent inception" means an age of the earth from 6,000 to 10,000 years and is based on the genealogy of the Old Testament using the rather astronomical ages assigned to the patriarchs (Gilkey and several of defendants' scientific witnesses); (5) Separate ancestry of man and ape focuses on the portion of the theory of evolution which Fundamentalists find most offensive. *Epperson* v. *Arkansas,* 393 U.S. 97 (1968).

20. "[C]oncepts concerning. . .a supreme being of some sort are manifestly religious. . .These concepts do not shed that religiosity merely because they are presented as philosophy or as a science. . ." *Malnak* v. *Yogi,* 440 F.Supp. 1284, 1322 (D.N.J. 1977); *aff'd per curiam,* 592 F.2d 197 (3d Cir. 1979).

21. See, e.g., Px 76, Morris, *et al., Scientific Creationism,* 203 (1980) ("If creation really is a fact, this means there is a *Creator,* and the universe is His creation.") Numerous other examples of such admissions can be found in the many exhibits which represent creationist literature, but no useful purpose would be served here by a potentially endless listing.

22. Morris, the Director of ICR and one who first advocated the two model approach, insists that a true Christian cannot compromise with the theory of evolution and that the Genesis version of creation and the theory of evolution are mutually exclusive. Px 31, Morris, *Studies in the Bible & Science,* 102–103. The two model approach was the subject of Dr. Richard Bliss's doctoral dissertation. (Dx 35). It is presented in Bliss, *Origins: Two Models—Evolution, Creation* (1978). Moreover, the two model approach merely casts in educationalist language the dualism which appears in all creationist literature—creation (i.e. God) and evolution are presented as two alternative and mutually exclusive theories. See, e.g., Px 75, Morris, *Scientific Creationism* (1974) (public school edition); Px 59, Fox, *Fossils: Hard Facts from the Earth.* Particularly illustrative is Px 61, Boardman, *et al., Worlds Without End* (1971) a CSRC publication: "One group of scientists, known as creationists, believe that God, in a miraculous manner, created all matter and energy. . .

"Scientists who insist that the universe just grew, by accident, from a mass of hot gases without the direction or help of a Creator are known as evolutionists."

23. The idea that belief in a creator and acceptance of the scientific theory of evolution are mutually exclusive is a·false premise and offensive to the religious views of many. (Hicks) Dr. Francisco Ayala, a geneticist of considerable reknown and a former Catholic priest who has the equivalent of a Ph.D. in theology, pointed out that

many working scientists who subscribed to the theory of evolution are devoutly religious.

24. This is so despite the fact that some of the defense witnesses do not subscribe to the young earth or flood hypotheses. Dr. Geisler stated his belief that the earth is several billion years old. Dr. Wickramasinghe stated that no rational scientist would believe the earth is less than one million years old or that all the world's geology could be explained by a worldwide flood.

25. "We do not know how God created, what processes He used, for *God used processes which are not now operating anywhere in the natural universe.* This is why we refer to divine creation as Special Creation. We cannot discover by scientific investigation anything about the creative processes used by God." Px 78, Gish, *Evolution? The Fossils Say No!,* 42 (3d ed. 1979) (emphasis in original).

26. The evolutionary notion that man and some modern apes have a common ancestor somewhere in the distant past has consistently been distorted by anti-evolutionists to say that man descended from modern monkeys. As such, this idea has long been most offensive to Fundamentalists. See *Epperson v. Arkansas,* 393 U.S. 97 (1968).

27. Not only was this point acknowledged by virtually all the defense witnesses, it is patent in the creationist literature. See, e.g., Px 89, Kofahl & Segraves, *The Creation Explanation,* 40: "The Flood of Noah brought about vast changes in the earth's surface, including vulcanism, mountain building, and the deposition of the major part of sedimentary strata. This principle is called 'Biblical catastrophism.' "

28. See n. 7, *supra,* for the full text of the CRS creed.

29. The theory is detailed in Wickramasinghe's book with Sir Fred Hoyle, *Evolution From Space* (1981), which is Dx 79.

30. Ms. Wilson stated that some professors she spoke with sympathized with her plight and tried to help her find scientific materials to support Section 4(a). Others simply asked her to leave.

31. Px 129, published by Zonderman Publishing House (1974), states that it was "prepared by the Textbook Committee of the Creation Research Society." It has a disclaimer pasted inside the front cover stating that it is not suitable for use in public schools.

32. Px 77, by Duane Gish.

33. The passage of Act 590 apparently caught a number of its supporters off guard as much as it did the school district. The Act's author, Paul Ellwanger, stated in a letter to "Dick," (apparently Dr. Richard Bliss at ICR): "And finally, if you know of any textbooks at any level and for any subjects that you think are acceptable to you and also constitutionally admissible, these are things that would be of *enormous* to these bewildered folks who may be caught, as Arkansas now has been, by the sudden need to implement a whole new ball game with which they are quite unfamiliar." (sic) (Unnumbered attachment to Ellwanger depo.)

Part 6
Some Thoughts on Genesis as Science

On Giving Equal Time to the Teaching of Evolution and Creation*

by John A. Moore

I believe that these are the inevitable conclusions that a science teacher and his students must reach if they stick to an entirely scientific analysis of biblical statements. Either the Bible is wrong or science is wrong, and very few educated persons in the modern world maintain the latter. Is this what the lawmakers and creationists desire? I doubt it.

On April 30, 1973, Senate Bill 394, having been passed by an overwhelming majority of both houses of the General Assembly of the state of Tennessee, became law. The new law, which to some extent replaced the antievolution law that was repealed only in 1967, reads in part:

Any biology textbook used for teaching in the public schools, which expresses an opinion of, or relates to a theory about origins or creation of man and his world shall be prohibited from being used as a textbook in such a system unless it specifically states that it is a theory as to the origin and creation of man and his world and is not represented to be scientific fact. Any textbook so used in the public education system which expresses an opinion or relates to a theory or theories shall give in the same text book and under the same subject commensurate attention to, and an equal amount of emphasis on, the origins and creation of man and his world as the same is recorded in other theories, including, but not limited to, the Genesis account in the Bible. . . . The teaching of all occult or satanical beliefs

*Based on a lecture given at a symposium entitled "The Role of Controversy in Science," Annual Meeting, American Association for the Advancement of Science, San Francisco, February 1974. Reprinted from *Perspectives in Biology and Medicine,* vol. 18 (March 1975), by permission of The University of Chicago Press. Copyright © 1980 by The University of Chicago Press. All rights reserved.

of human origins is expressly excluded from this act. . . . Provided
however that the Holy Bible shall not be defined as a textbook, but
is hereby declared to be a reference work, and shall not be required
to carry the disclaimer above provided for textbooks. . . . This
Act shall take effect upon becoming a law, the public welfare requiring it.[1]

Similar bills have been or are being considered by the legislatures
or departments of education of Georgia, Michigan, Washington, California, and Colorado, but only Tennessee's has become law.

When teachers of science are confronted with a situation of this
sort, a variety of responses might be expected. Some teachers might
welcome the possibility of being able to present their own religious
beliefs to their students. Others might avoid the problem by omitting
all references to scientific data and hypotheses about the origin and
evolution of the world and its inhabitants. This last course has been
widely adopted in the past; lots of problems never arise if one ignores
the topic. The Tennessee law does not require one to teach the accounts
of creation given in Genesis and elsewhere. It says only that if you do
include the scientific explanations, you have to include the religious
ones as well.

Still another response would be to abide by the law and give
"commensurate attention to" and "an equal amount of emphasis on"
the two conflicting points of view. This is the option that I plan to
discuss in this paper.

So let us assume that we will carry out the stipulations of the
Tennessee law as honestly and as competently as we can. Let us assume
also that we do this as teachers of science and not as advocates of some
religious doctrine or sect. That is, we will employ *only* the canons of
scientific and scholarly procedures in exploring the topic. Statements
and hypotheses will be evaluated solely on the basis of the scientific
evidence in their favor. Many accounts of creation, including Genesis,
are precise enough to be used as working hypotheses from which various deductions can be made. The deductions can be tested, again with
scientific data and procedures, and from the results the original hypothesis can be substantiated, made more probable, made less probable, or
rejected.

One might object at this point by saying that what I propose to do
is not what the Tennessee lawmakers had in mind. That may be, but
if I am asked to consider Genesis in a science course, and to treat it as
a scientific theory, how else am I expected to do it? Furthermore, as I
understand them, this is precisely what the most effective creationists
in the country are requesting. I am referring here to members of the
Creation Research Society and the Institute for Creation Research.

Their campaign in California was for equal time and emphasis to be given to biological evolution and creationism. Their theory of creation, which is now more often referred to as the "creation model," is derived from Genesis. The basis of their beliefs is given by the credo to which all members of the Creation Research Society ascribe. They "are committed to full belief in the Biblical record of special creation and early history as opposed to evolution, both of the universe and of the earth with its complexity of living forms." They believe, further, "that science should be realigned within the framework of Biblical creationism." More specifically:

> All members of the Society subscribe to the following statement of belief.
>
> 1. The Bible is the written Word of God, and because it is inspired throughout, all its assertions are historically and scientifically true in all the original autographs. To the student of nature this means that the account of origins in Genesis is a factual presentation of simple historical truths.
>
> 2. All basic types of living things, including man, were made by direct acts of God during the Creation Week described in Genesis. Whatever biological changes have occurred since Creation Week have accomplished only changes within the original created kinds.
>
> 3. The great Flood described in Genesis, commonly referred to as the Noachian Flood, was a historic event worldwide in its extent and effect. . . .[2]

It is important to note also that all regular voting members of the Creation Research Society must have an earned postgraduate degree (M.S., Ph.D., or the equivalent) in science.

Thus, for these influential creationists, at least, we would be complying with the Tennessee law if we concentrated on Genesis as an example of an account of creation. So, for the purposes of this paper, space being a limiting factor, I will suggest how the "equal time and emphasis" for creationism might be devoted to analyzing how adequately Genesis can account for the origin and diversity of living things.

First, it would be necessary to establish what is, in fact, said in Genesis. This is not a simple matter. There is a serious problem concerning what was originally written. Some students may need to be reminded that Genesis has not always existed in the language of the King James Version (KJV). The ultimate source is the ancient beliefs of the Jewish people, which were first written down at various times before the beginning of the Christian Era. The earliest may date to the second millennium B.C., though the oldest surviving Hebrew texts of Genesis are about 1,000 years old. Nevertheless, there is much evidence that the surviving Hebrew texts are highly accurate. That is, when it has been possible to compare the Hebrew Bible with ancient manuscripts, such as the Dead Sea Scrolls, the two are essentially identical.

A far more substantial problem is the adequacy of translation. Hebrew was almost a dead language even before the time of Christ. In fact, the sacred texts had become such a mystery that, in the days of Ptolemy II, the Jewish people of Alexandria engaged a group of 70 scholars to translate their sacred texts into Greek. Their product was the Septuagint—dating from the third century B.C. It is the oldest version of the Old Testament. The Septuagint was *the* Bible of the early church in the West and is the Bible of the Eastern church today. Nevertheless, there were many different versions and revisions. The difficulty of knowing what was *the* Word led Origen (A.D. 185–254) to prepare his Hexapla, which survived only in fragments. This consisted of six parallel columns, each with a different version of the sacred texts.

Early in the fifth century A.D., Jerome completed the Vulgate, which was to become the official Bible of the Western church. His was a translation from Hebrew to Latin, using the best Hebrew manuscripts that could be obtained at the time. It is to be noted, however, that he produced not a literal but an idiomatic translation. Jewish scholars continued to work on the problems of choosing the most accurate versions and the most probable readings of the ancient Hebrew words. By the end of the tenth century A.D., they completed what was to become the first official Hebrew text—the Massoretic text.

The Vulgate was translated into English in the fourteenth century by Nicholas of Hereford and John Purvey—their product generally known as the Wycliffe Bible. Early in the sixteenth century Tyndale translated much of the Bible from Hebrew. Various other versions—Coverdale (1535), the Great Bible (1560), the Geneva Bible (1560), and the Bishops' Bible—appeared in the sixteenth century.

What is often regarded as *the* Bible, namely the King James Version, was published in 1611. This was based on the Bishops' Bible, modified by reference to the best currently available Hebrew and Greek texts. Other revisions followed.

The New English Bible (NEB) of 1961 and 1970 will probably be the standard for some years. A. A. Macintosh has this to say about it: "The importance of the N.E.B. as a translation of the Old Testament lies in the fact that it is based upon the most up-to-date scholarship and that it is a *new* translation. This independence has made possible the maximum utilization of the results of modern research. The last century or so has seen a very considerable increase in our knowledge of the languages, customs and institutions of the ancient Near East, as well as of the history of the Old Testament text. The twentieth-century translators of the Old Testament are therefore able to make use of knowledge which was simply not available to their predecessors. . . ."[3]

He goes on to point out that many problems still remain—does an unintelligible word represent an ancient copyist's error, or is it a word for which the meaning is totally lost? Sometimes the problem can be tentatively resolved by reference to other Semitic languages. For example, a word thought to mean only "to know" in Hebrew means both "to know" and "to be tamed" in Arabic, suggesting that Judges 16:9, which is about Samson, should be translated, "And his strength was not tamed," instead of "So his strength was not known," as it has been rendered by previous translators.

Sometimes the new information suggests a wording that modifies the beauty of the King James Version. Take the case of the Twenty-third Psalm, "Yea, though I walk through the valley of the shadow of death." One hearing that statement for the first time might be very confused as to the possible meaning. What is the "shadow" of death? Is the speaker at the point of death? That would be one possibility. Most individuals familiar with the Twenty-third Psalm have no doubt treasured the King James translation for its poetic beauty—and have not worried too much about true meanings. The better understanding of ancient Hebrew, which has come in recent years, suggests that the word translated as "shadow of death" really means "darkest shadow." The modern translation becomes less ambiguous, therefore, even though possibly it becomes less beautiful.

Sometimes the results of biblical scholarship suggest changes that deeply affect church dogma. Consider, for example, the virginity of Mary. Isaiah 7:14, as translated from the Septuagint, and which would have been familiar to the compilers of the New Testament, can be rendered, "Behold a virgin shall conceive and bear a son and shall call his name Immanuel" (KJV). Matthew 1:22–23 refers to this as follows: "Now all this was done, that it might be fulfilled which was spoken of the Lord by the prophet, saying, Behold a virgin shall be with child, and shall bring forth a son, and they shall call his name Emmanuel." However, the official Hebrew Massoretic text speaks not of a virgin but of a young woman. Thus the NEB translates Isaiah as, "A young woman is with child, and she will bear a son, and (you) will call him Immanuel."

One could discuss the evolution of the Bible for a very long time. The amount of scholarship devoted to gaining a better understanding of the Bible is simply enormous. Many science teachers might find this a new and very interesting subject. In any event, they would soon gain the impression that the Bible is something more than the King James

Version, and that there still remains great uncertainty in understanding some of the ancient words and statements.

This problem is avoided by many fundamentalists who hold that the translators of the Bible were inspired by God and, therefore, that what they wrote must be correct. If this is so, we are left with the problem of which inspiration is correct. This would be a serious problem for the science teacher trying to fulfill the mandates of the Tennessee law. Neither should the teacher sidestep the problem. If the account of creation being discussed is given in the Bible, one has to evaluate the source, just as one is bound to evaluate the data of paleontology, genetics, etc., when dealing with biological evolution.

But let us go on and assume with the members of the Creation Research Society that "the account of origins in Genesis is a factual presentation of simple historical truths." We will assume, therefore, that the statements in Genesis are working hypotheses, and we will make deductions from the hypotheses and test them.

First, what are the statements? Here many individuals are in for a great surprise. Although the Bible may be the most widely read of all books for all time, few readers seem aware that Genesis has two accounts of creation. So the science class will have to investigate that problem before continuing the analysis.

The first chapter of Genesis plus the first four verses of the second chapter give what is generally considered *the* account of creation:

> On the first day, when the earth was dark, wet, and formless, light was created.
> On the second day the sky (heaven) separated waters above and below.
> On the third day, land and water were separated and plants created.
> On the fourth day, sun, moon, and stars were created.
> On the fifth day, aquatic creatures and flying creatures, the birds, were created.
> On the sixth day, terrestrial forms—mammals, reptiles, and man were created.
> On the seventh day God "ceased from all the work he had set himself to do."

Note the sequence of creation, as far as living creatures are concerned:

> first plants
> then aquatic creatures and birds
> finally reptiles and mammals, including man.

The second account of creation begins with the fifth verse of chapter two of Genesis. The order of creation is not described in days, but there is this sequence:

> We begin with a barren earth totally without plant life.
> Then the Lord God forms Adam from dust.
> Then the Garden of Eden was planted, which contained all the plants.
> Then the Lord God, noting that "It is not good for man to be alone," formed all the wild animals and birds out of dust.
> Finally, none of the wild animals being a satisfactory partner, one of Adams's ribs was removed to form woman.

Some theologians have interpreted the Scriptures as saying that all of this was done instantaneously—not in six days as before.

How is one to interpret these totally different accounts of creation? If we are to regard the statements in Genesis as working hypotheses, we face the problem that the two hypotheses are mutually exclusive. One or the other may be correct, but both cannot be correct. Remember, we are seeing how we can discuss creation as a scientific theory, so we are bound by accepted scientific procedures.

Some fundamentalists insist that there is no conflict whatsoever, but it is beyond my comprehension to understand how they arrive at their position. And, in my defense, it can be stated that the fathers of the church regarded this as a nearly insoluble problem. Andrew Dickson White, the famous historian, diplomat, and first president of Cornell, gives a fascinating account of how the early theologians sought to resolve the dilemma.[4]

In the minds and words of the fathers of the church, and in the art of the great cathedrals, Genesis was assumed to mean what was literally said. Creation was the *work* of God. This work was more than a moulding of matter: matter was first created, and then it was formed into the earth and its inhabitants and into the celestial bodies. Considerable difficulty arose when an attempt was made to understand the sequence of creation. Most early theologians accepted the first account of creation—in the first chapter of Genesis. Others, however, maintained that the account in the second chapter was more acceptable. Finally, it was agreed that both accounts must be accepted, since the Bible in its entirety was the Word of God. Saint Augustine, among others, maintained and encouraged this point of view. As White describes this problem: "Serious difficulties were found in reconciling these two views, which to the natural mind seem absolutely contradictory; but by ingenious manipulation of texts, by dexterous play upon phrases, and by the abundant use of metaphysics to dissolve away facts, a reconciliation was affected, and men came at least to believe that they believe in a creation of the universe instantaneous and at the same time extending through six days."[5] I wonder what would be the effect on a high school student's mind of recounting this bit of history?

Though Augustine and the other fathers of the church could not resolve the dilemma, more recent biblical scholarship can. In fact, the mystery of the two conflicting accounts of creation in Genesis was cleared up during the nineteenth century, a period during which the Bible was subjected to searching analysis.

It was observed, for example, that in the various parts of Genesis there are great differences in style and vocabulary. Sometimes the creator is referred to as Yahweh, at other times as Elohim. This is reflected in the English Bible, where Elohim is translated as God and Yahweh as Lord God. It so happens that the creator mentioned in the first Genesis account is Elohim, or God, whereas in the second account he is Yahweh, or the Lord God. A huge amount of scholarly detective work was done before it was clear, beyond a reasonable doubt, that the two accounts of creation included in Genesis had very different origins. In fact, by the 1880s it was established that Genesis and the other books of the Pentateuch represent a compilation of numerous ancient documents. As far as the first two chapters of Genesis are concerned, they are derived from what are called the P and J documents, but, according to *The Interpreter's Bible,* "both of them bear the marks of having been elaborated by writers other than their original authors."[6] The P (for Priestly) document is the youngest. It is thought to have been written after the Jews returned from exile in Babylonia (sixth century B.C.). The Priestly document refers to the creator as Elohim. Its account of creation relies heavily on the Babylonian creation myth, which the priests would have learned about during the exile if it was not already known to them.

The J (for Yahweh) manuscript is much more ancient. It probably was written about the tenth century B.C., presumably after a long period during which the traditions were transmitted orally. This manuscript derives from the beliefs of the southern tribes of Israel, with their fierce god, Yahweh.

This solution to the problem is no longer seriously debated by biblical scholars. There *are* two conflicting accounts of creation in Genesis. One recounts the ancient beliefs of the nomadic tribes of southern Israel; the other unites some of the beliefs of the Jews with Babylonian accounts of creation. The interval between the writing of P and J is roughly the same as between the Dark Ages and today. The fact that numerous conflicting narratives were included in the Pentateuch is interpreted by biblical scholars as an example of political compromise between conflicting groups of priests—of Hebron, Shechem, and Jerusalem. If you can't agree on a single point of view, give all.

Needless to say, this flowering of biblical scholarship in the nineteenth century produced a profound revolution in scripture interpretation. Whereas biblical scholars from the time of Augustine to the Enlightenment might make heroic efforts to believe two incompatible accounts of creation, scholars of the nineteenth and twentieth centuries accepted neither as "a factual presentation of simple historical truths." Biblical scholars, Jew and Gentile, Catholic and Protestant, are almost unanimous in placing the first two chapters of Genesis among the creation myths that form parts of the sacred traditions of nearly all primitive peoples. One would, in scholarly honesty, have to present this point of view to one's students.

It is often maintained that biblical statements, such as the accounts of creation given in Genesis, cannot be proven or disproven by scientific procedures. In some sense this is true. If one accepts an initial supernatural phenomenon, there are no restraints on invoking additional supernatural phenomena to explain away difficulties of interpretation. No doubt everyone has heard arguments of the sort that one need not accept the fossil data for evolution at all. It is conceivable, at least in metaphysics, that the earth, complete in its present form (including the fossils), was created 10 minutes ago, etc. But from the time of Francis Bacon, this approach has not proven to be a generally acceptable way of gaining an understanding of the natural world. We and all our works may be an illusion—but it is at least an internally consistent and satisfying illusion to a lot of people.

But we can agree to examine biblical statements as *scientific* statements, as the Tennessee law and its advocates are asking, and to see how they fare. And it must be emphasized again that, in our procedures, we cannot invoke supernatural phenomena to explain away the difficulties. That is, when the time comes to squeeze the creatures of the earth into the ark, we cannot decide to suspend their heterotrophicity or to miniaturize them. A scientific hypothesis must assume an ark with sufficient space for the creatures and for their food, and enough caretakers to control a situation that would make the Augean stables seem like a rose garden.

The key elements in biblical accounts of creation, which will be our hypothesis to be tested, are these: First, the earth and its inhabitants were created in essentially the same form in which we observe them today. We can ignore the differences between an instantaneous creation, suggested by J and a creation requiring six days, as in P. Well into the nineteenth century, scholars of all sorts assumed that all forms that *could be created were created,* and that all persist today. Ecclesiastes 3:14 was one basis: "I know that whatever God does lasts forever: to add to it or subtract from it is impossible" (NEB). Even so great an authority as Linnaeus maintained this view early in his career.[7] He believed that all species must have been created in the beginning; if not, this would imply that God was inefficient. Furthermore, none could be extinct—this would imply that God's products were defective.

Second, the time of creation was approximately 6,000 years ago. Bishop Ussher usually gets credit for having determined this date, but it was generally believed long before his time. The fifth and tenth chapters of Genesis give much of the data. Bishop Ussher was more precise and fixed the beginning of creation at 4004 B.C., and his dates for all biblical events were included in the KJV until quite recently. For many they became part of divine scripture. It was Dr. John Lightfoot, vice-chancellor of Cambridge and one of the most eminent Hebrew scholars of the seventeenth century, who fixed the time of creation more precisely as 9 A.M., October 23, 4004 B.C.[8]

Both of these elements of the Genesis creation hypothesis suggest deductions. The most obvious one from the hypothesis that life has been the same from the moment of creation to the present is this: If there is a record of past life, then, barring sampling errors, the record should show essentially identical faunas and floras throughout the period for which the record is available. For a test of this deduction one turns to the data of geology. There *is* a record going back about 3 billion years, but useful for this deduction for only about half a billion years. This record shows that the successive strata of the earth's crust contain different assemblages of organisms—the differences increasing with the distances between the strata.

With respect to the Genesis hypothesis of a young earth, we can make this deduction: If there are scientific methods for determining age, natural objects must be younger than, roughly, 6,000 years. Again we can turn to the physical sciences, where we find that various methods of determining age are available. These are of varying accuracy, all lacking the precision of Dr. Lightfoot's, but they do demonstrate that, beyond a reasonable doubt, the earth is extraordinarily old.

These two hypotheses, which can be tested readily by accepted scientific procedures, show that beyond a reasonable doubt the accounts of creation given in Genesis cannot be scientifically true. They may be of extraordinary religious, emotional, metaphysical, metaphorical, or literary importance, but they are not useful working hypotheses for science.

A point of even greater importance is that a science teacher would have to explain to the students why hypotheses based on the accounts of creation given in Genesis, or from other religious traditions, can *never* be useful in science. Natural phenomena are to be explained by a scientist only in terms of phenomena that he can observe and study. Supernatural explanations are not permitted. Thus science must ignore hypotheses that involve the creation of matter and energy ex nihilo. Thus, there are valid scientific and procedural grounds for rejecting the hypotheses of creation based on Genesis.

Yet there are many other statements in Genesis about events after creation that apparently involve no supernatural elements, and hence may be treated as hypotheses to be tested by scientific procedures. A few of these will be mentioned to illustrate how they might be developed in a classroom.

The problem of the continuity of human beings is a serious one if the biblical statements are to be taken literally. Only two human beings were created—one male and one female. Their first two children were males (Cain and Abel). Subsequently there were other males (Enoch and Seth). Very much later other males and females were produced by Adam and Eve. However, the first members of the F_1 generation consisted only of males. Current biological theory suggests that there could have been no F_2. Yet, according to Genesis, F_2 were produced in abundance.

Following the creation, the flood was by far the most important event for living creatures. The account given in the sixth through ninth chapters of Genesis is a combination of both J and P manuscripts—which accounts for the contradictory statements. Both seem to be based on the Babylonian story of the flood given in the Gilgamesh Epic. The essential points of the Genesis account are these:

1. Every living thing perished. As Genesis 7:23 gives it, "God wiped out every living thing that existed on earth," except for those on the ark.

2. The waters covered the entire earth reaching a height of 15 cubits (a cubit is the distance from the elbow to the end of the middle finger), or about 7 meters above the highest mountains.

3. The flood was due to rain water according to J, and to rainwater plus subterranean water according to P.

4. The duration of the flood was 40 days according to J and 150 according to P. J and P also differ on the time before the waters dried up, but, in any event, they did.

Thus, all life subsequent to the flood was descended from the animals and plants that Noah had taken into the ark. The ark, therefore, becomes a bottleneck, and numerous biological questions can be asked about it: how was it filled, and what was the history of the organisms once they were released from the ark, etc? Once again, these matters must be dealt with in a scientific manner—that is, we cannot invoke supernatural phenomena to explain difficulties that may arise. A host of problems present themselves. Some of the more obvious ones are (of course these are not new questions—they sorely beset theologians of earlier times):

1. What was the mechanism that caused the animals to migrate from their homelands to the Near East? Did the giant earthworms of Australia have a premonition of the flood and a nervous system complex enough for them to take the necessary precautionary steps?

2. By what route did all the animals, especially those with very limited means of dispersal, get to the Middle East to board the ark? This would seem to have been especially difficult for all organisms of the New World and essentially impossible for those in Australia (and all remote islands).

3. How did Noah obtain plants or their seeds from areas distant from the site of the construction of the ark?

4. What so modified the patterns of behavior of the animals that they were able to exist together for the duration of the voyage?

5. How could the roughly 2,000,000 species of organisms known to inhabit the earth, including terrestrial, fresh-water, and marine forms, plus food to last for about a year, be domiciled in an ark which we are told was about 150 meters long, 25 meters wide, and 15 meters high?

6. If, as Genesis says, all living things not in the ark were destroyed, how could the dove sent out in search of dry land return with a freshly plucked olive leaf?

7. When the ordeal was finally over and the ark door opened, how did the organisms reach the localities where we now find them? They would have the same problems as they did in coming to the ark, except for an additional major disadvantage: the flood has sterilized the earth of all living creatures. What would have served as food for the animals?

One could continue this sort of scientific exegesis and hermeneutics, but more than likely enough has been given to allow us to reach some obvious, though important, conclusions.

The first is that, if one is to subject Genesis to the sort of analysis that the law of Tennessee and some of the more prominent creationists are demanding, the Genesis account is demolished from a scientific point of view.

The second point is that if one gets out on the fundamentalist's limb of maintaining that *all* biblical statements must be true, and one demonstrates that part cannot be scientifically true, then the entire opus becomes questionable.

I believe that these are the inevitable conclusions that a science teacher and his students must reach if they stick to an entirely scientific analysis of biblical statements. Either the Bible is wrong or science is wrong, and very few educated persons in the modern world maintain the latter.

Is this what the lawmakers and the creationists desire? I doubt it. Yet, unless one makes the improbable assumption that they seek to hold religion up to ridicule or to destroy it, I cannot imagine that they truly desire a critical and scientific evaluation of Genesis. During the past century biblical scholars and scientists have independently reached the same conclusion: the ancient Hebrew accounts of creation, as recorded in Genesis, cannot be accepted as "a factual presentation of simple historical truths."

Those ancient Hebrews left a rich legacy to the world—but this legacy was singularly lacking in scientific accomplishments. One looks in vain for a single Hebrew scientist in the long ages down to the Roman destruction of Jerusalem (A.D. 70). I do not know of a single scientific discovery that is credited to the ancient Hebrews. Seemingly they put little store in such matters, for how else is one to explain the inclusion in Genesis of that part of the creation myth that has light created before there was a sun, or that the race was continued with only males, or any of the other numerous notions that must have been obviously false to the Hebrews by the time they finally began to assemble the Bible. It makes far more sense to me to believe that these ancient scholars simply were not enough interested in natural or scientific matters to think it necessary to expunge their ancient traditions of obvious errors. No doubt all races live happily with intellectual skeletons. No one today, at a time when genetics has reached such glorious heights, is upset if we speak of "our blood relations." Somehow that sounds more comfortable than "sharing the same genetic code."

I think that the most probable explanation of the creationists' demands is that neither they nor the Tennessee lawmakers have thought out the consequences of those demands. Had they done so, surely they would not wish science teachers to deal with these questions. To give "equal time and emphasis" to creation myths and to the biological theory of evolution must lead to the destruction of the former.

Quite possibly the creationists would say that I have not developed the topic along the lines that they wish. No doubt this is so. Their main activity for the past century has been to advance the creationist point of view, not by developing a creationist hypothesis, but by attacking the biological theory. Somehow they seem to work on the supposition that there are only two explanations, and that if you can cast sufficient doubt on one, the other is thereby established as true. There were uncertainties in Darwinian theory in 1859, and there are uncertainties today. Nevertheless, there has been a steady progress in understanding what all, with even a partially open mind, must admit. Creationism, on the other hand, has become ever more bankrupt as an explanatory hypothesis. More than a century ago Herbert Spencer remarked: "Those who cavalierly reject the theory of evolution as not adequately supported by facts seem quite to forget that their own theory is supported by no facts at all. Like the majority of men who are born to a given belief, they demand the most rigorous proof of any adverse belief, but assume that their own needs none."[9]

But we must remember that creationists have a strange relationship with what everyone else regards as facts. I have recently surveyed the creationists' arguments of a century ago and compared them to the present time. For the most part the same objections are being raised now as then to the biological theory of evolution. Seemingly the discoveries in the biological and physical sciences of the past century have made no impression. Each discovery of new evidence of the age of the earth, of fossil remains that give improved understanding of lineages, and of experiments dealing with the components of the evolutionary process is ignored or rejected. Seemingly there is no amount of data that will convince a creationist if he does not wish to be convinced. Not infrequently they behave as though they were adhering to the advice of Robert Owen—"Never argue: repeat your assertion."[10]

But, to a very limited extent, the creationists do more than argue. Recently the *New York Times* reported that the Institute for Creation Research is mounting an expedition to Mount Ararat to search for remnants of Noah's ark.[11] Previous attempts to secure the approval of the Turkish government had been unsuccessful, but now, hopefully, permission will be granted. The eight-man expedition is to be led by the son of the director of the Institute for Creation Research. The plan is to search for the remains of the ark near the 14,500 foot level of the mountain. I should like to offer a helpful suggestion: even the most elementary computations will show that, if the ark did what Genesis demands, it must have been so huge that Mount Ararat could easily rest on it, rather than it on Mount Ararat. Thus, I suggest that the expedition should look, not at the 14,500-foot level, but underneath the mountain.

REFERENCES

1. In September 1974 the Nashville Chancellery Court found this law to be unconstitutional [Editor's Note: See Jerry P. Lightner, "Tennessee 'Genesis Law'

Ruled Unconstitutional," *NABT News and Views* vol. 19, no. 2 (April, 1975). The Tennessee Law is included in this volume; see p. 385.]

2. The quotations are from a leaflet, "Creation Research Society."

3. A. A. Macintosh in *The Cambridge Bible Commentary on the New English Bible, Introductory Volume, The Making of the Old Testament* (Cambridge: Cambridge University Press, 1972), p. 162.

4. A. D. White, *A History of the Warfare of Science with Theology* (New York: Appleton, 1898).

5. White, 1898, vol. 1, p. 6.

6. G. A. Buttrick, ed., *The Interpreter's Bible* (New York: Abingdon, 1952), vol. 1, p. 465.

7. See G. H. Daniels, *Science in American Society: A Social History* (New York: Knopf, 1971), p. 98.

8. White, 1898, vol. 1, p. 9.

9. Quoted in J. R. Tompkins, ed., *D-Days at Dayton: Reflections on the Scopes Trial* (Baton Rouge, LA: Louisiana State University Press, 1965), p. 154.

10. Quoted in G.B.S., *Saturday Review* (April 29, 1950), p. 9.

11. Report published in Riverside Press, January 16, 1974.

Six "Flood" Arguments
Creationists Can't Answer

by Robert J. Schadewald

In pseudoscience, hypotheses are erected as defenses against the facts. Pseudoscientists frequently offer hypotheses flatly contradicted by well-known facts which can be ignored only by well-trained minds. Therefore, to demonstrate that creationists are pseudoscientists, one need only carry some creationist hypotheses to their logical conclusions.

Some years ago, NASA released the first deep-space photographs of the beautiful cloud-swirled blue-green agate we call Earth. A reporter showed one of them to the late Samuel Shenton, then president of International Flat Earth Research Society. Shenton studied it for a moment and said, "It's easy to see how such a picture could fool the untrained eye."

Well-trained eyes (and minds) are characteristic of pseudoscientists. Shenton rejected the spherical earth as conflicting with a literal interpretation of the Bible, and he trained his eyes and his mind to reject evidence which contradicted his view. Scientific creationists must similarly train their minds to reject the overwhelming evidence from geology, biology, physics and astronomy which contradicts their interpretation of the Bible. In a public forum, the best way to demonstrate that creationism is pseudoscience is to show just how well-trained creationist minds are.

Pseudoscience differs from science in several fundamental ways, but most notably in its attitude toward hypothesis testing. In science, hypotheses are ideas proposed to explain the facts, and they're not considered much good unless they can survive rigorous tests. In pseudoscience, hypotheses are erected as defenses against the facts. Pseudoscientists frequently offer hypotheses flatly contradicted by well-known facts which can be ignored only by well-trained minds. Therefore, to demonstrate that creationists are pseudoscientists, one need only carry some creationist hypotheses to their logical conclusions.

1. The Karroo Formation. Scientific creationists interpret the fossils found in the earth's rocks as the remains of animals which perished in the Noachian Deluge. Ironically, they often cite the sheer number of fossils in "fossil graveyards" as evidence for the Flood. In particular, creationists seem enamored of the Karroo Formation in Africa, which is estimated to contain the remains of 800 billion vertebrate animals.[1] As pseudoscientists, creationists dare not test this major hypothesis that all of the fossilized animals died in the Flood.

Robert E. Sloan, a paleontologist at the University of Minnesota, has studied the Karroo Formation. He told me that the animals fossilized there range from the size of a small lizard to the size of a cow, with the average animal perhaps the size of a fox.[2] A minute's work with a calculator shows that, if the 800 billion animals in the Karroo Formation could be resurrected, there would be 21 of them for every acre of land on earth. Suppose we assume (conservatively, I think) that the Karroo Formation contains 1% of the vertebrate fossils on earth. Then when the Flood began there must have been at least 2100 living animals per acre, ranging from tiny shrews to immense dinosaurs. To a noncreationist mind, that seems a bit crowded.

I sprang this argument on Duane Gish during a joint appearance on WHO Radio in Des Moines, Iowa, on October 21st, 1980. Gish did the only thing he could: he stonewalled by challenging my figures, in essence calling me a liar. I didn't have a calculator with me, but I duplicated the calculation with pencil and paper and hit him with it again. His reply? Creationists can't answer everything. It's been estimated that there are 100 billion billion herring in the sea. How did *I* account for *that?!* Later, I tried this number on a calculator and discovered that it amounts to about 27,000 herring *per square foot* of ocean surface. I concluded (a) that all of the herring are red, and (b) that they were created *ex nihilo* by Duane Gish on the evening of October 21st, 1980.

2. Marine fossils. The continents are, on average, covered with sedimentary rock to a depth of about one mile. Some of the rock (chalk, for instance) is essentially 100% fossils and many limestones also contain high percentages of marine fossils. On the other hand, some rock is barren. Suppose that, on average, marine fossils comprise .1% of the volume of the rock. If all of the fossilized marine animals could be resurrected, they would cover the entire planet to a depth of at least 1.5 feet. What did they eat?

Creationists can't appeal to the tropical paradise they imagine existed below the pre-Flood canopy because the laws of thermodynamics prohibit the earth from supporting that much animal biomass. The first law says that energy can't be created, so the animals would have to get their energy from the sun. The second law limits the efficiency with which solar energy can be converted to food. The amount of solar energy available is not nearly sufficient.

3. The Green River Formation. The famous Green River formation covers tens of thousands of square miles. In places, it contains about 20 million varves, each varve consisting of a thin layer of fine light sediment and an even thinner layer of finer dark sediment. According to the conventional geologic interpretation, the layers are sediments laid down in a complex of ancient freshwater lakes. The coarser light sediments were laid down during the summer, when streams poured run-off water into the lake. The fine dark sediments were laid down in the winter, when there was less run-off. (The process can be observed in modern freshwater lakes.) If this interpretation is correct, the varves of the Green River formation must have formed over a period of 20 million years.

Creationists insist that the earth is no more than 10,000 years old, and that the geologic strata were laid down by the Flood. Whitcomb and Morris therefore attempt to attribute the Green River varves to "a complex of shallow turbidity currents . . ."[3] Turbidity currents, flows of mud-laden water, generally occur in the ocean, resulting from underwater landslides. If the Green River shales were laid down during the Flood, there must have been 40 million turbidity currents, alternately light and dark, over about 300 days. A simple calculation (which creationists have avoided for 20 years) shows that the layers must have formed at the rate of about three layers every two seconds. A sequence of 40 million turbidity currents covering tens of thousands of square miles every two-thirds of a second seems a bit unlikely.

Henry Morris apparently can't deal with these simple numbers. Biologist Kenneth Miller of Brown University dropped this bombshell on him during a debate in Tampa, Florida, on September 19th, 1981, and Morris didn't attempt a reply. Fred Edwords used essentially the same argument against Duane Gish in a debate on February 2, 1982. In rebuttal, Gish claimed that some of the fossilized fishes project through several layers of sediment, and therefore the layers can't be semiannual. As usual, Gish's argument ignores the main issue, which is the alleged formation of millions of distinct layers of sediment in less than a year. Furthermore, Gish's argument is false, according to American Museum of Natural History paleontologist R. Lance Grande, an authority on the Green River Formation. Grande says that while bones or fins of an individual fish may cut several layers, in general each fish is blanketed by a single layer of sediment.[4]

4. Disease Germs. For numerous communicable diseases, the only known "reservoir" is man. That is, the germs or viruses which cause these diseases can survive only in living human bodies or well-equipped laboratories. Well-known examples include measles, pneumococcal pneumonia, leprosy, typhus, typhoid fever, small pox, poliomyelitis, syphillis and gonorrhea. The scientific creationists insist on a *completed* creation, where the creator worked but six days and has been resting ever since. Thus, between them, Adam and Eve had to have been created with every one of these diseases. Later, somebody must have carried them onto Noah's Ark.

Note that the argument covers *every* disease germ or virus which can survive only in a specific host. But even if the Ark was a floating pesthouse, few of these diseases could have survived. In most cases, only two animals of each "kind" are supposed to have been on the Ark. Suppose the male of such a pair came down with such a disease shortly after the Ark embarked. He recovered, but passed the disease to his mate. She recovered, too, but had no other animal to pass the disease to, for the male was now immune. Every disease for which this cycle lasts less than a year should therefore have become extinct!

Creationists can't pin the blame for germs on Satan. If they do, the immediate question is: How do we know Satan didn't create the rest of the universe? That has frequently been proposed, and if Satan can create one thing, he can create another. If a creationist tries to claim germs are mutations of otherwise benign organisms (degenerate forms, of course), he will actually be arguing for evolution. Such hypothetical mutations could only be considered favorable, since only the mutated forms survived.

5. Fossil Sequence. At all costs, creationists avoid discussing how fossils came to be stratified as they are. Out of perhaps thousands of pages Henry Morris has written on creationism, only a dozen or so are devoted to this critical subject, and he achieves that page count only by recycling three simple apologetics in several books. The mechanisms he offers might be called victim habitat, victim mobility, and hydraulic sorting. In practice, the victim habitat and mobility apologetics are generally combined. Creationists argue that the Flood would first en-gulf marine animals, then slow lowland creatures like reptiles, etc., while wily and speedy man escaped to the hilltops. To a creationist, this adequately explains the order in which fossils occur in the geologic column. A scientist might test these hypotheses by examining how well they explain the fact that flowering plants don't occur in the fossil record until early in the Cretaceous era. A scenario with magnolias (a primitive plant) heading for the hills, only to be overwhelmed along with early mammals, is unconvincing.

If explanations based on victim habitat and mobility are absurd, the hydraulic sorting apologetic is flatly contradicted by the fossil record. An object's hydrodynamic drag is directly proportional to its cross sectional area and its drag coefficient. Therefore when objects with the same density and the same drag coefficient move through a fluid, they are sorted according to size. (Mining engineers exploit this phenomena in some ore separation processes.) This means that all small trilobites should be found higher in the fossil record than large ones. That is not what we find, however, so the hydraulic sorting argument is immediately falsified. Indeed, one wonders how Henry Morris, a hydraulic engineer, could ever have offered it with a straight face.

6. Overturned Strata. Ever since George McCready Price, many creationists have pointed to overturned strata as evidence against conventional geology. Actually, geologists have a good explanation for overturned strata, where the normal order of fossils is precisely reversed. The evidence for folding is usually obvious, and where it's not, it can be inferred from the reversed fossil order. But creationists have no explanation for such strata. Could the Flood suddenly reverse the laws of hydrodynamics (or whatever)? All of the phenomena which characterize overturned strata are impossible for creationists to explain. Well-preserved trilobites, for instance, are usually found belly down in the rock. If rock strata containing trilobites are overturned, we would expect to find most of the trilobites belly up. Indeed, that is what we do find in overturned strata. Other things which show a geologist or paleontologist which way is up include worm and brachiopod burrows, footprints, fossilized mud cracks, raindrop craters, graded bedding, etc. Actually, it's not surprising that creationists can't explain these features when they're upside down; they can't explain them when they're right side up, either.

Each of the six preceding arguments subjects a well-known creationist hypothesis to an elementary and obvious test. In each case, the hypothesis fails miserably. In each case, the failure is obvious to anyone not protected from reality by a special kind of blindness.

Studying science doesn't make one a scientist any more than studying ethics makes one honest. The studies must be applied. Forming and testing hypotheses is the foundation of science, and those who refuse to test their hypotheses cannot be called scientists, no matter what their credentials. Most people who call themselves creationists have no scientific training, and they cannot be expected to know and apply the scientific method. But the professional creationists who flog the public with their doctorates (earned, honorary, or bogus) have no excuse. Because they fail to submit their hypotheses to the most elementary tests, they fully deserve the appellation of pseudoscientist.

REFERENCES

1. See John C. Whitcomb and Henry M. Morris, *The Genesis Flood* (Philadelphia: Presbyterian and Reformed Publishing Co., 1961), p. 160; and Duane T. Gish, *Evolution: The Fossils Say No!* (San Diego, CA: Creation-Life Publishers, 1978), p. 61.

2. Robert E. Sloan, personal communication, 1980, 1982.

3. Whitcomb and Morris, 1961, p. 427.

4. R. Lance Grande, personal communication, 1982.

Creationism and Evolution: The Real Issues

by N. Patrick Murray and Neal D. Buffaloe

The dispute is not really over biology *or* faith, but rather is essentially about *Biblical interpretation*. At issue here are two irreconcilable viewpoints regarding the characteristics of Biblical literature, and more comprehensively the nature of Biblical authority. The doctrine of 'Scientific Creationism' has its roots in the Fundamentalist movement, with its absolutely central dogma of Biblical literalism.

A heated controversy has arisen in recent years centering around the theory of evolution, an explanatory model widely taught and accepted in biological science. The famous Scopes trial in Tennessee in 1925 by no means laid this issue to rest. After a period of relative calm, the rise of a more militant religious and political conservatism has once again thrust the name of Charles Darwin and the subject of evolution to the forefront of discussion. From Virginia to California the question of biological origins once again commands front-page and prime-time attention. Current intellectual trends suggest that the conflict is not likely to abate in the foreseeable future, but that it may escalate into one of the far-reaching sociocultural debates of our day.

In widely separated areas school boards wrestle with the question of whether to require the teaching of so-called Creationism alongside the standard biological orthodoxy represented by evolutionary theory. (From here on in this essay, the term "Creationism" refers to the viewpoint that the literal Biblical account of creation is the correct explanation for the origin of the earth and its living forms.) In March, 1981, a case reached the California Superior Court in which the plaintiff argued that the teaching of evolution violates the constitutional rights of certain religious groups who believe the "Biblical account" of creation. The writer of a recent *Time* article makes the observation that school textbook publishers, fearing jeopardy to lucrative markets, are drastically curtailing the space devoted to explanation of evolutionary theory, fossil formation, and even Darwin's life.[1]

In March, 1981, the Arkansas Legislature passed a bill (Act 590) requiring "balanced treatment of creation-science and evolution-science in public schools."[2] Thus Arkansas became the first state to enact such a law, although proponents of Creationism are waging a vigorous campaign to have similar laws passed by other states.

This issue is a potentially explosive one that carries grave implications for education and religion in the United States. It seems likely that many rational people may be further alienated from religion as they react against a new wave of anti-intellectualism. It is therefore a matter of urgency that the elements in this controversy be identified and clarified as fully as possible.

WHAT THE ISSUE IS *NOT*

The Issue Is Not Biology

There is no revival of debate about the scientific validity of the evolutionary model within any mainstream scientific community. Darwin's thesis, which orginated in the 19th century, has certainly undergone modification since he proposed it. However, it is accepted by virtually the entire scientific community. By contrast, so-called "Scientific Creationism" is the intellectual product of a few isolated "research centers" and "institutes" that are totally without standing in recognized scientific circles. They simply are not taken seriously by the overwhelming majority of biologists. In fact, the "research" conducted at these centers is directed toward the discovery of inconsistencies in the data supporting evolution and the publication of arguments for the Creationist viewpoint. It is not research in the traditional scientific sense; it involves no direct observation and experimentation and is not subjected to peer review by the entire scientific community.

In his epochal book, *Origin of Species* (1859), Darwin sought to establish two major tenets: (1) *that* evolution has occurred and is occurring, and (2) *how* evolution occurs. He was highly successful with regard to the first of these two points; his arguments in the *Origin* carried the weight of a boulder within the scientific community of his day. His explanation for the second point was that the major factor in evolution is natural selection. Certainly, Darwin's "how" has been modified with further research. However, there is no real disagreement among modern biologists concerning Darwin's "that." In fact, continuing research has strengthened the original hypothesis immeasurably. The evidence is overwhelming that present-day organisms are the product of long ages of descent with modification. Of course, Darwin was a pioneer who made a pioneer's first approximations in his statements and conclusions. But modern evolutionary theory, far from repudiating Darwin, is actually an extended footnote to it. The celebrated geneticist Theodosius Dobzhansky has stated the prevailing consensus succinctly: "Nothing in biology makes sense except in the light of evolution."[3]

The major thrust of Creationist argument is a questioning of the validity of evidences for evolution from the various fields of science. Fossil evidences, the reliability of radioactive dating methods, and the nature of the applicability of the second law of thermodynamics to living systems are questioned most particularly. Such questioning stems from a theological bias, which most Creationists readily admit: A desire (1) to discredit the hypothesis that life has existed on earth for more than a few thousand years, and (2) to show that *Homo sapiens* came into existence instantaneously. This theological bias stems, in turn, from a literal interpretation of the Genesis account of creation, about which we shall later comment more fully.

No attempt will be made here to review critically the Creationist arguments. This has been done elsewhere in detail.[4] Suffice it to say that the scientific community as a whole regards the evidences of Creationists for their major tenets as quaint, at best, ranking approximately with the evidences that the Flat Earth Society sets forth for its viewpoint. Creationist literature is heavily dependent on scientific data and theories that have long been discredited by the mainstream scientific community, and almost universally betrays a serious misunderstanding of the scientific issues under consideration.

The Issue Is Not Faith in God

Though largely unexplored, there exists a wide body of doctrinal agreement between Creationists and Theists who accept evolutionary theory. The theological affirmations of the two groups, while no doubt different in wording, overlap on many basic points.

First, both certainly affirm that the universe is not self-explanatory, but owes its origin to a transcendent Creator, who is the single source of all being. This could, in fact, stand as a reasonably adequate definition of "Theism," the religious outlook common to Christianity, Judaism, and Islam. The existence of God, the Creator, is in no way at issue here. Not all of these groups, of course, make use of traditional creeds, but few Theists of whatever stripe would quibble with the opening words of the Nicene Creed: "I believe in one God the Father Almighty, Maker of heaven and earth, and of all things visible and invisible."

Second, all Theistic groups attribute a place of special significance to the creation of man. The Biblical tradition from beginning to end consistently places man at the summit of creation, describing his divinely assigned role in a multitude of ways. The concept of the "image of God" voiced in the Genesis account has certainly been one of the most significant, producing a vast literature of theological reflection over the centuries. However interpreted, it certainly establishes common ground among Theists regarding the *discontinuity* between the human species and all other orders of creation. Man is a product of God's special intention and purpose, having qualities not paralleled in any other aspect of the created order. This unique kinship with God also has been expressed as man's having an "eternal soul," possessing the gift of reason, being endowed with a sense of moral duty, and many others.

Theists who accept the evolutionary explanation of man's biological origins make no compromise with this basic tenet. Whatever the "mechanics" of human origins, they affirm that man stands as the "crown of creation," the highest fulfillment of God's creative purpose, and is the creature called into an absolutely distinctive fellowship with God. This doctrine, like that of God's existence, *is unaffected by the theory of evolution as interpreted from the perspective of Theistic belief.*

Perhaps these two examples will suffice to make the point under consideration here, namely, that the current debate about "evolution vs. creation" is not primarily concerned with faith in God's existence and his creation of the world. The main tenets of Theism are not at stake, and in fact, the disputants in this particular issue share a fundamental agreement about the creedal essentials that traditionally define Theistic faith.

WHAT THE CENTRAL ISSUE *IS*

The dispute is not really over biology *or* faith, but rather is essentially about *Biblical interpretation.* At issue here are two irreconcilable viewpoints regarding the characteristics of Biblical literature, and more comprehensively the nature of Biblical authority. The doctrine of "Scientific Creationism" has its roots in the Fundamentalist movement, with its absolutely central dogma of Biblical literalism. This movement has its origins in the numerous conservative splinter groups of the Reformation, many of which moved rapidly in the 16th and 17th centuries to the substitution of another absolute authority (the Bible) for the lately-repudiated papal authority of Roman Catholic tradition. The resulting dictum is that every word of the Bible is equally divinely inspired and the Biblical record is therefore to be taken with uniform literalness throughout. Furthermore, the Biblical writings stand as absolute authority in all matters on which they touch, whether moral, religious, political, historical, or scientific. When this literalist doctrine of Biblical interpretation is applied to the early creation stories in Genesis 1–3, the elements of "Creationism" emerge.

The Three Basic Elements of "Scientific Creationism"

(1) *The Age of Creation.* The earth is declared to be only a few thousand years old, probably not more than 10,000, as compared with the modern scientific view that it is approximately 4.5–4.7 billion years old. Fundamentalists arrive at their figures by totaling up the life-spans of the ancient Hebrew patriarchs, whom they regard as historical figures, as recorded in the early pages of Genesis. By contrast, the mainstream of Biblical scholarship rejects the literal historicity of the Genesis stories prior to Chapter 12, and finds the literature of parable and symbol in the early chapters of Genesis.

(2) *The Time of Creation.* Included in the Fundamentalist concept, of course, is a literal interpretation of the "seven days" of creation as found in the first chapter of Genesis. Although some Creationist interpreters may go so far as to concede that the seven days were not necessarily 24-hour periods, they insist, nonetheless, on the scientific accuracy of the *order* of creation in these stories. Even conceding seven "epochs" instead of days, Creationists would not extend the period required for the fossil records. For God to have delayed the creative process so extensively would have been wasteful and unnecessary, given the premise of God's direct creative acts at each stage.

(3) *The Mode of Creation.* The literalist view requires the notion of numerous direct or "special" creative acts of God, so that there is no necessary continuity even between the seven periods or "days." In fact, discontinuity would be expected, so as to demonstrate more clearly God's immediate and direct intervention. Naturally, these interpreters exploit to the fullest any existing gaps or discrepancies among fossil records and methods of dating.

By contrast, Theists who embrace the evolutionary hypothesis posit a creator who initiates the order of nature and invests it with a limited autonomy. He could be described, perhaps, as the Overseer of the long, natural evolutionary spiral, making use of its semi-autonomous products in a kind of creative interplay. To Theists who hold this viewpoint, God's creation is just as remarkable and just as deserving of awe and reference as if he had created living forms directly and instantaneously.

The Fundamental Issue

Now we can isolate the essential element that separates Creationists from other Theists. That element is the different way in which they treat the ancient literary forms in the early chapters of Genesis. The literalist view regards these ancient narratives as *scientific* documents that actually describe the way in which the physical order originated. They are seen as *competing* with contemporary scientific accounts, and a choice between the two is required. The ancient Genesis stories and modern scientific explanations are regarded as having the same basic purpose: to *describe* how the universe came about. Since these two descriptions stand in marked contrast, both cannot be true—hence the court cases and school board controversies about "equal time" for Creationist views alongside evolutionary concepts.

The motives behind this movement by the conservative religious groups are undoubtedly well-intentioned. Their concern about evolution reflects an evident desire to counteract the powerful materialistic and secularistic currents of our day. They apparently hope to establish respectability for the "Biblical view" by amassing as many negative data as possible against evolutionary hypotheses, and, in addition, by arguing as forcefully as possible for the scientific plausibility of the Genesis story as a valid description of natural and human origins. While respecting the sincerity of those laboring so diligently toward this end, we must at the same time lament the profound misapprenhension upon which it all rests. The effort is not only foredoomed to failure, since the weight of scientific knowledge against the Creationist notion is truly overwhelming, but it also threatens to damage much painstakingly-wrought dialogue between the scientific and religious communities. It will appear to many informed persons as a new wave of anti-intellectualism with the likely result of a serious discreditation of religious faith.

AN ALTERNATIVE TO LITERALISM

Why, then, are mainstream Christians and other Theists not threatened by the modern expression of evolutionary theory, and why are they convinced that the current crusade against evolution is entirely wrongheaded? The answer lies in the fundamentally different approach to Biblical interpretation characteristic of the mainstream groups. We have observed that the "scientific" arguments of the Creationist groups are rejected by virtually all biological scientists, and that these arguments founder against the shoals of any objective analysis of modern biological and paleontological evidence. Similarly, the method of Biblical interpretation represented by the Creationists is not accepted among reputable Biblical scholars and theological schools in the mainstream of either Protestantism, Roman Catholicism, or Judaism.

Advances in Biblical scholarship, principally through the study of archeology, comparative literature, and ancient languages, have produced during approximately the last century a much better understanding of the nature and origins of many Biblical writings. This body of objective data has come to be universally accepted in standard theological circles. The first chapter of Genesis, as one important example of these advances, has been shown to have a completely different style and vocabulary from the creation story found in Chapters 2 and 3. Because the Genesis 1 story is believed to have been written by the priestly scholars of late Judaism around 500–400 B.C., it is generally referred to as the "Priestly," or "P," account. Beginning in Chapter 2 (2:4) a different writer takes over, probably dating from around 950 B.C. This passage is often called the "Yahwist," or "J," account, because of the writer's consistent preference for the Hebrew term "Yahweh" as the name for God. In other words, Genesis contains two different creation accounts, and the "J" account is believed to antedate the "P" account by several centuries, despite the order of their appearance in our Old Testament. We will devote our attention here to the Priestly account in Genesis 1, because it is the one that contains the famous "seven-day" description of creation.

It can easily be shown that the Priestly account is closely related to an earlier Babylonian creation account, known as the *Enuma elish.* The general order in which things were created, for example, is the same in the two accounts. However, there are very dramatic and significant differences between the ancient Babylonian story and the Hebrew version, and it is clear how the Hebrew writer adapted the well-known Middle Eastern story for his purposes. The Babylonian tale is thoroughly polytheistic, depicting the creation as emerging from violent conflict among the gods. The Hebrew writer has greatly refined the account, and of course, has adapted it to strictly monotheistic theology. Bernard M. Anderson, a leading Old Testament scholar whose textbook is one of the most widely used in American colleges and seminaries, has summarized clearly the significance of the Genesis version:

> But the Priestly account is not a treatise on scientific origins. Here the poetry of faith speaks of something that lies behind or beyond human experience and scientific inquiry: the origination and ordering of all that is by the sovereign, initiating will of the Creator (Job 38:4–7). Unlike ancient polytheistic myths, which depicted the birth of the gods out of the intermingling waters of chaos, this liturgy

affirms the holy transcendence of the Creator, who originated the cosmos in the beginning and upon whose sovereign will all creatures, terrestrial and celestial, are dependent for their being.[5]

In other words, the aim of the ancient Hebrew writers was to establish and nurture monotheistic faith in the one supreme God who is Creator of all that exists, and to fire a broadside against the degraded polytheistic rivals to Hebrew faith in the surrounding cultures. This is accomplished with compelling literary and theological purity. Arthur Smethurst, an authority on Hebrew writing whose *Modern Science and Christian Beliefs* is something of a classic, comments on the early chapters of Genesis as follows:

> These chapters are creation *epics,* or *poems.* They were written as, and intended to be interpreted as, poetry. They were imaginative creations, in poetic language, and no Oriental people would attempt to interpret them any other way. Great mischief has been caused by the misrepresentation of them by totally unimaginative and literal English and Western minds . . . the purpose of those who wrote them was never to give a precise scientific account of the Creation, but rather to enshrine in splendid language and rich symbolism the profound doctrine of God, the Creator.[6]

A helpful contribution to our understanding of the Genesis creation stories was made some years ago by B. D. Napier, formerly Professor of Old Testament at Yale University and more recently at Stanford University. Dr. Napier offers an approach to these accounts that sets them free once again to speak compellingly to the modern scene. Quoting the German novelist, Thomas Mann, Dr. Napier says of the type of literature found in early Genesis, "It is, it always is, however much men try to say it was." In other words, the Genesis poems are significant not because they tell us how things *were,* or the way things happened long ago. Rather, they are talking about man's situation *now*—the eternal importance of man's relationship to God, and the primordial disruption of that fellowship that lies at the root of human nature and history. When we read the ancient Hebrew accounts of the creation—Adam and Eve, the Garden of Eden, man's "fall" by listening to the seductive words of a serpent, and God's Sabbath rest —we must understand, says Napier, that "these things never were, but always are . . . the stories are told and retold, recorded and read and reread not for their *wasness* but for their *isness.*"[7]

In other words, the Hebrew account is a beautiful devotional and theological poem addressed to faith in all ages. It seeks to establish a relationship between God and man at the level of feeling and hope by showing that the life of mankind and the realm of nature in which humanity lives are not blind confluences of indifferent forces, but purposeful designs of a loving God with whom humans can have fellowship. The Hebrew poet had a far greater purpose in mind than merely giving information about the "mechanics" of how nature originated. He could not even have conceived of his writing in terms of modern scientific theories. This pre-scientific account is an *interpretation of the essential and universal meaning of human existence,* not a technical treatise whose aim is to delineate the origins, structure, and properties of the physical order of nature. If that simple yet absolutely pivotal point were accepted by Creationists, the whole unfortunate controversy would evaporate overnight. Instead Creationists harness Theistic faith both to a hopelessly discredited methodology of Biblical interpretation and to a similarly absurd attempt to compete with modern scientific knowledge. The upshot of this effort is not an enhancement of the Biblical message in the eyes of contemporary culture, but an obscuring of just those features of the ancient accounts that have a universal and permanent significance.

To the literalist, it is profoundly threatening to Biblical authority if different strata and types of literature are found in the Biblical record. The interpretation of the early Genesis stories as the literature of saga and parable is regarded as demeaning their power and relevance. The watchword of the literalist viewpoint is formalized in the slogan, "To take the Bible seriously is to take it literally." The application of this naive and misguided premise, however, is absolutely fatal to the power and authority of the overall Biblical message. Ironically, what is intended as a recovery of the appeal of Biblical faith issues finally in a kind of myopia with regard to the essential message of these magnificent passages.

SCIENCE AND WORLDVIEWS

As is true of all scientific knowledge, the concept of evolution is theologically neutral. It has to be *interpreted*—that is, put into some larger framework of meaning and explanation, as do all the raw data of experience. For example, if a person smiles at us, that is simply a physical event that occurs at the level of *observation.* At the level of *interpretation,* however, we have to assess the *meaning* of the smile. We do this by calling upon some wider frame of reference, *not directly given in the smile-event itself,* about the person smiling and the situation in which the event occurs. If we have just entered a store, we may interpret the smile as meaning merely that the approaching salesperson is eager to sell us something. But if the salesperson also happens to be someone whom we regard as a friend, we will probably interpret and respond to the smile quite differently. In either case, we must necessarily draw upon a *larger* framework of understanding, accurately or not, to assess the *meaning* of the smile. So it is with interpreting scientific data.

In themselves, scientific discoveries about the operations of nature neither affirm nor deny the existence of God. They merely describe how physical things work. On the one hand, they can be interpreted as manifesting the designful handiwork of God. The early astronomer Johannes Kepler (1571–1630) said as he formulated the mathematical rules describing interrelationships among the stars, "I am but thinking God's thoughts after him." Charles Darwin acknowledged a Creator, and closed the *Origin of Species* with this statement:

> There is grandeur in this view of life, with its several powers, having been originally breathed by the Creator into a few forms or into one; and that, whilst this planet has gone cycling on according to the fixed law of gravity, from so simple a beginning endless forms most beautiful and most wonderful have been, and are being evolved.

On the other hand, astronomy, biology, atomic theory, or any field of scientific inquiry whatever can be interpreted non-theistically as merely the tabulation of the blind principles of nature which emerged over eons of time by sheer random chance. Perhaps the most eloquent and impassioned spokesman in modern times for this latter view is Bertrand Russell (1872–1970):

> That man is the product of causes which had no prevision of the end they were achieving; that his origin, his growth, his hopes and fears, his loves and his beliefs, are but the outcome of accidental collocations of atoms; that no fire, no heroism, no intensity of thought and feeling can preserve an individual life beyond the grave; that all the labours of all the ages, all the devotion, all the inspiration, all the noonday brightness of human genius are destined to extinction in the vast death of the solar system, and that the whole temple of man's achievement must inevitably be buried beneath the debris of a universe in ruins—all these things, if not quite beyond dispute, are yet so nearly certain that no philosophy which rejects them can hope to stand.[8]

Obviously, there is quite a wide distance between Kepler and Darwin on the one hand and Russell on the other, representing the polarity between Theism and Atheism. Each man has posited a *worldview,* an overall interpretation of the ultimate meaning of life in the universe. Worldviews, in the nature of things, cannot finally be proven or disproven by the accumulation of facts or information, because bare facts are not self-interpreting. Worldviews are holistic "leaps of faith," to use an often-misconstrued phrase, and all three of these men— Kepler, Darwin, and Russell—have made such an interpretive leap.

Widely divergent worldviews have persisted and have been held by sincere and rational people throughout the ages. It is difficult to analyze just how and why we formulate our respective faith-perspectives as we do. Why is one person such a profoundly convinced Theist, and another an equally avid Materialist or Secularist? We cannot adequately answer, but it is abundantly clear that two equally intelligent and informed persons can stand before the same body of data and interpret its worldview-meaning in opposite ways.

Nowhere does this seem to be more clearly borne out than in the evolution controversy. Some persons examine the pattern of evolution and conclude that living species, including mankind, emerged through blind, mechanistic forces totally devoid of any direction or purpose. Evolutionary principles are perceived as operating in total indifference to any values—aesthetic, moral, or otherwise—and without the guidance of any conscious or rational principle whatsoever. The crucial point in the current controversy, however, is that evolutionary theory does not *have* to be so interpreted. There is no inherent logical or epistemological principle in evolutionary data that requires a leap into a non-theistic worldview. One of our main purposes here, in fact, is to show that the vast mainstream of Theistic interpretation has long ago assimilated the concept of evolution into its faith-perspective, along with modern astronomy, the atomic theory, and other scientific findings.

Thus, no matter what the advances of science *as such,* there is no threat to faith. Of course, many modern scientists, as well as non-scientists, continue to *interpret* our vastly expanded knowledge of the workings of nature in non-theistic ways. Theism naturally must remain in constant critical dialogue with these opposing *worldviews,* seeking always to expose what it believes to be their inadequacy to interpret the whole story. Tension will probably always exist between proponents of these divergent worldviews, as has been well expressed by the theologian Reinhold Niebuhr:

> . . . the religious symbols of ultimate meaning are poetic rather than exact and scientific, and the fearfully pious are always tempted to buttress their validity by a frantic adhesion to some outmoded science, against the challenge of a marching science, which always has immediate truth on its side but which always threatens to construct a scientific world picture in which no meaning can be found for man in his grandeur and his misery.[9]

With regard to the evolution debate, the enlightened theist will surely be heard finally to say, "What is all the fuss about? If God chose to use the process of evolution over a span of billions of years, isn't that his prerogative as sovereign Creator?" As more and more is learned through scientific research, biological or otherwise, the mainstream Theist sees increasing affirmation of the grandeur and awesome complexity of the Creator's design. Science continues only and ever to affirm what the Hebrew Psalmist said: "The heavens declare the glory of God, and the firmament showeth his handiwork." (Psalm 19:1; see also Job 38–39) Thus the findings of evolution, in striking contrast to the ways Creationists react to them, can even become a subject for devotional reflection. This is seen in the words of a Benedictine devotional writer, Maria Boulding:

> Creation is the more glorious in our eyes now that we know that God did not produce things ready-made but empowered them to make themselves and transcend their origins, and that by these processes he created the mind of man with responsibility to shape the world.[10]

Far from undermining the canons of Theistic faith, the long eons of evolutionary meandering can be viewed as a kind of expression of the creative "playfulness" of God. We are led

> into perpetual wonder at the huge array of God's creative ideas embodied in beings of all shapes and sizes and colours and aptitudes, with their ingenious methods of locomotion, feeling, fighting, reproducing themselves, caring for their young and simply enjoying life. When to the contemporary spectacle is added what we know of earlier sports of creative joy now left behind in the evolutionary process, creation appears as a very surprising affair.[11]

Rather than posing a threat to faith, then, evolution becomes for most modern Theists but another way of "thinking God's thoughts after him."

WHAT BIBLICAL LITERALISM LEADS TO

When consistently applied, the literalist approach to Biblical interpretation leads to a maze of difficulties. One of the best ways to demonstrate this is to examine the "blueprint," or "model," of the universe that is found in Genesis 1 and throughout the Bible. This concept of how the universe is built was common to all ancient peoples and was simply taken for granted by the Hebrews, who undoubtedly adopted it from their Middle Eastern cultural environment. In fact, it was the standard way of viewing the universe in Western culture until Copernicus and Galileo challenged it in the 16th century. It is quite clearly outlined in verses 6–10 of Chapter 1:

Then God said, "Let there be a dome in the middle of the waters, to separate one body of water from the other." And so it happened: God made the dome, and it separated the water above the dome from the water below it. God called the dome "the sky." Evening came, and morning followed—the second day. Then God said, "Let the water under the sky be gathered into a single basin, so that the dry land may appear." And so it happened: the water under the sky was gathered into its basin, and the dry land appeared. God called the dry land "the earth," and the basin of water he called "the sea." (New American Bible)

Here we see plainly set forth the basic structure of the pre-scientific view of the universe: the earth is essentially a flat plain, partly covered by water, and over the earth is a great dome, the sky or "heavens." The Genesis account explains that there is a vast reservoir of water collected above the dome ("firmament"), which of course is how ancient peoples accounted for rainfall. This picture of the universe is presupposed throughout the Bible.

Though it does not appear in the Genesis account, the concept of an underworld or nether region is an additional important feature of the ancient model. This region under the earth was a kind of repository for the "shades," or ghosts, of the dead. The Hebrew term used to describe the underworld was *Sheol,* and it corresponds with the Greek concept of *Hades,* which has a prominent place in Greek mythology. (Since this notion of an underground place of the dead has no counterpart in modern thought, and hence no English term to translate it, the words *Sheol* and *Hades* are commonly transliterated in modern Bible translations, a practice much to be commended for its accuracy. In no case should these terms be translated "Hell," an entirely different concept of later origin.)

The ancient Biblical picture of the world is commonly termed the "three-story" or "three-tiered" view of the universe. (For readers who wish to explore the numerous references to this model, any good annotated Bible will supply cross-references, or refer to "Sheol," "Hades," and "Heavens" in a modern concordance. See, for example, Genesis 28:12, Exodus 20:4, I Kings 8:35, Job 11:7–8, Psalms 78:23–24, 139:8, 148:4, Isaiah 7:10, 14:13–15, Ezekiel 26:19–20, Amos 9:2, Matthew 11:23, Romans 10:6–7, Philippians 2:10.) According to this model, the universe consists essentially of the sky-dome or "heavens" above; the flat earth stretched out beneath; and the underworld, pictured something like underground caverns. The ancients envisioned this whole world-structure, finally, as floating in a vast ocean. An added touch was that the heavenly bodies—sun, moon, and stars—rolled across the underside of the sky-dome, being attached to it in some fashion. (In the art and story of ancient cultures, considerable ingenuity was expressed in explaining how the sun traveled from west to east under the earth overnight.)

The Hebrew mind naturally had its own application of the ancient model. The region above the sky-dome, the "heavens," was regarded as the dwelling place of God, who created and placed man upon the great plain of earth, originally a beautiful garden ("Eden"), to nurture, enjoy, and replenish it. Man was to live in fellowship with God, who viewed his activities from his lofty vantage point and interacted quite directly with man on occasion, particularly on the heights of mountains, viewed as man's closest access to God. (The "Tower of Babel" story in Genesis 11 is one of the Hebrew parables depicting the "fall of man, who tries to build a great mountain-tower of access to the third tier of the universe for the purpose of invading the heavenly kingdom and usurping God's power.)

Our purpose in the foregoing discussion is to establish a point that is beyond any reasonable question: The Bible expounds a version of the ancient pre-scientific view of the universe common to all people of the period in which the Bible was written (ca. 1100 B.C.–150 A.D.). Now, the critical question emerges: What is the modern Biblical interpreter to do with this charming, but absolutely outmoded, view of the world? If the literalist is to remain consistent, he now faces the insurmountable difficulty of having to accept it. But surely the Creationists, so bent on propounding the "seven-day" doctrine of creation, cannot swallow *also* the three-storied picture of the universe that appears along with it! Yet, by the literalist's own methodology, one cannot "pick and choose" in Biblical interpretation; one must treat it all as uniformly authoritative. There simply is no logical escape-route for the literalist interpreter. By what conceivable principle can he accept the seven-day creation, yet reject the three-story cosmology that is intertwined with it?

Thus, Creationists are in the position of having accepted at least one modern scientific viewpoint in clear contradiction to the ancient view of the universe set forth in the Bible. But how can this be, given their literalist principle that Biblical accounts are to be regarded as scientific descriptions of the origin of things, and must be placed into direct competition with modern discoveries? This exposes a radical inconsistency in the Creationist treatment of the Bible and science. The incorporation of evolutionary theory into modern Theistic faith does not differ in principle from the acceptance of the post-Copernican model of the universe. Perhaps it is relevant here to recall that dogmatic religious forces once opposed the findings of Copernicus and Galileo with the same vehemence that has been displayed over the last century against Darwin. The Copernican "revolution" was perceived originally by the entrenched religious authorities as absolutely destructive to the foundations of Christian faith. After all, it overturned the earth-centered, three-story view of the universe that had prevailed for many centuries, and no one had ever thought this way before. Four centuries later, the Copernican model is commonplace. One can only hope that we shall not have to wait so long for the cessation of the similarly fruitless and needless debate that has been generated over evolution.

DOES EVOLUTION "EXPLAIN AWAY" THE CREATOR?

We have seen that the theory of evolution contradicts a particular type of Biblical interpretation, termed "literalism." We have shown why this approach to Biblical understanding is unsound and how the arbitrary doctrine of Creationism is derived from it. When a sound method of Biblical interpretation is adopted, the need for Creationism vanishes, so far as "protecting" the Bible from science is concerned. The Bible is no longer seen as competing with evolutionary theory. However, we must not ignore the fact that evolution is often misinterpreted as inherently contradictory to Theistic belief, aside from the issues of Biblical interpretation involved. We must now examine the reasons for this rather widespread position, because it also provides a source of fuel for the fires of Creationism.

The original "sting of Darwinism," as it has been called, is not difficult to identify. Darwin's theory seemed to raise questions about a centuries-old line of argument for the existence of a Creator-God, which runs like this: Whenever we observe something that has order, plan, pattern, or design in it, we know logically that there has to be a conscious intelligence or mind behind it capable of producing this plan or design. To illustrate this idea, let us imagine a pilot flying over a presumably uninhabited island. On the beach below he observes a pile of debris—sticks, rocks, and brush. He pays it little notice because he assumes that it was merely washed there by tides or heavy rains, perhaps abetted by the winds. Since the rubble shows no discernible pattern or structure, there is no reason to assume that anyone deliberately stacked it that way or intended it for any purpose.

Now imagine a different circumstance in which the debris is arranged in a definite pattern that spells out the word "HELP." This observation would undoubtedly cause the pilot to react very differently, perhaps trying to land or radio for a search party. Even if he should not know English, he would not reasonably assume that the pattern displayed by the letters happened by sheer accident, because the probability of natural forces spelling out a word in this fashion seems very small.

This line of reasoning is classically translated into an argument for the existence of a Creator: The world around us shows the most intricate, complex, and sometimes awesome forms of design. From the amazing structure of a molecule to the motion of the planets, the universe demonstrates order, regularity, and mind-boggling complexity. Design on such a scale requires a cause sufficient to explain its existence. (If an arrangement of sticks and stones spelling out a simple word cannot be explained without some causal intelligence, wouldn't it be much more unreasonable to believe that the cosmic order came about by the random interplay of mechanistic forces or blind "natural law"?) Hence a Cosmic Designer, God, must exist. Historically, this line of reasoning has been called the Teleological Argument, or the Argument from Design.

Within this framework of reasoning, which was universally respected in the 19th century, we do not have to look far to see the impact of Darwin's researches. His principles of natural selection, survival of the fittest, and progressive adaptability of organisms to their environment suddenly seemed to provide purely *natural* causes for what had formerly been attributed to God. The small gradations of difference between one generation of a species and the next, for example, could account for vast changes when magnified over eons of time. The adaptability of organisms through natural selection and mutation seemed to account for the emergence of higher complex organisms from primitive origins. All this could be conceived as moving blindly and randomly through the vast successions of time, eventually spinning out products (including man) purely by accident. By such a process nothing could be categorized as "purposeful" or designful. No great Intelligence seemed necessary to explain even the complexity of the most advanced species. To many people, God seemed excluded—a "needless hypothesis." We shall now address this challenge to Theism, not by an arbitrary appeal to Biblical authority as in Creationism, but by showing an alternative way of interpreting evolution.

One of the most thorough and helpful attempts to do just this was made some years ago by Professor F. R. Tennant, a theologian and philosopher at Cambridge University.[12] First, Professor Tennant readily concedes that Darwinian research described in a new and far more comprehensive manner the natural causes for the existence of the various species of organisms, much as advances in physics and astronomy, for example, have helped to explain the operation of our planetary system.

However, there is nothing *inherent* in Darwinian theory that excludes the possibility of an overall "directivity" in the evolutionary process. "The survival of the fittest," observes Professor Tennant, "presupposes the arrival of the fit, and throws no light thereupon." In other words, Darwinian science merely describes some of the principles or patterns by which the innumerable variations and adaptations in organisms occur. It does not and certainly cannot explain *why* they occur. Once again, the familiar axiom that science explains *how,* but not *why,* is relevant here. So far as evolutionary science is concerned, "room is left for the possibility that variation is externally predetermined or guided, so that not only the general trend of the organic process, but also its every detail, may be preordained or divinely controlled." Professor Tennant's language may sound extravagant here, but this should not be allowed to obscure his central point: Nothing in the nature of evolution precludes the *possibility* of divine direction.

Once again, we confront the issue raised earlier in our comparison of scientific knowledge and "worldviews." Interpreting the *meaning* of the evolutionary process is not a matter that is accessible to science; rather it is a worldview question. Modern biological science has made enormous strides in explaining the physical origin and development of organisms through the evolutionary process. But why is there an evolutionary process at all? Is there any purpose for it? Does it have any overall meaning? Was it God-directed? The answers to such questions necessarily depend on one's *interpretation* of the whole meaning of human existence. This is the most elementary thing to understand about the evolution controversy. Although the *facts* of evolution are not seriously at issue, what these facts *mean* represents one of the most basic philosophical and theological issues.

There is a second line of reasoning by which Professor Tennant seeks to incorporate evolution into a Theistic worldview. It might be termed the "ecological" basis for belief in a Designer-God. The term "ecological" was probably unknown to Tennant at the time of his writing (1930), but modern advances in this field can only be seen as strengthening his argument. Tennant asks us to reflect on the remarkable way in which the *inorganic* realm of nature assumed the extremely complex arrangement necessary to support *organic* life: Clearly, he says, there exists a "continuity of apparent purposiveness between the two realms." We know that "the fitness of our world to be the home of living beings depends upon certain primary conditions, astronomical, thermal, chemical, etc., and on the coincidence of qualities apparently not causally connected with one another, the number of which would doubtless surprise anyone wholly unlearned in the sciences. . . ." Everything in modern science argues, moreover, that the inorganic world took form *before* the organic. Therefore, it seems entirely reasonable to interpret the inorganic world as *itself* being purposeful, moving toward "an end to which the inorganic processes were means," namely, the support of living things.

In fact, Tennant sees a special force in this "ecological" line of argument. If we accept purposiveness in the inorganic realm, it would seem clearly to be unexplainable by any *internal* causal principle. After all, the inorganic order lacks even a semblance of the innate consciousness or intelligent adaptiveness that many organisms enjoy. Whatever purposiveness is found at the inorganic level, then, cannot possibly be "explained away" by such principles as natural selection and survival of the fittest, as some have tried to do at the organic level. Curiously, therefore, the "lower" inorganic order argues all the more forcefully for an *external* cause of its designful properties. In Theistic terms, inorganic nature calls all the more loudly for God. There simply is no *intrinsic* causal explanation for its highly intricate preparedness to support life.

Whatever one's assessment of this reasoning, it should not be allowed to obscure Tennant's central thrust: The life-supporting ecological balance in the *inorganic* realm—atmosphere, temperature, moisture, soil, fertility, and much more—demonstrates design in its own way perhaps just as remarkably as the physiological marvels of the organic realm. In fact, it seems to present an even greater challenge to mechanistic worldviews than does the organic world. And so, concludes Professor Tennant, unless divine direction is invoked, "the intricate adaptations that have been mentioned must be referred ... to a mechanically controlled concourse of atoms." This he sees as an offense to "common sense reasonableness," which rejects as "infinitesimally small" the possibility that mere chance prepared the inorganic world to be a "theatre of life."

There is one additional strand to be woven into the cable of Professor Tennant's argument, and that is an appeal to the fact that the evolutionary process did after all result in *Homo sapiens*—a species capable of bearing moral, aesthetic, and spiritual values. In other words, the evolutionary process must be given the "credit" it is due in producing man's marvelous biological make-up. Leaving aside momentarily the issue of a spiritual dimension, man's *physical* nature clearly shows its linkage with the whole natural process that preceded it. In Tennant's words, "We can affirm that man's body, with all its conditioning of his mentality, his sociality, knowledge and morality, is 'of a piece' with Nature; and that, insofar as he is a phenomenal being, man is organic to Nature." And so man's higher capabilities, it could be said, had to await the arrival of a fit biological product as their instrument. This, evolution eventually provided.

Whether or not one attributes this result of the evolutionary process to divine action, it nonetheless seems reasonable to say that evolution has been "purposeful" in at least this sense. It is simply a fact that it produced that wonder which we know as the human species. Whatever one's worldview-commitment, surely it cannot be further argued that evolution is *of necessity* at enmity with the Theistic concept of a purposive Creator. Tennant's aim, in fact, is to show that it is more reasonable than not to regard evolution as harmonious with this concept. The goal of the natural process would then be seen as that of bringing the human creature, in Tennant's words, to the "threshold of spirit." Whereas "naturalism once preached that Darwin had put an end to the assumption that man occupies an exceptional position on our planet," it is now possible in Tennant's view to make a distinct reversal on that notion: "But if we judge the tree by its fruits, Darwin may rather be said to have restored man to the position from which Copernicus seemed to have ousted him, in making it possible to regard man as not only the last term and crown of Nature's long upward effort, but also as its end or goal." The process of evolution, supported by the collaboration of the inorganic realm, after all, "has produced moral beings, is instrumental to moral life . . . and is modifiable by operative moral ideals—or, rather, by moral agents pursuing ideas." Thus the central question is focused once again: Is one willing to interpret such a seemingly purposeful "achievement" of nature as "a chance product of mindless agency"? Theism is not willing; Naturalism is.

It is not our claim (nor Tennant's) that we have produced a conclusive *proof* for the Theistic view. Such proof is not possible in the realm of worldviews. Insofar as the term "proof" has been applied to the classical arguments for the existence of God, it has been misleading, and should be rejected as a troublesome misnomer. (Absolute proof is not possible even in the empirical sciences—only varying degrees of probability. In this respect, the opponents of evolution are correct in saying that it has not been "proven." No scientific theory has been, in the absolute sense. The accumulation and convergence of evidence, however, eventually renders some hypotheses so highly probable that continued attempts at refuting them border on irrationality.) Rather, as we have stated, our purpose has been to provide an alternative way of interpreting evolution. We have sought to show that evolution is not *in itself* the enemy of Theism, as the Creationists mistakenly assume, but rather can reasonably be interpreted as providing support for the doctrine of divine creation. In fact, Theists believe that when the whole picture is viewed impartially, this interpretation commends itself as *more* reasonable than the alternatives. This will never become a matter of coercive proof, however, for when one enters the realm of worldviews, one must pass through the doorway of choice.

CONCLUSION

Our purpose in this essay has been two-fold: (1) to expose the inadequacies of the Creationists' method of Biblical interpretation, and (2) to demonstrate that evolution does not have to be interpreted as an enemy of Theism. Perhaps a brief summary of the path we have taken will be useful.

(1) Creationism rests squarely on a literalist approach to Biblical interpretation and collapses without this foundation. For this reason, we reject as categorically false the opinion reflected in the aforementioned Arkansas Act 590, which states: "Creation-science is an alternative scientific model of origins and can be presented from a strictly scientific standpoint without any religious doctrine just as evolution-science can. . . ."[13] In actual fact, Creationism cannot possibly be extricated from the principles of Biblical literalism upon which it depends. Literalism is the central dogma of Fundamentalism, the extreme conservative wing of Protestantism, which has its counterpart in Orthodox Judaism, also a strong supporter of Creationism.

Since Creationism regards the Bible as a scientific authority, it feels called upon to defend the Bible against any modern scientific knowledge that "conflicts" with the Biblical accounts. We have argued that the purpose of the Bible is to set forth monotheistic faith, and thus is not in competition with modern scientific knowledge.

Of course, a hallmark of the modern age is the currency of many nontheistic philosophies, variously called Materialism, Naturalism, Humanism, Secularism, and so on. The Bible most assuredly is in opposition to these worldviews, and on this level, fruitful debate can and must continue. The battle between the Bible and science, however, is an artificial one, based on mistaken premises, and is an unfortunate waste of time and resources.

(2) Our second purpose has been to show that the findings of modern evolutionary science pose no threat to Theistic faith unless they are *interpreted* that way. The raw data of evolution, or of any other branch of science, are not self-interpreting. We always assess the meaning of these data in light of whatever larger framework of assumptions we hold about the overall nature of life and reality. This framework is often referred to as one's "worldview," "faith," or "philosophy of life." By whatever name, it is these fundamental assumptions and convictions that determine how we construe the *meaning* of what we learn from science, history, or any other branch of knowledge.

We have seen that the Copernican "revolution" of the 16th century, which overthrew the geocentric, flat-earth model, was initially viewed as a monumental threat to Theistic faith. The Copernican view has so completely supplanted the old three-story notion, however, that

FIGURE 1. A summary of the major differences between Theistic Evolution and Creationism.

Issue	Theistic Evolution	Creationism
1. Existence of God	1. God is the sole creator of the universe—purposefully brought it into existence	1. (Same)
2. The Place of Man	2. Man bears the "image of God"—is called into special relationship with his Creator.	2. (Same)
3. Age of the World	3. 4.5–4.7 billion years	3. Not more than 10,000 years
4. Mode of Creation	4. God initiated the principles of nature and allowed them to take a gradual, unfolding course, which is continuing.	4. God directly created all things in seven days or periods, following the sequence in Genesis 1.
5. Origin of Man	5. Man's biological nature emerged from more primitive origins, until he became capable of bearing a spiritual nature.	5. Man was instantaneously created by God's direct act, physically and spiritually.
6. Biblical Interpretation	6. Genesis accounts of creation are pre-scientific literature of parable and saga, not in competition with modern science.	6. Genesis accounts are to be interpreted as literal, scientific descriptions.

even the Creationists seem unwilling to fight for the pre-Copernican model, despite the fact that it is clearly taught throughout the Bible. In fact, it is intertwined with the seven-day account of creation in Genesis 1, which is the cornerstone of Creationism. This kind of inconsistency cannot fail to raise serious questions about the coherency of the Creationist viewpoint.

We have also shown that evolution has long ago been appropriated by what may be termed "mainstream" Theism, and has even become the basis for a deeper reverence for God's creative power. Evolution is viewed by these Theists as the means by which the Creator brought about the biological order of nature, culminating in a species capable of bearing the very "image of God."

We have observed, of course, that evolution can also be interpreted nontheistically. With the currency of non-theistic worldviews in Western society, it is hardly surprising to find this interpretation so common. However, this perspective is not confined to *biological* science. The findings of modern astronomy, for example, tracing the vast expanses of inter-galactic space, can likewise be interpreted as pushing God out of nature. The awesome advances of modern atomic physics, disclosing the fundamental structure of all material things, can be viewed as explaining how the world came to be without any need for God. Since biological science engages more directly the issue of man, however, it seems to be a more attractive target for those predisposed to see science and religion as enemies.

This observation focuses our last and possibly most important consideration. One could say that the central issue raised by this whole discussion has been, What is man? It is essential to remember that evolution, after all, addresses only the *biological* side of this question. To say that the findings of modern biology can be assimilated in Theism does not mean that biology can replace theology. A bedrock claim of Theism, lying entirely beyond the power of biology to evalute, is that man is more than an animal and is distinguished from the lower orders by possession of a "spiritual" dimension. We have seen the one way the ancient Hebrew writers sought to express this was by the concept of the "image of God" in man (Genesis 1:26–27).

Man's physical nature is seen in Theism as only the *instrument* of his higher, spiritual nature, which has both an origin and destiny not ultimately reducible to a material source or explanation. Do Creationists somehow fail to understand that even when evolution and all its ramifications are accepted, it can do no more than trace man's *biological* origins? The methods of science are confined to exploration of the physical order, and claims that man's essence is more than physical lie outside the realm of scientific investigation. The Lutheran theologian Helmut Thielicke has stated this point memorably:

> I can ask where man comes from biologically, and receive the answer that he sprang from animal forms. Or I can ask why he is here, what is his destiny, what is the "role" assigned to him. If I put the question this way, the answer I get is that he was designed to be a child of God . . .
>
> I must not mix up these two questions. Once one sees this, then the ill-famed question of "Faith and Science" will look quite different . . .
>
> I have no objection, even as a Christian, to your deriving man from previous animal forms and declaring that the monkey is his grandfather and the tadpole his great-great-grandfather. Why should I? That is something for science to inquire into.
>
> But I have objections to something else. I object to your saying that therefore the nature of man, that your nature and my nature is like that of a tadpole. No, if you are going to define the mystery of man, if you are going to define what God has in mind for man and what he breathed into him, then you *cannot* say: "He is only a little more

than a tadpole." *Then* you must say: "He is a little less than God." In other words, you cannot define man on the basis of his biological origin; you must define him in the light of his destiny, his goal.[14]

With this we rest our case, in the hope that this discussion has contributed something to the clarification of the *real* issues in the present conflict. For until that task is accomplished, human beings cannot hope to make intelligent responses to any of the questions that life poses.

REFERENCES

1. *Time* (March 16, 1981), p. 81.

2. Ark Stat. Ann., § 80-1663, et seq. (1981 Supp.). Included in this volume.

3. Theodosius Dobzhansky, "Nothing in Biology Makes Sense Except in the Light of Evolution." Included in this volume.

4. See, for example, *A Compendium of Information on the Theory of Evolution and the Evolution-Creationism Controversy* (Reston, VA: National Association of Biology Teachers, 1978).

5. Bernard M. Anderson, *Understanding the Old Testament,* 3d ed. (Englewood Cliffs, NJ: Prentice-Hall, 1975), p. 428.

6. Arthur Smethurst, *Modern Science and Christian Belief* (New York: Abingdon Press, 1955), pp. 109-10.

7. B. D. Napier, *Come Sweet Death* (Philadelphia: United Church Press, 1967), p. 12.

8. Bertrand Russell, *Mysticism and Logic* (London: Allen and Unwin, 1917), p. 47.

9. Reinhold Niebuhr, "Christianity and Darwin's Revolution," in *A Book That Shook the World* (Pittsburgh, PA: University of Pittsburgh Press, 1958), p. 33.

10. Maria Boulding, *Prayer: Our Journey Home* (Ann Arbor, MI: Servant Books, 1979), p. 61.

11. Boulding, p. 59.

12. Our discussion draws upon Volume II, Chapter 4, of Tennant's work, *Philosophical Theology* (Cambridge: Cambridge University Press, 1930). See also John Hick, ed., *Classical and Contemporary Readings in the Philosophy of Religion,* 2d ed. (Englewood Cliffs, NJ: Prentice-Hall, 1970), pp. 247-70, for this portion of Tennant's writing.

13. Ark, Stat. Ann., § 80-1663, et seq. (1981 Supp.), section 7 (j).

14. Helmut Thielicke, *How the World Began* (Philadelphia, PA: Muhlenberg Press, 1961), pp. 79-84.

BIBLIOGRAPHY

BOOKS

Adams, Frank Dawson. *The Birth and Development of the Geological Sciences.* New York: Dover (1938) 1950.

Albritton, Claude C. *The Abyss of Time: Changing Conceptions of the Earth's Antiquity after the Sixteenth Century.* San Francisco, CA: Freeman, Cooper, 1980.

Alexander, Richard D. *Darwinism and Human Affairs.* Seattle, WA: University of Washington Press, 1980.

Allen, Leslie H. *Bryan and Darrow at Dayton.* New York: Russell and Russell, 1925.

Anderson, B. W. *Creation versus Chaos.* New York: Association Press, 1967.

Anderson, Bernard M. *Understanding the Old Testament.* 3d ed. Englewood Cliffs, NJ: Prentice-Hall, 1975.

Appleman, Philip, ed. *Darwin: A Norton Critical Edition.* 2d ed. New York: W. W. Norton, 1979.

Austin, W. H. *The Relevance of Natural Science to Theology.* London: Macmillan, 1976.

Aveling, Edward. *The Religious Views of Charles Darwin.* London: Freethought Publishing Company, 1883.

Averill, Lloyd J. *American Theology in the Liberal Tradition.* Philadelphia, PA: Westminster Press, 1967.

Awbrey, Frank, and Thwaites, William. *Evolution vs. Creation.* San Diego, CA: Aztec Lecture Notes, San Diego State University, n.d.

Badash, Lawrence. *Radioactivity in America: Growth and Decay of a Science.* Baltimore, MD: Johns Hopkins University Press, 1979.

Bailey, E. B. *James Hutton: The Founder of Modern Geology.* New York: Elsevier, 1967.

Bales, James D. *The Genesis Account and a Scientific Test, or, The Predictive Value of Genesis One through Three.* Searcy, AK: Privately published, 1975.

Bannister, Robert C. *Social Darwinism: Science and Myth in Anglo-American Social Thought.* Philadelphia, PA: Temple University Press, 1979.

Barash, D. P. *Sociobiology and Behavior.* New York: Elsevier, 1976.

Barbour, Ian G. *Issues in Science and Religion.* New York: Harper and Row, 1966.

———. *Science and Religion: New Perspectives on the Dialogue.* New York: Harper and Row, 1968.

Barnes, Thomas G. *Origin and Destiny of the Earth's Magnetic Field.* San Diego, CA: Creation-Life Publishers, 1973.

Barnett, S. A., ed. *A Century of Darwin.* Cambridge, MA: Harvard University Press, 1959.

Baylo, J. G. *Creation and Evolution.* Chicago: The Regular Baptist Press, 1961.

Beckner, M. *The Biological Way of Thought.* New York: Columbia University Press, 1959.

Benz, Ernst. *Evolution and Christian Hope.* Translated by H. G. Frank. London: Victor Gollancz, 1968.

Bergman, Jerry. *Teaching about the Creation/Evolution Controversy.* Bloomington, IN: Phi Delta Kappa Educational Foundation, 1979.

Blacker, C., and Loewe, M., eds. *Ancient Cosmologies.* London: Allen and Unwin, 1975.

Bliss, Richard, *Origins: Two Models Evolution Creation.* San Diego, CA: Creation-Life Publishers, 1976.

Boardman, William W., Jr.; Koontz, Robert F.; and Morris, Henry M. *Science and Creation.* San Diego, CA: Creation-Science Research Center, 1973.

Boller, Paul F. *American Thought in Transition: The Impact of Evolutionary Naturalism, 1865–1900.* Chicago: Rand, McNally and Company, 1969.

Brachtell, D. F. *The Impact of Darwinism: Texts and Commentary Illustrating Nineteenth Century Religious, Scientific, and Literary Attitudes.* London: Avebury, 1980.

Brent, Peter. *Charles Darwin.* New York: Harper and Row, 1981.

Broom, R. *The Mammal-like Reptiles of South Africa and the Origin of Mammals.* London: H. F. and G. Witherby, 1932.

Brown, Ira V. *Lyman Abbott: Christian Evolutionist.* Cambridge, MA: Harvard University Press, 1953.

Brush, Stephen G. *The Temperature of History: Phases of Science and Culture in the Nineteenth Century.* New York: Burt Franklin and Company.

Burchfield, Joe. *Lord Kelvin and the Age of the Earth.* New York: Science History Publications, 1975.

Buttrick, George Arthur. *The Interpreter's Bible.* Vol. 1. New York: Abingdon Press, 1952.

Campbell, John P. *Biological Teaching in the Colleges of the United States.* Washington, DC: Government Printing Office, 1891.

Cloud, Preston. *Cosmos, Earth, and Man.* New Haven, CT: Yale University Press, 1978.

Colbert, Edwin H. *Evolution of the Vertebrates: A History of the Back-boned Animals through Time.* 2d ed. New York: John Wiley and Sons, Inc., 1969.

Coleman, W. R. *George Cuvier, Zoologist: A Study in the History of Evolution Theory.* Cambridge, MA: Harvard University Press, 1964.

Collingwood, R. G. *The Idea of Nature.* New York: Oxford University Press (1945) 1960.

Cook, Melvin A. *Prehistory and Earth Models.* London: Max Parrish, 1966.

Coppedge, James F. *Evolution: Possible or Impossible?* Grand Rapids, MI: Zondervan Publishing House, 1973.

Cubie, W. "A Comparative Analysis of the Objectives and Content of Biology Instruction in the Secondary Schools in Three Periods as Revealed by Representative Textbooks in the Field during Those Periods." Ph.D. diss., Indiana University, 1958.

Cummings, Violet M. *Noah's Ark: Fact or Fable?* San Diego, CA: Creation-Science Research Center, 1972.

Custance, Arthur. *Genesis and Early Man.* Grand Rapids, MI: Zondervan Publishing House, 1975.

Daniels, George, ed. *Darwin Comes to America.* Waltham, MA: Blaisdell, 1968.

———. *Science in American Society: A Social History.* New York: Knopf, 1971.

Darwin, Charles. *The Autobiography of Charles Darwin, 1809–1882, with Original Omissions Restored.* Edited by Nora Barlow. New York: W.W. Norton, 1969.

———. *The Collected Papers of Charles Darwin.* Edited by Paul H. Barrett. Chicago: University of Chicago Press, 1977.

———. *Life and Letters of Charles Darwin.* Edited by Francis Darwin. New York: Johnson Reprint Corporation, (1888) 1969.

———. *On the Origin of Species* (1859). A Facsimile of the first edition, with an introduction by Ernst Mayr. Boston: Harvard University Press, 1964.

———. *The Origin of Species: A Variorum Text.* Edited by Morse Peckham. Philadelphia, PA: University of Pennsylvania Press, 1959.

Dawkins, R. *The Selfish Gene.* New York: Oxford University Press, 1976.

de Beer, Gavin. *Charles Darwin: Evolution by Natural Selection.* Westport, CT: Greenwood Press, 1976.

de Camp, L. Sprague. *Darwin and His Great Discovery.* New York: Macmillan, 1972.

———. *The Great Monkey Trial.* Garden City, NY: Doubleday, 1968.

Deely, John N., and Raymond J. Nogar. *The Problem of Evolution.* New York: Appleton, Century, Crofts, 1973.

Did Man Get Here by Evolution or by Creation? New York: Watchtower Bible and Tract Society, 1967.

Dillenberger, John. *Protestant Thought and Natural Science: A Historical Interpretation.* Nashville, TN: Abingdon Press, 1960.

Dobzhansky, Theodosius. *Genetics and the Origin of Species.* New York: Columbia University Press, 1951.

———. *Genetics of the Evolutionary Process.* New York: Columbia University Press, 1970.

Dobzhansky, Theodosius, et al. *Evolution.* San Francisco, CA: W. H. Freeman, 1977.

Dobzhansky, Theodosius; Hecht, M. K.; and Steere, W. C., eds. *Evolutionary Biology.* New York: Appleton, Century, Crofts, 1969.

Drake, E. T. *Evolution and Environment.* New Haven, CT: Yale University Press, 1968.

Draper, John William. *History of the Conflict between Religion and Science.* New York: Appleton, 1874.

Eaton, Theodore H., Jr. *Evolution.* New York: W. W. Norton, 1970.

Ehrlich, Paul; Holm, Richard W.; Parnell, Dennis R. *The Process of Evolution.* 2d ed. New York: McGraw-Hill, 1974.

Eicher, Don L. *Geologic Time.* 2d ed. Englewood Cliffs, NJ: Prentice-Hall, 1976.

Eiseley, Loren C. *Darwin and the Mysterious Mr. X: New Light on the Evolutionists.* New York: E. P. Dutton, 1979.

———. *Darwin's Century: Evolution and the Men Who Discovered It.* Garden City, NY: Doubleday and Company, 1958.

Eldredge, Niles. *The Monkey Business: A Scientist Looks at Creationism.* New York: Pocket Books, Inc., 1982.

Elmendorf, R. G. *How to Scientifically Trap, Test, and Falsify Evolution.* Bairdford, PA: Association of Western Pennsylvania, 1978.

Erickson, Lonni R. *The Teaching of Evolution in Public Schools: A Comparison of Evolution and Special Creation.* Lyons, CO: Scandia Publishers, 1980.

Farrington, Benjamin. *What Darwin Really Said.* New York: Schocken Books, Inc., 1967 (reprint).

Fisher, David E. *The Creation of the Universe.* Indianapolis, IN: Bobbs-Merrill, 1977.

Fisher, R. A. *The Genetical Theory of Natural Selection.* 2d ed. New York: Dover, 1958.

Ford, E. B. *Ecological Genetics.* 4th ed. London: Chapman and Hall, 1975.

Fothergill, Philip G. *Evolution and Christians.* London: Longmans, 1961.

Futuyma, D. J. *Evolutionary Biology.* Sunderland, MA: Sihaner, 1979.

Gallant, Roy A. *How Life Began: Creation vs. Evolution.* New York: Four Winds Press, 1975.

Gardner, Martin. *Fads and Fallacies in the Name of Science.* 2d ed. New York: Dover, 1957.

Gaustad, Edwin Scott. *Religious Issues in American History.* New York: Harper and Row, 1968.

Gatewood, Willard B., Jr. *Preachers, Pedagogues, and Politicians: The Evolution Controversy in North Carolina.* Chapel Hill, NC: University of North Carolina Press, 1966.

Gay, Peter. *The Enlightenment: An Interpretation.* New York: Vintage Books, 1966.

Geikie, Archibald. *Charles Darwin as Geologist.* New York: Gordon Press Publishers, 1977.

———. *The Founders of Geology.* New York: Dover (1905) 1962.

Ghiselin, Michael. *The Triumph of the Darwinian Method.* Berkeley, CA: University of California Press, 1969.

Gilkey, Landon. *Maker of Heaven and Earth: A Study of the Christian Doctrine of Creation.* Garden City, NY: Doubleday and Company, 1959.

Gillespie, Neal C. *Charles Darwin and the Problem of Creation.* Chicago: University of Chicago Press, 1979.

Gillispie, Charles Coulston. *Genesis and Geology: A Study of the Relations of Scientific Thought, Natural Theology, and Social Opinion in Great Britain, 1790–1850.* New York: Harper and Row, 1951.

Ginger, Ray. *Six Days or Forever? Tennessee v. John Thomas Scopes.* Boston: Beacon Press, 1958.

Gish, Duane T. *Evidence against Evolution.* Wheaton, IL: Tyndale House Publications, 1972.

———. *Evolution: The Fossils Say No!* 3d ed. San Diego, CA: Creation-Life Publishers, 1979.

———. *Have You Been Brainwashed?* Seattle, WA: Life Messengers, 1974.

———. *Speculations and Experiments Related to Theories on the Origin of Life.* ICR Technical Monograph No. 1. San Diego, CA: Institute for Creation Research, 1972.

Glass, B., Temkin, O., and Straus, W. L. Jr., eds. *Forerunners of Darwin: 1745–1859.* Baltimore, MD: Johns Hopkins Press, 1968.

Glick, Thomas F., ed. *The Comparative Reception of Darwinism.* Austin, TX: University of Texas Press, 1972.

Godfrey, Laurie, ed. *Scientists Confront Creationism.* New York: W. W. Norton, 1982.

Gould, Stephen J. *Ever Since Darwin: Reflections in Natural History.* New York: W. W. Norton, 1977.

Grant, Verne. *Plant Speciation.* New York: Columbia University Press, 1971.

Gray, A. P. *Mammalian Hybirds.* Franham Royal, England: Commonwealth Agricultural Bureau, 1971.

Greene, John C. *Darwin and the Modern World View.* Baton Rouge, LA: Louisiana State University Press, 1961.

————. *The Death of Adam: Evolution and Its Impact on Western Thought.* Ames, IA: Iowa State University Press, 1959.

Gruber, Howard E. *Darwin on Man: A Psychological Study of Scientific Creativity.* Chicago: University of Chicago Press, 1981.

Gruber, Howard E., and Paul H. Barrett. *Darwin and Man.* London: Wildwood House, 1974.

Haber, Francis C. *The Age of the World: Moses to Darwin.* Baltimore, MD: Johns Hopkins Press, 1957.

Haldane, J. B. S. *The Causes of Evolution.* Ithaca, NY: Cornell University Press, 1966.

Hall, George M. *Farewell to Darwin, Unified Field Theory of Physics, the Genetic Process, and Psychology.* St. Louis, MO: Green, n.d.

Hallam, A. *A Revolution in the Earth Sciences: Continental Drift to Plate Tectonics.* Oxford: Clarendon Press, 1973.

Hamilton, T. H. *Process and Pattern in Evolution.* New York: Macmillan, 1967.

Hardin, Garrett. *Nature and Man's Fate.* New York: Holt, Rinehart, and Winston, 1959.

Hefley, James C. *Textbooks on Trial.* Wheaton, IL: Victor Books, 1976.

Heim, Karl. *Christian Faith and Natural Science.* New York: Harper and Row, 1953.

Hick, John, ed. *Classical and Contemporary Readings in the Philosophy of Religion.* 2d ed. Englewood Cliffs, NJ: Prentice-Hall, 1970.

Hofstadter, Richard. *Anti-Intellectualism in American Life.* New York: Vintage Books, 1963.

————. *Social Darwinism in American Thought.* Boston: Beacon Press, 1955.

Hofstadter, Richard, and Matzger, Walter P. *The Development of Academic Freedom in the United States.* New York: Columbia University Press, 1955.

Holton, Gerald, and Blanpied, William A., eds. *Science and Its Public: The Changing Relationship.* Boston: D. Reidl, 1976.

Hooykaas, R. *Natural Law and Divine Miracle: A Historical Critical Study of the Principle of Uniformity in Geology, Biology, and Theology.* 2d ed. Leiden, Holland: E. J. Brill, 1963.

————. *Religion and the Rise of Modern Science.* Grand Rapids, MI: William B. Eerdmans, 1972.

House, G. F. *Speak to the Earth: Creation Studies in Geo Science.* Nutley, NJ: Presbyterian and Reformed Publishing Company, n.d.

Howard, Jonathan. *Darwin.* New York: Hill and Wang, 1982.

Hoyle, Fred. *The Cosmogony of the Solar System.* Hillside, NJ: Enslow Publishers, 1979.

Hull, David. *Darwin and His Critics: The Reception of Darwin's Theory of Evolution by the Scientific Community.* Cambridge, MA: Harvard University Press, 1973.

————. *Philosophy of Biological Science.* Englewood Cliffs, NJ: Prentice-Hall, 1974.

Huxley, Julian. *Evolution: The Modern Synthesis.* New York: Harper and Row, 1943.

Huxley, Thomas Henry. *Science and the Hebrew Tradition.* New York: Appleton, 1895.

Irvine, William. *Apes, Angels, and Victorians: The Story of Darwin, Huxley, and Evolution.* New York: McGraw-Hill, 1955.

Jeddeloh, Kenneth, ed. *Journal of the Minnesota Science Teachers Association,* vol. 1 (1979). Entire issue devoted to position papers on the Minnesota Creationism Bill (1978).

Johanson, Donald C., and Edey, Maitland A. *Lucy: The Beginnings of Humankind.* New York: Simon and Schuster, 1981.

Johnson, L. "The Evolution Controversy during the 1920's." Ph.D. diss., New York University, 1953.

Karp, Walter. *Charles Darwin and the Origin of Species.* New York: Harper and Row, 1968.

Kennedy, Gail, ed. *Evolution and Religion: The Conflict between Science and Theology in Modern America.* Lexington, MA: D. C. Heath and Company, 1957.

Kenyon, D., and Steinman, G. *Biochemical Predestination.* New York: McGraw-Hill, 1969.

Klaaren, Eugene M. *Religious Origins of Modern Science.* Grand Rapids, MI: William B. Eerdmans, 1977.

Klotz, John W. *Genes, Genesis, and Evolution.* St. Louis, MO: Concordia Publishing House, 1970.

Kofahl, Robert E. *Handy Dandy Evolution Refuter.* San Diego, CA: Beta Books, 1977.

Kofahl, Robert E., and Segraves, K. L. *The Creation Explanation.* Wheaton, IL: Harold Shaw Publishers, 1975.

Komreich, Y., ed. *A Science and Torah Reader.* New York: Union of Orthodox Jewish Congregations of America, 1970.

Lack, D. *Evolutionary Theory and Christian Belief.* London: Methuen, 1957.

LaHaye, Tim, and Morris, John D. *The Ark on Ararat.* San Diego, CA: Creation-Life Publishers, 1976.

Lammerts, Walter E., ed. *Scientific Studies in Special Creation.* Grand Rapids, MI: Baker Book House, 1971.

———. *Why Not Creation?* Grand Rapids, MI: Baker Book House, 1970.

Leach, Maria. *The Beginning: Creation Myths around the World.* New York: Harper and Row, 1956.

Le Gros Clark, W. E. *The Fossil Evidence for Human Evolution.* Revised ed. Chicago: University of Chicago Press, 1964.

Lemmon, Richard M. *Review of Science and Creation Series.* Sacramento, CA: California State Department of Education, 1975.

Levitt, Z. *Creation: A Scientist's Choice.* Wheaton, IL: Victor Books, 1971.

Lewontin, R. C. *The Genetic Basis of Evolutionary Change.* New York: Columbia University Press, 1974.

Lightener, J., ed. *A Compendium of Information on the Theory of Evolution and the Evolution-Creationist Controversy.* Reston, VA: National Association of Biology Teachers, 1977.

Lillegraven, J. A.; Keilan-Jaworowska, Z.; and Clemens, W. A., eds. *Mesozoic Mammals.* Berkeley, CA: University of California Press, 1979.

Lindsell, Harold. *The Battle for the Bible.* Grand Rapids, MI: Zondervan Publishing Company, 1976.

Loewenberg, Bert J. *Darwinism Comes to America: 1859–1900.* Philadelphia, PA: Fortress Press, 1969.

Lovejoy, Arthur O. *The Great Chain of Being.* Cambridge, MA: Harvard University Press, 1936.

Lovtrup, S. *Epigenetics: A Treatise on Theoretical Biology.* London: John Wiley and Sons, Inc., 1974.

Lyell, Charles. *Principles of Geology: Being an Inquiry How Far the Former Changes of the Earth's Surface Are Referable to Causes Now in Operation.* London: John Murray, 1830–33.

Lyttleton, R. A. *Mysteries of the Solar System.* Oxford: Clarendon Press, 1968.

Macbeth, Norman. *Darwin Retried: An Appeal to Reason.* Boston: Gambit, 1971.

MacLagan, D. *Creation Myths.* London: Thames and Hudson, 1977.

Mascall, E. L. *Christian Theology and Natural Science: Some Questions on Their Relations.* London: Longmans, Green, and Company, 1956.

Mayfield, J. C. *Using Modern Knowledge to Teach Evolution in the High School as Seen by the Participants in the High School Conference of the Darwin Centennial Celebration at the University of Chicago, November 24–25, 1958.* Chicago: The Graduate School of Education of the University of Chicago, 1960.

Mayr, Ernst. *Animal Species and Evolution.* Cambridge, MA: Harvard University Press, 1963.

———. *Population, Species, and Evolution.* Cambridge, MA: Harvard University Press, 1970.

———. *Systematics and the Origin of Species from the Viewpoint of a Zoologist.* New York: Columbia University Press, 1942.

Mayr, Ernst, and Provine, William B., eds. *The Evolutionary Synthesis: Perspectives on the Unification of Biology.* Cambridge, MA: Harvard University Press, 1980.

McKinney, H. Lewis, ed. *Lamarck to Darwin: Contributions to Evolutionary Biology, 1809–1859.* Lawrence, KS: Coronado Press, Inc., 1971.

McPherson, T. *The Argument from Design.* London: Macmillan, 1972.

Medawar, P. W. *The Art of the Soluble.* London: Methuen, 1967.

Mellor, E. B., ed. *The Cambridge Bible Commentary on the New English Bible, Introductory Volume, The Making of the Old Testament.* Cambridge: Cambridge University Press, 1972.

Millhauser, Milton. *Just Before Darwin.* Middletown, CT: Wesleyan University Press, 1959.

Mixter, R. *Creation and Evolution.* Wheaton, IL: American Scientific Affiliation, 1949.

Montgomery, John Warwick. *The Quest for Noah's Ark.* Minneapolis, MN: Bethany Fellowship, Inc., 1972.

Moore, John N. *Questions and Answers on Creation/Evolution.* Grand Rapids, MI: Baker Books, 1976.

Moore, John N., and Slusher, Harold, eds. *Biology: A Search for Order in Complexity.* Developed by the Creation Research Society. Grand Rapids, MI: Zondervan Publishing Company, 1970.

Moorehead, Alan. *Darwin and the Beagle.* New York: Harper and Row, 1969.

Moorhead, Paul S., and Kaplan, Martin M., eds. *Mathematical Challenges to the Neo-Darwinian Interpretation of Evolution.* A symposium held at The Wistar Institute of Anatomy and Biology, April 25–26, 1966. Philadelphia, PA: Wistar Institute Press, 1967.

Morison, William James. "George Frederick Wright: In Defense of Darwinism and Fundamentalism, 1838–1921." Ph.D. diss., Vanderbilt University, 1971.

Morris, Henry M. *Biblical Cosmology and Modern Science.* Grand Rapids, MI: Baker Book House, 1970.

———. *Evolution and the Modern Christian.* Grand Rapids, MI: Baker Book House, 1978.

———. *King of Creation.* San Diego, CA: Creation-Life Publishers, 1980.

———. *The Remarkable Birth of Planet Earth.* San Diego, CA: Creation-Life Publishers, 1972.

———. *The Scientific Case for Creation.* San Diego, CA: Creation-Life Publishers, 1977.

———, ed. *Scientific Creationism.* General and public school editions. San Diego, CA: Creation-Life Publishers, 1974.

———. *The Troubled Waters of Evolution.* San Diego, CA: Creation-Life Publishers, 1974.

———. *The Twilight of Evolution.* Grand Rapids, MI: Baker Book House, 1963.

Morris, Henry M.; Boardman, William W.; and Koontz, Robert F. *Creation Science Teachers Handbook.* San Diego, CA: Creation Science Research Center, 1971.

Morris, Henry M., and Gish, Duane T., eds. *The Battle for Creation.* San Diego, CA: Creation-Life Publishers, 1976.

Morris, John D. *Adventure on Ararat.* San Diego, CA: Institute for Creation Research, 1973.

———. *Tracking Those Incredible Dinosaurs and the People Who Knew Them.* San Diego, CA: Creation-Life Publishers, 1980.

Nagel, E. *The Structure of Science.* New York: Harcourt, Brace, and World, 1961.

Nelkin, Dorothy. *Science Textbook Watchers and the Politics of Equal Time.* Cambridge, MA: Massachusetts Institute of Technology Press, 1977.

Nelson, Byron C. *The Deluge Story in Stone.* Minneapolis, MN: Bethany Fellowship, Inc., 1968.

Newman, J. R., ed. *What is Science?* New York: Simon and Schuster, 1955.

Noorbergen, Rene. *The Ark File.* Mountain View, CA: Pacific Press Publishing, 1974.

————. *Secrets of the Lost Races.* New York: The Bobbs-Merrill Company, Inc., 1977.

O'Brien, Charles F. *Sir William Dawson: A Life in Science and Religion.* Philadelphia, PA: American Philosophical Society, 1971.

Ong, Walter, J. ed., *Darwin's Vision and Christian Perspectives.* New York: Macmillan, 1960.

Overman, Richard H. *Evolution and the Christian Doctrine of Creation.* Philadelphia, PA: Westminster Press, 1967.

Patten, D., Ed. *A Symposium on Creation.* Grand Rapids, MI: Baker Book House, 1969.

Patterson, Colin. *Evolution.* London: British Museum (Natural History), 1978.

Peacocke, A. R. *Creation and the World of Science: The Bampton Lectures, 1978.* Oxford: Clarendon Press, 1979.

Persons, Stow, ed. *Evolutionary Thought in America.* New Haven, CT: Yale University Press, 1950.

Price, Derek de Solla. *Little Science, Big Science.* New York: Columbia University Press, 1963.

Provine, William B. *The Origins of Theoretical Population Genetics.* Chicago: University of Chicago Press, 1971.

Quebedeaux, Richard. *The Worldly Evangelicals.* New York: Harper and Row, 1978.

Raven, Charles E. *Natural Religion and Christian Theology.* Cambridge: Cambridge University Press, 1953.

Rehwinkle, A. *The Flood.* St. Louis, MO: Concordia Publishing Company, 1951.

Riddle, O. *The Teaching of Biology in the Secondary Schools of the United States.* Washington, DC: Union of American Biological Societies, 1942.

————. *The Unleashing of Evolutionary Thought.* New York: Vantage Press, 1954.

Rogers, Jack B., and McKim, Donald K. *The Authority and Interpretation of the Bible.* New York: Harper and Row, 1979.

Romer, Alfred Sherwood. *Vertebrate Paleontology.* 3d ed. Chicago, IL: University of Chicago Press, 1966.

Rudwick, M. J. *The Meaning of Fossils.* London: Macdonald, 1972.

Ruffa, Anthony. *Darwinism and Determinism: The Role of Direction in Evolution.* Boston: Branden Press, 1981.

Ruse, Michael. *Darwinism Defended: A Guide to the Evolution Controversies.* Reading, MA: Addison-Wesley Publishing Company, 1982.

Russett, Cynthia E. *Darwin in America: The Intellectual Response, 1865–1912.* San Francisco, CA: W. H. Freeman and Company, 1976.

Rust, Eric C. *Evolutionary Philosophies and Contemporary Theology.* Philadelphia, PA: Westminster Press, 1969.

Rutten, M. *The Origin of Life by Natural Causes.* New York: Elsevier, 1971.

Sahlins, M. *The Use and Abuse of Biology.* Ann Arbor, MI: University of Michigan Press, 1976.

Sandeen, Ernest R. *The Origins of Fundamentalism.* Philadelphia, PA: Fortress Press, 1968.

———. *The Roots of Fundamentalism.* Chicago: University of Chicago Press, 1970.

Sayers, Dorothy L. *The Mind of the Maker.* Westport, CT: Greenwood Press, 1941.

Schrödinger, E. *What Is Life?* New York: Macmillan, 1945.

Schropf, T. J. M., ed. *Models in Paleobiology.* San Francisco, CA: Freeman, Cooper, and Company, 1972.

Scientific American Books. *Evolution.* San Francisco, CA: W. H. Freeman and Company, 1978. (Originally published as articles in September 1978 issue of *Scientific American.*)

Scopes, John T., and Presley, James. *Center of the Storm: Memoirs of John T. Scopes.* New York: Holt, Rinehart, and Winston, 1967.

Segraves, Kelly L. *The Great Dinosaur Mistake.* San Diego, CA: Beta Books, 1975.

Segraves, N. J. *The Creation Report.* San Diego, CA: Creation-Science Research Center, 1977.

Shepherd, J. J. *Experience, Inference, and God.* London: Macmillan, 1975.

Simpson, George Gaylord. *The Geography of Evolution.* New York: Capricorn Books, 1965.

———. *Life of the Past.* New Haven, CT: Yale University Press, 1953.

———. *The Major Features of Evolution.* New York: Columbia University Press, 1953.

———. *The Meaning of Evolution: A Study of the History of Life and Its Significance for Man.* 2d ed. New Haven, CT: Yale University Press, 1967.

———. *Tempo and Mode in Evolution.* New York: Columbia University Press, 1944.

Simpson, J. Y. *Landmarks in the Struggle between Science and Religion.* London: Hodder and Stoughton, 1925.

Skoog, Gerald. "The Topic of Evolution in Secondary School Biology Textbooks: 1900–1968." Ph.D. diss., University of Nebraska, 1969.

Slusher, Harold S. *Age of the Cosmos.* ICR Technical Monograph No. 9. San Diego, CA: Institute for Creation Research, 1980.

————. *Critique of Radiometric Dating.* ICR Technical Monograph No. 2. San Diego, CA: Institute for Creation Research, 1981.

————. *The Origin of the Universe.* San Diego, CA: Institute for Creation Research, 1978.

Smethurst, Arthur. *Modern Science and Christian Belief.* New York: Abingdon Press, 1955.

Smith, John Maynard. *The Theory of Evolution.* 3d ed. Baltimore, MD: Penguin Books, 1975.

Spilsbury, Richard. *Providence Lost: A Critique of Darwinism.* Oxford: Oxford University Press, 1975.

Stacey, Frank D. *Physics of the Earth.* 2d ed. John Wiley and Sons, Inc., 1977.

Stebbins, G. Ledyard. *Darwin to DNA: Molecules to Humanity.* San Francisco, CA: W. H. Freeman, 1982.

————. *Processes of Organic Evolution.* 3d ed. Englewood Cliffs, NJ: Prentice-Hall, 1977.

————. *Variation and Evolution in Plants.* New York: Columbia University Press, 1950.

Steidl, Paul M. *The Earth, the Stars, and the Bible.* Grand Rapids, MI: Baker Book House, 1979.

Stevens, L. Robert. *Charles Darwin.* Boston: Twayne Publishers, 1978.

Suppe, F. *The Structure of Scientific Theories.* Urbana, IL: University of Illinois Press, 1974.

Tax, Sol, and Callender, Charles, eds. *Evolution after Darwin.* Vol. 1: *The Evolution of Life.* Vol. 2: *The Evolution of Man.* Vol. 3: *Issues in Evolution.* Chicago,: University of Chicago Press, 1960.

Thielicke, Helmut. *How the World Began.* Philadelphia, PA: Muhlenberg Press, 1961.

Tompkins, J. R., ed. *D-Days at Dayton: Reflections on the Scopes Trial.* Baton Rouge, LA: Louisiana State University Press, 1965.

Toulmin, Stephen, and Goodfield, June. *The Discovery of Time.* London: Hutchinson, 1956.

Turner, Frank Miller. *Between Science and Religion: The Reaction to Scientific Naturalism in Late Victorian England.* New Haven, CT: Yale University Press, 1974.

Twenty-One Scientists Who Believe in Creation. San Diego, CA: Creation-Life Publishers, 1977.

Van Wylen, G. J., and Sonntag, R. E. *Fundamentals of Classical Thermodynamics.* 2d. ed. New York: John Wiley and Sons, Inc., 1973.

Volpe, E. Peter. *Understanding Evolution.* 3d ed. Dubuque, IA: William C. Brown, 1977.

Vorzimmer, P. J. *Charles Darwin: The Years of Controversy.* Philadelphia, PA: Temple University Press, 1970.

Wallace, A. R. *Natural Selection and Tropical Nature.* London: Macmillan, 1891.

Wallace, Bruce. *Chromosomes, Giant Molecules, and Evolution.* New York: W. W. Norton, 1977.

Weinberg, Stanley. *Biology.* Boston, MA: Allyn and Bacon, 1977.

Wells, D. F., and Woodbridge, D. J., eds. *The Evangelicals: What They Believe, Who They Are, Where They Are Changing.* Nashville, TN: Abingdon Press, 1975.

Wendt, Herbert. *In Search of Adam.* Westport, CT: Greenwood Press, 1955.

Whitcomb, J. *The Early Earth.* Grand Rapids, MI: Baker Book House, 1972.

Whitcomb, John C. *The Origin of the Solar System.* Nutley, NJ: Presbyterian and Reformed Publishing Company., 1977.

————. *The World that Perished.* Grand Rapids, MI: Baker Book House, 1973.

Whitcomb, John C., and Morris, Henry M. *Genesis Flood.* Philadelphia, PA: Presbyterian and Reformed Publishing Company, 1961.

White, Andrew Dickson. *A History of the Warfare of Science with Theology.* New York: Appleton, 1898.

White, Edward A. *Science and Religion in American Thought.* Stanford, CA: Stanford University Press, 1952.

White, M. J. D. *Modes of Speciation.* San Francisco, CA: W. H. Freeman and Company, 1978.

Whitehouse, W. A. *Creation, Science, and Theology: Essays in Response to Karl Barth.* Grand Rapids, MI: William B. Eerdmans, 1981.

Whitney, D. *The Case for Creation.* 5 parts. Malverne, NY: Christian Evidence League, 1946.

Wiener, Philip P. *Evolution and the Founders of Pragmatism.* Cambridge, MA: Harvard University Press, 1949.

Wikke, Peter. *Christianity Challenges the University.* Downers Grove, IL: InterVarsity, 1981.

Williams, E. L., ed. *Thermodynamics and the Development of Order.* Norcross, GA: Creation Research Society Books, 1982.

Wilson, Edward O. *Sociobiology: The New Synthesis.* Cambridge, MA: Harvard University Press, 1975.

Wilson, Leonard C. *Sir Charles Lyell's Scientific Journals on the Species Question.* New Haven, CT: Yale University Press, 1970.
Wilson, R. J., ed. *Darwinism and the American Intellectual: A Book of Readings.* Homewood, IL: Dorsey Press, 1967.
York, Derek, and Farquhar, Ronald M. *The Earth's Age and Geochronology.* New York: Pergamon Press, 1972.
Yourgrau, W. and Breck, A. D., eds. *Cosmology, History, and Theology.* New York: Plenum, 1977.
Zimmerman, P., ed. *Darwin, Evolution, and Creation.* St. Louis, MO: Concordia Publishing House, 1959.

ARTICLES, REPORTS AND PAMPHLETS

Alexander, Richard D. "The Evolution of Social Behavior." *Annual Review of Ecology and Systematics* 5 (1974): 325–83.
———. "Natural Selection and the Analysis of Human Sociality." In *Changing Scenes in the Natural Sciences,* edited by C. E. Goulden. Academy of Natural Sciences Special Publication 12, Philadelphia, PA: Academy of Natural Sciences, 1977.
———. "The Search for a General Theory of Behavior." *Behavioral Science* 20 (1975): 77–100.
———. "The Search for an Evolutionary Philosophy of Man." *Proceedings of the Royal Society of Victoria, Melbourne* 84 (1971): 99–119.
Anderson, V. Elving. "Evangelicals and Science: Fifty Years after the Scopes Trial." In *The Evangelicals: What They Believe, Who They Are, and Where They Are Changing,* edited by D. F. Wells and J. D. Woodbridge. Nashville, TN: Abingdon Press, 1975.
Armstrong, H. L. "Creation Research Society." *Creation Research Society Quarterly* 18 (September 1981): 135.
Aulie, Richard P. "The Doctrine of Special Creation." (2 parts.) *American Biology Teacher* 34 (April 1972): 191–93; and (May 1972): 261–68, 81.
———. "The Post Darwinian Controversies." *Journal of the American Scientific Affiliation* 34 (1982): 24–29.
Awbrey, Frank. "Evidence of the Quality of Creation Science Research." *Creation/Evolution* 2 (Fall 1980): 40–42.
———. "Yes, Virginia, There *Is* a Creation Model." *Creation/Evolution* 1 (Summer 1980): 1.
Awbrey, Frank T., and Thwaites, William M. "A Closer Look at Some Biochemical Data that 'Support' Creationism." *Creation/Evolution* 7 (Winter 1982): 14–17.

Badash, Lawrence. "Rutherford, Boltwood, and the Age of the Earth: The Origin of Radioactive Dating Techniques." *Proceedings of the American Philosophical Society* 112 (1968): 157–69.

"Balanced Treatment for Scientific Creationism and Evolution Act." House Bill 107, Florida House of Representatives, 1980.

Ball, Howard G., and Nordmann, Terrance J. *Scientific Creationism in our Public School Curricula.* U.S. Alabama, 1981. ERIC document ED 205 381.

Ball, Ian R. "Nature and Formulation of Biogeographical Hypotheses." *Systematic Zoology* 24 (1975): 407–30.

Barker, A. "An Approach to the Theory of Natural Selection." *Philosophy* 44 (1969): 271–90.

Barker, F. D. "Evolution and the University of Nebraska." *Science* 63 (1926): 71.

Barnes, Thomas G., and Upham, R. J. "Another Theory of Gravitation: An Alternative to Einstein's General Theory of Relativity." *Creation Research Society Quarterly* 12 (1976): 194–97.

Bassi, J. "Review of: *The Moon: Its Creation, Form and Significance.*" *Journal of the American Scientific Affiliation* 33 (1981): 249.

Baum, Rudy. "Battle Brewing over Arkansas Creationism Law." *Chemical and Engineering News* 59 (July 1981): 25–26.

———. "Science Confronts Creationist Assault." *Chemical and Engineering News* 60 (March 1982): 12–26.

Beckner, Morton. "Aspects of Explanation in Biological Theory." In *Philosophy of Science Today,* edited by S. Morgenbesser. New York: Basic Books, 1967.

Bernhard, R. "Heresy in the Halls of Biology: Mathematicians Question Darwin." *Scientific Research* 2 (November 1967): 59–66.

Bethell, T. "Darwin's Mistake." *Harper's Magazine* 252 (February 1976): 70–72, 74–75.

Betts, John Richard. "Darwinism, Evolution, and American Catholic Thought." *Catholic Historical Review* 45 (July 1959): 161–85.

Birch, L. C., and Ehrlich, P. R. "Evolutionary History and Population Biology." *Nature* 214 (April 1967): 349–52.

Bird, Wendell R. "Creationism and Evolution: A Response to Gerald Skoog." *Educational Leadership* 38 (November 1980): 157.

———. "Freedom from Establishment and Unneutrality in Public School Instruction and Religious School Regulation." *Harvard Journal of Law and Public Policy* 2 (1979): 125–205.

———. "Freedom of Religion and Science Instruction in Public Schools." *Yale Law Journal* 87 (January 1978): 515–70.

———. "Resolution for Balanced Presentation of Evolution and Scientific Creationism." *ICR Impact Series,* no. 71 (1979).

Black, D. "The Creationists Are Coming—Again." *Next* 2 (March/ April 1981): 64.

Bock, Walter J. "Philosophical Foundations of Classical Evolutionary Classification." *Systematic Zoology* 22 (1973): 375–92.

Bowen, Francis. "On the Origin of Species." *North American Review* 90 (1860): 474–506.

Boylan, D. R. "Process Constraints in Living Systems." *Creation Research Society Quarterly* 15 (1978): 133–38.

Brady, R. H. "Natural Selection and the Criteria by which a Theory is Judged." *Systematic Zoology* 28 (1979): 600–21.

Brice, William R. "Bishop Ussher, John Lightfoot, and the Age of Creation." *Journal of Geological Education* 30 (January 1982): 18–24.

Broad, William J. "Creationists Limit Scope of Evolution Case," *Science* 211 (March 1981): 1331–32.

———. "Louisiana Puts God into Biology Lessons." *Science* 213 (August 1981): 628–29.

Brown, J. H. "The Teaching of Evolution." *Science* 56 (1922): 448–49.

Brown, Robert H. "Can We Believe Radiocarbon Dates?" *Creation Research Society Quarterly* 12 (1975): 66–68.

Bruce, E. J., and Ayala, F. J. "Humans and Apes Are Genetically Very Similar." *Nature* 276 (1978): 264–65.

———. "Phylogentric Relationships between Man and the Apes: Electrophoretic Evidence." *Evolution* 33 (1979): 1040–56.

Brush, Stephen G. "Creationism/Evolution: The Case Against 'Equal Time'." *Science Teacher* 48 (April 1981): 29–33.

———. "A Geologist among Astronomers: The Rise and Fall of the Chamberlin-Moulton Cosmogony." *Journal for the History of Astronomy* 9 (1978): 1–41, 77–104.

———. "Should the History of Science be Rated X?" *Science* 183 (1974): 11164–72.

Bube, Richard H. "Creation and Evolution in Science Education." *Journal of the American Scientific Affiliation* 25 (1973): 69–70.

Bush, C. "The Teaching of Evolution in Arkansas." *Science* 64 (October 1926): 356.

Calbreath, D. G. "The Challenge of Creationism: Another Point of View." *American Laboratory* (November 1980): 8, 10.

Caldwell, O. W., and Weller, F. "High School Biology Content as Judged by Thirty College Biologists." *School Science and Mathematics* 32 (1932): 411–24.

Callaghan, Catherine A. "Evolution and Creationist Arguments." *American Biology Teacher* 42 (October, 1980): 422–25, 27.

Carroll, R. L. "The Ancestry of Reptiles." *Philosophical Transactions of the Royal Society of London* (B) 257 (1970): 267–308.

———. "The Earliest Reptiles," *Journal of the Linnean Society, Zoology* 45 (1964): 61–83.

———. "Problems of the Origin of Reptiles." *Biology Review* 44 (1969): 393–432.

Carson, Hampton L. "The Genetics of Speciation at the Diploid Level." *American Naturalist* 109 (1975): 83–92.

Carter, L. J. "Creationists Sue to Ban Museum Evolution Exhibits." *Science* 204 (1979): 925.

Chambers, Bette. "Why a Statement Affirming Evolution?" *The Humanist* 37 (January/February 1977): 4–6, 23–24.

Cloud, Preston. "Evolution Theory and Creation Mythology." *The Humanist* 37 (November/December 1977): 53–55.

Cole, John R. "Film Review: Theories of the Origin of Life." *Creation/Evolution* 5 (Summer 1981): 39.

———. "Misquoted Scientists Respond." *Creation/Evolution* 6 (Fall 1981): 34–44.

Conrad, Ernest C. "Tripping Over a Trilobite: A Study of the Meister Tracks." *Creation/Evolution,* Issue VI (Fall 1981): 30–33.

Cook, M. A. "Do Radiological 'Clocks' Need Repair?" *Creation Research Society Quarterly* 5 (1968): 69–77.

Cory, Lawrence R. "Creationism and the Scientific Method." *American Biology Teacher* 35 (1973): 223–25.

Cowen, J. "Evolution: Equal Time for God." *Technology Review* 81 (June/July 1979): 10–11, 19.

Cramer, J. A. "General Evolution and the Second Law of Thermodynamics." In *Origins and Change,* edited by David L. Willis. Elgin, IL: *American Scientific Affiliation,* 1978.

"Creation-Science Challenge." *Academe* 67 (October 1981): 298.

Crompton, A. W., and Parker, P. "Evolution of the Mammalian Masticatory Apparatus." *American Scientist* 66 (1978): 192–201.

Cuffey, Roger J. "Evidence for Evolution from the Fossil Record." *Journal of the American Scientific Affiliation* 23 (1971): 158–59.

———. "Transitional Fossils Well-Known." *Journal of the American Scientific Affiliation* 23 (1971): 38.

Dalrymple, G. Brent. "Radiometric Dating, Geologic Time, and the Age of the Earth: A Reply to 'Scientific Creationism'." (Preprint, 1982.)

Darwin, Charles, and Wallace, Alfred. "On the Tendency of Species to Form Varieties; and on the Perpetuation of Varieties and Species by Natural Means of Selection." *Journal of the Linnean Society of*

London, Zoology 3 (1858): 45–62. Reprinted in John A. Moore, *Principles of Zoology.* New York: Oxford University Press, 1957.

Deadman, J. A., and Kelly, P. J. "What Do Secondary School Boys Understand about Evolution and Heredity before They Are Taught the Topics." *Journal of Biological Education* 12 (March 1978): 7–15.

Dean, Dennis R. "The Age of the Earth Controversy: Beginnings to Hutton." *Annals of Science* 38 (1981): 435–56.

"Decide: Evolution and Creation—One or Both in Public School Science?" Creation Science Research Society pamphlet, March 19, 1980.

Dexter, John Smith. "Anti-Evolution Propaganda in Georgia." *Science* 62 (October 1925): 399.

Dobzhansky, Theodosius. "Darwinian Evolution and the Problem of Extraterrestrial Life." *Perspectives in Biology and Medicine* 15 (1972): 157–75.

———. "Evolution at Work." *Science* 127 (1958): 1091–98.

Dodge, R. A. "Divine Creation: A Theory?" *American Institute of Biological Science Education Review* 2 (April, 1973): 29–30.

Duns, John. "On the Origin of Species." *North British Review,* American edition 27 (1860): 245–63.

Dutch, Steven I. "A Critique of Creationist Cosmology." *Journal of Geological Education* 30 (January, 1982): 27–33.

———. "Notes on the Nature of Fringe Science." *Journal of Geological Education* 30 (January 1982): 6–13.

Edwords, Frederick. "Victory in Arkansas: The Trial, Decision, and Aftermath." *Creation/Evolution* 7 (Winter 1982): 33–45.

———. "Why Creationism Should Not be Taught as Science: The Educational Issues." *Creation/Evolution* 3 (Winter 1981): 6–36.

Eglin, Paula G., and Graham, Mildred W. "Creationism Challenges Geology: A Retreat to the Eighteenth Century." *Journal of Geological Education* 30 (January 1982): 14–17.

Eigen, M., and Gradiner, W. "The Origins of Genetic Information." *Scientific American* 244 (April 1981): 88–118.

Eldredge, Niles. "Creationism Isn't Science." *The New Republic* 184 (April 1981): 15–18.

———. "Do Gaps in the Fossil Record Disprove Descent with Modification?" *Creation/Evolution,* Issue IV (Spring 1981): 17–19.

———. "Evolution and Prediction." *Science* 212 (May 1981): 737.

Eldredge, Niles, and Gould, Stephen J. "Punctuated Equilibria: An Alternative to Phyletic Gradualism." In *Models in Paleobiology,* edited by T. J. M. Schopf. San Francisco, CA: Freeman, Cooper, 1972).

Elmendorf, R. G. "$5000 Reward and a Challenge to Evolution." *Creation/Evolution* 4 (Spring 1981): 1–4.

"Evolving to the Beat of a Different Theory." *Science News* 120 (July 1981): 52.

Faul, Henry. "A History of Geologic Time." *American Scientist* 66 (1978): 159–65.

Fenner, Peter. "Classrooms and Creation." *Geotimes* 17 (November 1972): 13.

Fowler, Dean R. "Biology Texts Under Fire: A Case Study in Science and Religion." *Perspectives in Religious Studies* 8 (Fall 1981): 231–44.

Frank, O. D. "Data on Textbooks in the Biological Sciences Used in the Middle West." *School Science and Mathematics* 16 (1916): 354–57.

Freske, Stanley. "Evidence Supporting a Great Age for the Universe." *Creation/Evolution* 2 (Fall 1980): 34–38.

Gerlovich, Jack. *Creation, Evolution, and Public Education: The Position of the Iowa Department of Public Instruction.* Des Moines, IA: Iowa State Department of Public Instruction, 1980. ERIC document ED 199 074.

Gish, Duane T. "A Challenge to Neodarwinism." *American Biology Teacher* 32 (1970): 495–97.

―――. "Creation, Evolution, and the Historical Evidence." *American Biology Teacher* 35 (March 1973): 132–40.

―――. "A Decade of Creationist Research." *Creation Research Society Quarterly* 12 (June 1975): 34–46.

―――. "The Scopes Trial in Reverse." *The Humanist* 37 (November/December 1977): 50–51.

Gish, Duane T., and Asimov, Isaac. "The Genesis War." *Science Digest* 89 (October 1981): 82–87.

Glass, H. Bentley. "The Centrality of Evolution in Biology Teaching." *American Biology Teacher* 29 (December 1967): 705–15.

Godfrey, Laurie R. "An Analysis of the Creationist Film, *Footprints in Stone.*" *Creation/Evolution,* 6 (Fall 1981): 23–29.

―――. "The Flood of Antievolutionism: Where is the Science in 'Scientific Creationism'?" *Natural History* 90 (June 1981): 4, 6, 9–10.

―――. "Science and Evolution in the Public Eye." *Skeptical Inquirer* 4 (1979): 21–32.

Gorman, James. "Creationists vs. Evolution." *Discover* 2 (May 1981): 32–33.

Gould, Stephen J. "Evolution as Fact and Theory." *Discover* 2 (May 1981): 34–37.

―――. "Evolution's Erratic Pace." *Natural History* 86 (May 1977): 12, 14, 16.

―――. "The Piltdown Conspiracy." *Natural History* 89 (August 1980): 8–28.

―――. "The Return of Hopeful Monsters." *Natural History* 86 (1977): 22–30.

Grabiner, Judith, and Miller, Peter D. "Effects of the Scopes Trial: Was It a Victory for Evolutionists?" *Science* 185 (1974): 832–37.

Gray, Asa. "Review of Darwin's Theory on the Origin of Species." *American Journal of Science and Arts,* 2nd series, 29 (1860): 158–84.

Greene, John C. "Reflections on the Progress of Darwinian Studies." *Journal of the History of Biology* 8 (1975): 243–73.

Grobman, A. "National Science Foundation's Role in the Improvement of Science Education in American Schools." *Biological Sciences Curriculum Study Newsletter* 27 (1965): 7–10.

Haas, J. W., Jr. "Biogenesis: Paradigm and Presupposition." In *Origins and Change,* edited by David L. Willis. Elgin, IL: American Scientific Affiliation, 1978.

Halstead, Beverly. "Popper: Good Philosophy, Bad Science?" *New Scientist* 87 (July 1980): 215–17.

Hamilton, W. D. "The Genetical Evolution of Social Behavior." *Journal of Theoretical Biology* 7 (1964): 1–52.

Hammond, Allen, and Margulis, Lynn. "Creationism as Science." *Science '81* (December 1981): 55–57.

Hardin, Garrett. "Ambivalent Aspects of Evolution." *American Biology Teacher* 35 (1973): 15–19.

Harris, Leon. C. "An Axiomatic Interpretation of the Neo-Darwinian Theory of Evolution." *Perspectives in Biology and Medicine* 18 (1975): 179–84.

Hatfield, L. "Educators against Darwin." *Science Digest,* special edition (Winter 1979–80): 94.

Helgeson, S. L. et al. *The Status of Pre-college Science, Mathematics, and Social Science Education. I. Science Education: 1955–1975.* Columbus, OH: Ohio State University, Center for Science and Mathematics Education, 1977. ERIC document ED 153 876.

Henig, Robin Marantz. "The Battle Continues: Evolution Called a 'Religion,' Creationism Defended as a 'Science'." *Bioscience* 29 (September 1979): 513–16.

Herrman, Robert L. "Implications of Molecular Biology for Creation and Evolution." In *Origins and Change,* edited by David L. Willis. Elgin, IL: American Scientific Affiliation, 1978.

Holmes, S. J. "Proposed Laws against the Teaching of Evolution." *Bulletin of the American Association of University Professors* 13 (December 1927): 549–54.

Hopson, J. A. "The Origin and Adaptive Radiation of Mammal-like Reptiles and Non-therian Mammals." *Annals of the New York Academy of Sciences* 167 (1969): 199–216.

Hull, David L. "Charles Darwin and Nineteenth-Century Philosophies of Science." In *Foundations of Scientific Method,* edited by R. N. Giere and R. S. Westfall. Bloomington, IN: Indiana University Press, 1973.

———. "The Metaphysics of Evolution." *British Journal for the History of Science* 3 (December 1967): 309–37.

"Human-Like Tracks in Stone Are Riddle to Scientists." *Science News Letter* (October 1938): 278–79.

Hutton, James. "Theory of the Earth, or an Investigation of the Laws Observable in the Composition, Dissolution, and Restoration of the Land upon the Globe." *Transactions of the Royal Society of Edinburgh* 1 (1788): 209–304.

Huxley, Julian. "Evolution and Genetics." In *What Is Science?,* edited by James R. Newman. New York: Simon and Schuster, 1955.

Huxley, Thomas Henry. "Darwin on the Origin of Species." *Westminster Review,* American edition 73 (1860): 295–310.

———. "Time and Life: Mr. Darwin's 'Origin of Species'." *Macmillan's Magazine* 1 (1859): 142–48.

"The ICR Scientists." *ICR Impact Series,* no. 86 (August, 1980).

Ingram, T. Robert. "The Theological Necessity of a Young Universe." *Creation Society Research Quarterly* 12 (1975): 32–33.

Institute for Creation Research. "Distinction between Scientific Creationism and Biblical Creationism." *Acts and Facts* (December 1978): 4–8.

Jacob, Francois. "Evolution and Tinkering." *Science* 196 (June 1977): 1161–66.

Jenkins, Fleeming. "The Origin of Species." *North British Review* 46 (June 1867): 277–318.

Jones, D. Gareth. "Evolution: A Personal Dilemma." In *Origins and Change,* edited by David L. Willis. Elgin, IL: American Scientific Affiliation, 1978.

Kaminsky, M. "The Biochemical Evolution of the Horse." *Comparative Biochemical Physiology* 63b (1979): 175–78.

Kauffman, E. G. "Plate Tectonics: Major Force in Evolution." *Science Teacher* 43 (March 1976): 13–17.

Kettlewell, H. B. D. "Darwin's Missing Evidence." *Scientific American* 200 (March 1959): 48–53.

King, M. C., and Wilson, A. C. "Evolution at Two Levels in Humans and Chimpanzees." *Science* 188 (1975): 107–16.

Kitts, David B. "Paleontology and Evolutionary Theory." *Evolution* 28 (September 1974): 458–72.

Kofahl, Robert E. "The Bombardier Beetle Shoots Back." *Creation/ Evolution* (Summer 1981): 12–14.

———. "Could the Flood Waters Have Come from a Canopy or Extraterrestrial Source?" *Creation Research Quarterly* 13 (1977): 202–06.

Kranz, R. A. "Karl Popper's Challenge." *Creation Social Science and Humanities Quarterly* 2 (1980): 20–23.

Kraus, David. "The New York Creation Battle." *Creation/Evolution* 2 (Fall 1980): 8–9.

Kube-McDowell, Michael P. *The Scientific Creationist Challenge to the Treatment of Evolution in the Public School Curriculum.* U.S. Indiana, 1981. ERIC document ED 202 786.

Kyle, William C., Jr. "Should 'Scientific' Creation and the Science of Evolution Be Taught with Equal Emphasis?" *Journal of Research in Science Teaching* 17 (November 1980): 519–27.

Le Clercq, Frederic S. "The Constitution and Creationism." *American Biology Teacher* 36 (March 1974): 139–45.

———. "The Monkey Laws and the Public Schools: A Second Consumption?" *Vanderbilt Law Review* 27 (March 1974): 211–42.

Lee, K. "Popper's Falsifiability and Darwin's Natural Selection." *Philosophy* 44 (1969): 291–302.

Leith, Thomas H. "Explanation, Testability, and the Theory of Evolution." (2 parts.) *Journal of the American Scientific Affiliation* 32 (1980): 13–18; 156–163.

———. "Knowing and Understanding in Earth History." (2 parts.) *Christian Scholar's Review* 8 (1979): 30–50; 131–51.

———. "On Understanding the Geological Record." *Christian Scholar's Review* 5 (1975): 99–118.

———. "Some Logical Problems with the Thesis of Apparent Age." (3 parts.) *Journal of the American Scientific Affiliation* 17 (1965): 118–22; 18 (1966): 61–63, 125–26; 19 (1967): 62–63.

Levin, Florence S., and Lindbeck, Joy S. "An Analysis of Selected Biology Textbooks for the Treatment of Controversial Issues and Biosocial Problems." *Journal of Research in Science Teaching* 16 (May 1979): 199–203.

Lewin, Roger. "A Response to Creationism Evolves." *Science* 214 (November 1981): 635–36, 38.

Lewis, Ralph W. "Why Scientific Creationism Fails to Meet the Criteria of Science." *Creation/Evolution* 5 (Summer 1981): 7–11.

Lewontin, R. C. "Biological Models." In *Dictionary of the History of Ideas,* edited by Philip R. Wiener. New York: Scribners, 1973 (vol. 1, pp. 242–46).

———. "Selection in and of Populations." In *Ideas in Modern Biology.* Garden City, NY: Natural History Press, 1965.

Lightner, Jerry P. "Tennessee 'Genesis Law' Ruled Unconstitutional." *National Association of Biology Teachers News and Views* 19 (April 1975).

Linton, E. "Note on the Scientific Method and Authority." *Science* 64 (1926): 526–27.

Little, John. "Evolution: Myth, Metaphysics, or Science?" *New Scientist* 87 (September 1980): 708–09.

Lubenow, Marvin L. "Does a Proper Interpretation of Scripture Require a Recent Creation?" *ICR Impact Series,* nos. 65 and 66 (1978).

Lucas, E. C., et al. "Creationism and Evolutionism." *Science* 179 (1973): 953–54, 56.

Lyons, Gene. "Repealing the Enlightenment: The Biggest Thing in Arkansas since Creation—And When Was that, by the Way?" *Harper's* (April 1982): 38–40, 73–78.

Macrae, Barbara. "Creationism: The Nonsense of Nonscience." *Et Cetera* 38 (Summer 1981): 137–43.

Mandelbaum, Maurice. "Darwin's Religious Views." *Journal of the History of Ideas* 19 (1958): 363–78.

Marsh, F. L. "The Genesis Kinds in the Modern World." In *Scientific Studies in Special Creation,* edited by W. E. Lammerts. Grand Rapids, MI: Baker Book House, 1971.

Mayer, William V. "The Emperor's New Clothes—Sold Again." *The Humanist* 37 (November/December 1977): 52–53.

———. "Evolution and the Law." *American Biology Teacher* 35 (1973): 144–45, 62.

———. "Evolution: Theory or Dogma." Paper presented at the National Convention of the National Science Teachers Association, March 30–April 3, 1973. ERIC document ED 093 572.

———. "Evolution: Yesterday, Today, Tomorrow." *The Humanist* 37 (January/February 1977): 16–22.

Mayer, William V., and Roth, Ariel A. "Creation Concepts Should/Should Not Be Taught in Public Schools." *Liberty* 73: 3–7, 24–27.

Mayr, Ernst. "Agassiz, Darwin, and Evolution." *Harvard Library Bulletin* 13 (Spring 1959): 165–94.

———. "Darwin and Natural Selection—How Darwin May Have Discovered His Highly Unconventional Theory." *American Scientist* 65 (1977): 321–27.

————. "The Nature of the Darwinian Revolution." *Science* 176 (June 1972): 981–89.

————. "Theory of Biological Classification." *Nature* 220 (1968): 545–48.

McDaniel, Thomas R. "Special Creation and Evolution in the Classroom: Old Wine in New Wineskins." *School Science and Mathematics* 77 (January 1977): 47–52.

McKenzie, D. P. "Plate Tectonics and Its Relationship to the Evolution of Ideas in the Geological Sciences." *Daedalus* 106 (1977): 97–124.

McKown, Delos B. "Creationism and the First Amendment." *Creation/Evolution* 7 (Winter 1982): 24–32.

Miller, D. A. "Evolution of Primate Chromosomes." *Science* 198 (1977): 1116–24.

Miller, Kenneth R. "Special Creation and the Fossil Record: The Central Fallacy." *American Biology Teacher* 44 (February 1982): 85–89.

Milne, David H. "How to Debate with Creationists—and 'Win'." *American Biology Teacher* 43 (May 1981): 235–45, 66.

Moore, James R. "Evolutionary Theory and Christian Faith: A Bibliographic Guide to the Post-Darwinian Controversies." *Christian Scholar's Review* 4 (1975): 211–30.

Moore, John A. "Countering the Creationists." *Academe* (1982) (forthcoming).

————. "Creationism in California." *Daedalus* 103 (Summer 1974): 173–90.

————. "The Evolution-Creationism Debate." Sigma Xi Lecture, University of California, April 7, 1982.

————. "The Perils of Equal Time." Paper presented at the Annual Conference of the Association for Supervision and Curriculum Development, March 21, 1982.

————. "Teaching Creation Science." Paper presented at the National Convention of the National Association of Biology Teachers, October 23, 1981.

Moore, John N. "Evolution, Creation, and the Scientific Method." *American Biology Teacher,* 35 (1973): 23–26, 34.

————. "Evolution, Creation, Scientific Method, and Legislation." *Michigan Science Teachers Bulletin* 21 (1974): 6.

————. "Evolution—Required or Optional in a Science Course?" *Journal of the American Scientific Affiliation* 22 (1970): 82–87.

Moore, John N., and Cuffey, Roger J. "Dialogue on Paleontological Evidence and Organic Evolution." In *Origins and Change,* edited by David L. Willis. Elgin, IL: American Scientific Affiliation, 1978.

Moore, Robert A. "Arkeology: A New Science in Support of Creation?" *Creation/Evolution* 6 (Fall 1981): 6–15.

Morris, Henry M. "Evolution, Creation, and the Public Schools." *ICR Impact Series,* no. 1 (1973).

———. "Introducing Creationism into the Public Schools." *ICR Impact Series,* no. 20 (1974).

———. "Resolution for Equitable Treatment of Both Creation and Evolution." *ICR Impact Series,* no. 26 (1975).

———. "The Tenets of Creationism." *ICR Impact Series,* no. 85 (1980).

Morris, John. "The Paluxy River Tracks." *ICR Impact Series,* no. 35 (1976).

Moyer, Wayne A. "Arguments for Maintaining the Integrity of Science Education." *American Biology Teacher* 43 (October 1981): 380–81.

———. "The Challenge of Creationism." *American Laboratory* (August 1980): 12–14.

Myers, R. H., and Shafer, D. A. "Hybrid Ape Offspring of a Mating of Gibbon and Siamang." *Science* 205 (1979): 308–10.

Nelkin, Dorothy. "Science or Scripture: The Politics of 'Equal Time'." In *Science and Its Public: The Changing Relationship,* edited by Gerald Holton and William A. Blanpied. Boston: D. Reidl, 1976.

———. "The Science-Textbook Controversies." *Scientific American* 234 (April 1976): 33–39.

Nevins, Stuart E. "Evolution: The Ocean Says No!" *ICR Impact Series,* no. 8 (1973).

Newell, Norman D. "Evolution under Attack." *Natural History* 83 (April 1974): 32–39.

Newfeld, Berney. "Dinosaur Tracks and Giant Men." *Origins* 2 (1975): 64–76.

Niebuhr, Reinhold. "Christianity and Darwin's Revolution." In *A Book that Shook the World.* Pittsburgh, PA: University of Pittsburgh Press, 1958.

Oakley, Kenneth P., and Weiner, J. S. "Piltdown Man." *American Scientist* 43 (October 1955): 573–83.

Orlich, D. C.; Ratcliff, J. L.; and Stronk, D. R. "Creationism in the Science Classroom." *The Science Teacher* 42 (May 1975): 43–45.

Ost, David H. "Teacher-Training Use of ABT 'Creationism' Controversy." *American Biology Teacher* 34 (1972): 349.

Oxnard, Charles E. "Human Fossils: New Views of Old Bones." *American Biology Teacher* 41 (1979): 264–76.

————. "The Place of the Australopithecines in Human Evolution: Grounds for Doubt?" *Nature* 258 (December 1975): 389–95.

Parker, Barbara. "Creation vs Evolution: Teaching the Origin of Man." *American School Board Journal* 167 (March 1980): 25–34.

Parker, Barry. "The Age of the Universe." *Astronomy* 9 (1981): 66–71.

Parker, Franklin. *Why the Evolution/Creation Battle Rages: What Educators Can Do.* U.S. West Virginia, 1981. ERIC document ED 207 904.

Peckham, Morse. "Darwinsim and Darwinisticism." *Victorian Studies* 3 (September 1959): 19–40.

Peter, W. G. "Fundamentalist Scientists Oppose Darwinian Evolution." *Bioscience* 20 (1970): 1067–69.

Peters, R. H. "Predictable Problems with Tautology in Evolution and Biology." *American Naturalist* 112 (1978): 759–62.

————. "Tautology in Evolution and Ecology." *American Naturalist* 110 (1976): 1–12.

Phillips, Perry G. "Meteoritic Influx and the Age of the Earth." In *Origins and Change,* edited by David L. Willis. Elgin, IL: American Scientific Affiliation, 1978.

Pilbeam, David. "Rearranging Our Family Tree." *Human Nature* (June 1978): 38–45.

Pipho, Chris. "Scientific Creationism." *Compact* 14 (Winter 1981): 32.

————. "Scientific Creationism: A Case Study." *Education and Urban Society* 13 (February 1981): 219–33.

Pratt, Donald L. *Evolution vs. Creationism: 1973–81.* U.S. Florida, 1981. ERIC document ED 209 086.

Price, Robert. "The Return of the Navel, the 'Omphalos' Argument in Contemporary Creationism." *Creation/Evolution* 2 (Fall 1980): 26–33.

————. "Scientific Creationism and the Science of Creative Intelligence." *Creation/Evolution* 7 (Winter 1982): 18–23.

Protsch, R. "The Absolute Dating of Upper Pleistocene SubSaharan Fossil Hominids and Their Place in Human Evolution." *Journal of Human Evolution* 4 (1975): 297–322.

Ralph, E. K., and Michael, N. H. "Twenty-five Years of Radiocarbon Dating." *American Scientist* 62 (1974): 553–560.

Randall, John Herman. "The Changing Impact of Darwin on Philosophy." *Journal of the History of Ideas* 22 (1961): 435–62.

Rice, E. L. "Darwin and Bryan: A Study in Method." *Science* 61 (March 1925): 243–50.

Richards, O. W. "The Present Status of Biology in the Secondary Schools." *School Review* 31 (1923): 143–46.

Robinson, James T. "The Incommensurability of Evolution and Special Creation." *American Biology Teacher* 33 (December 1971): 535–38, 45.

Rogers, James Allen. "Darwinism and Social Darwinism." *Journal of the History of Ideas* 33 (1972): 165–80.

Root-Bernstein, Robert. "Views on Evolution, Theory, and Science." *Science 212* (June 1981): 1446–48.

Ruse, Michael. "Charles Darwin and Artificial Selection." *Journal of the History of Ideas* 36 (1975): 339–50.

———. "Karl Popper's Philosophy of Biology." *Philosophy of Science* 44 (1977): 638–61.

Schadewald, Robert J. "Equal Time for Flat-Earth Science." *Creation/ Evolution* 3 (Winter 1981): 37–41.

———. "Moon and Spencer and the Small Universe." *Creation/Evolution,* 4 (Spring 1981): 20–22.

Schweber, S. S. "Genesis of Natural Selection—1838: Some Further Insights." *Bioscience* 28 (May 1978): 321–27.

Schweinsberg, John, "The Alabama Creation Battle." *Creation/Evolution* 5 (Summer 1981): 31–32.

Scriven, M. "Explanation and Prediction in Evolutionary Theory." *Science* 130 (1959): 477–82.

Shea, James H. "Scientific Creationism and the Future of Geological Education—An Editorial." *Journal of Geological Education* 30 (January 1982): 4–5.

Siegel, Harvey. "Creationism, Evolution, and Education: The California Fiasco." *Phi Delta Kappan* 63 (October 1981): 95–101.

Simpson, George Gaylord. "One Hundred Years without Darwin Are Enough." *Teachers College Record* 62 (May 1961): 617–26.

———. "The World into which Darwin Led Us." *Science* 131 (April 1960): 966–74.

Simpson, Ronald D., and Anderson, Wyatt W. "Same Song, Second Verse—A Review of *Biology: A Search for Order in Complexity,* Revised Edition." *Science Teacher* 42 (May 1975): 40–42.

Skoog, Gerald. "The De-emphasis of Evolution in the Secondary School Biology Textbooks of the 1970's." Paper presented at the annual meeting of the National Science Teachers Association, April 7–11, 1978. ERIC document ED 156 539.

———. "Does Creationism Belong in the Biology Curriculum?" *American Biology Teacher* 40 (January 1978): 23–26, 29.

———. "Legal Issues Involved in Evolution vs. Creationism." *Educational Leadership* 38 (November 1980): 154–56.

———. "The Textbook Battle over Creationism." *The Christian Century* 105 (October 1980): 974–76.

Skow, John. "Creationism as Social Movement: The Genesis of Equal Time." *Science '81* (December 1981): 54, 57, 60.

Slusher, Harold S. "Clues Regarding the Age of the Universe." *ICR Impact Series,* no. 19 (1974).

———. "Some Astronomical Evidence for a Youthful Solar System." *Creation Research Society Quarterly* (June 1971): 55–57.

Sonleitner, Frank J. "Creationists Embarrassed in Oklahoma." *Creation/Evolution* 4 (Spring 1981): 23–26.

Spears, L. M. "Creationists Win Right to Revise Science Texts." *National Observer* 30 (December 1972).

"Statements by Scientists in the California Textbook Dispute." *American Biology Teacher* 34 (October 1972): 411–15.

Stebbins, G. Ledyard. "The Evolution of Design." *American Biology Teacher* 35 (1973): 57–61.

Strahler, Arthur N. "Creationists Change Their Strategy." *Journal of Geological Education* 30 (January 1982): 24–26.

Suppes, Patrick. "The Desirability of Formalization in Science." *Journal of Philosophy* 65 (1968): 651–64.

Tattersall, I., and Eldredge, N. "Fact, Theory, and Fantasy in Human Paleontology." *American Scientist* 65 (1977): 204–11.

"The Tennessee Antievolution Law." *Science* 65 (January 1927): 57.

Thoday, J. M. "Non-Darwinian Evolution and Biological Progress." *Nature* 255 (1975): 675.

Thwaites, William. "Another Favorite Creationist Argument: 'The Genes for Homologous Structures Are Not Homologous'." *Creation/Evolution* 2 (Fall 1980): 43–44.

———. "Book Review: *Biology: A Search for Order in Complexity.*" *Creation/Evolution* 1 (Summer 1980): 38.

Thwaites, William, and Awbrey, Frank. "Biological Evolution and the Second Law." *Creation/Evolution* 4 (Spring 1981): 5–7.

Toscano, Paul James. "A Dubious Neutrality: The Establishment of Secularism in the Public Schools." *Brigham Young University Law Review* 5 (1979): 181.

Tribus, Myron, and McIrvine, Edward C. "Energy and Information." *Scientific American* 225 (September 1971): 179–84, 86, 88.

Trivers, R. L. "The Evolution of Reciprocal Altruism." *Quarterly Review of Biology* 46 (1971): 35–57.

Urlich, Donald C., et al. "Creationism in the Science Classroom." *Science Teacher* 42 (May 1975): 43–45.

Van de Fliert, J. R. "Fundamentalism and the Fundamentals of Geology." In *Origins and Change,* edited by David L. Willis. Elgin, IL: American Scientific Affiliation, 1978.

Wade, N. "Creationists and Evolutionists: Confrontation in California." *Science* 178 (November 1972): 724–29.

———. "Evolution: Tennessee Picks a New Fight with Darwin." *Science* 182 (November 1973): 696.

Weber, Christopher George. "The Bombardier Beetle Myth Exploded." *Creation/Evolution* 3 (Winter 1981): 1–5.

———. "Common Creationist Attacks on Geology." *Creation/Evolution* 2 (Fall 1980): 10–25.

———. "The Fatal Flaws of Flood Geology." *Creation/Evolution* 1 (Summer 1980): 24–37.

———. "Paluxy Man: The Creationist Piltdown." *Creation/Evolution* 6 (Fall 1981): 16–22.

———. "Response to Dr. Kofahl." *Creation/Evolution* 5 (Summer 1981): 15–17.

Weinberg, Stanley. "Reactions to Creationism in Iowa." *Creation/Evolution* 2 (Fall 1980): 1–7.

———. "Response to Zuidema." *Creation/Evolution* 5 (Summer 1981): 22–30.

———. "Two Views on the Textbook Watchers." *American Biology Teacher* 40 (December 1978): 541–45, 60.

Welch, Claude A. "Evolution Theory and the Nature of Science." *Science Teacher* 39 (January 1972): 26–28.

Wells, B. W. "Fundamentalism in North Carolina." *Science* 64 (July 1926): 17–18.

Wenner, Adrian M. "Adam and Eve in Science." *American Biology Teacher* 35 (May 1973): 278–79.

Wetherill, G. W. "Formation of the Terrestrial Planets." *Annual Review of Astronomy and Astrophysics* 18 (1980): 77–113.

Whitcomb, John C., Jr. "The Creation of the Heavens and the Earth." In *Scientific Studies in Special Creation.* Nutley, NJ: Presbyterian and Reformed Publishing Company, 1971.

———. "The Ruin Reconstruction Theory of Genesis 1:2." In *Scientific Studies in Special Creation.* Nutley, NJ: Presbyterian and Reformed Publishing Company, 1971.

Wieland, C. "The Case for Teaching Special Creation as a Competing Model of Origins." *South Australian Science Teachers Journal* (December 1977): 34–39.

Wilberforce, Samuel. "On the Origin of Species." *Quarterly Review* 108 (1860): 225–64.

Wiley, E. O. "Karl R. Popper, Systematics, and Classification: A Reply to Walter Bock and Other Evolutionary Taxonomists." *Systematic Zoology* 24 (1975): 233–43.

Williams, Mary B. "Deducing the Consequences of Evolution: A Mathematical Model." *Journal of Theoretical Biology* 29 (1970): 343–85.

——. "Falsifiability Predictions of Evolutionary Theory." *Philosophy of Science* 40 (1973): 518–37.

Willis, David L. "Creation and/or Evolution." In *Origins and Change,* edited by David L. Willis. Elgin, IL: American Scientific Affiliation, 1978.

Wilson, E. O. "The Queerness of Social Evolution." *Bulletin of the Entomology Society of America* 19 (1973): 20–22.

Wonderly, Daniel E. "Non-Radiometric Data Relevant to the Question of Age." In *Origins and Change,* edited by David L. Willis. Elgin, IL: American Scientific Affiliation, 1978.

Woodmorappe, John. "The Essential Nonexistence of the Evolutionary-Uniformitarian Geologic Column: A Quantitative Assessment." *Creation Research Society Quarterly* 18 (June 1981): 46–71.

——. "Radiometric Geochronology Reappraised." *Creation Research Society Quarterly* 16 (September 1979): 102–29.

Young, Robert M. "Darwin's Metaphor: Does Nature Select?" *The Monist* (1971): 442–503.

Yunis, J.; Sawyer, J. R.; and Dunham, K. "The Striking Resemblance of High Resolution G-Banded Chromosomes of Man and Chimpanzee." *Science* 208 (1980): 1145–48.

Zuidema, Henry P. "Genetics and Genesis: The New Biology Textbooks that Include Creationism." *Creation/Evolution* 5 (Summer 1981): 18–21.

——. "The Scientific Creationists." *Liberty* 70 (March 1975): 3.

——. "A Survey of Creationist Field Research." *Creation/Evolution* 6 (Fall 1981): 1–5.

Index

Compiled by Frederick Ramey

3